# THE LIFE OF
## ISAMBARD KINGDOM
# BRUNEL
## CIVIL ENGINEER

ISAMBARD BRUNEL

NONSUCH

First published 1870
Copyright © in this edition 2006
Nonsuch Publishing Ltd

Nonsuch Publishing Limited
The Mill, Brimscombe Port, Stroud, Gloucestershire, GL5 2QG
www.nonsuch-publishing.com

Nonsuch Publishing Ltd is an imprint of Tempus Publishing Group

British Library Cataloguing in Publication Data.
A catalogue record for this book is available from the British Library.

ISBN 1-84588-341-1 (hardback)
ISBN 1-84588-031-5 (paperback)

Typesetting and origination by Nonsuch Publishing Limited
Printed in Great Britain by Oaklands Book Services Limited

THE LIFE OF
ISAMBARD KINGDOM
# BRUNEL
CIVIL ENGINEER

# CONTENTS

# Introduction to the Modern Edition

Isambard Kingdom Brunel was arguably one of the greatest engineers Britain has seen. His designs have stood the test of time and have ensured that his name is still recognised and respected today. Brunel was not a conservative engineer: his designs were flamboyant, majestic and often defied convention. Projects such as the Clifton Suspension Bridge and Maidenhead Bridge were criticised and doubted by fellow engineers and onlookers who were unable to believe that his designs would work and the structures would stay up. But on both occasions Brunel was triumphant, even employing daring and elaborate stunts to prove the credibility of his designs. He demonstrated to the world that he was not afraid to test the limits of what was believed possible.

The breadth of Brunel's engineering achievements is truly remarkable: bridges, tunnels and viaducts, railways, buildings, a hospital, ships and even, albeit unsuccessfully, locomotives all formed part of his design portfolio. The Great Western Railway was one of his most outstanding, but at the same time controversial, achievements. The adoption of the 7ft broad gauge as opposed to the 4ft 8½in. standard gauge showed a surprising short-sightedness in believing that his railway would remain isolated from the rest of the country's network. But Brunel's conviction in the superiority of his designs and the standard to which these designs were realised, earned it the nickname 'God's Wonderful Railway'.

The period of time in which Brunel was working was one of dramatic change and progression. Along with his contemporaries Brunel helped push forward the boundaries of science, engineering and technology, and played an important part in establishing Britain as the engineering giant of the world. In order to better understand the true significance of his achievements it is important to consider them in the context of the world for which he was designing.

By the 1830s Britain was in the grip of the Industrial Revolution. Advances in science and technology transformed traditional cottage industries into major manufacturing enterprises. Factories and cotton mills sprang up in towns and cities throughout the country, producing goods in large volumes, which in turn stimulated both internal and overseas trade. Advances in the use of steam power were not only used to drive machinery in the factories, but were quickly harnessed for use in transportation, with the development of railways and the first steam-ships. From the mid-1840s, railway mania saw railway lines spread across the length and breadth of the country. The development of faster forms of transport, including transatlantic shipping, allowed for the faster movement of goods and trade, and also led to the introduction of more efficient postal and telegraphic systems. The Great Exhibition of 1851 marked a high point in the Industrial Revolution, becoming a symbol of British manufacturing and ingenuity, and sealing the country's status as the 'workshop of the world.'

Geographically and demographically, Britain in the nineteenth century began to change. A move from an agricultural economy to an industrial one meant that people began to leave the country in search of work and by the middle of the century Britain was no longer a predominantly rural nation. With the workforce moving to be near their places of work, and in turn new factories and workshops being built in locations where there were convenient facilities and workforce, villages grew into towns, towns into cities. However, this influx of population to urban areas placed huge demands on the infrastructure of many towns and cities, and brought with it a wide range of social and political problems.

Rapid expansion meant that housing for the working classes was often built in haste, and was overcrowded and lacking in basic amenities. Some companies tried to lead by example and set about building facilities for their workers. Saltaire in Yorkshire and Brunel's own Railway Village in Swindon were among the earliest examples of planned industrial housing, but even they could not keep pace with demand. As such, conditions in urban areas were often unsanitary and dangerous. Advances in medicine and increasing awareness of public health issues were tools in highlighting these conditions, but it took political will to really begin to address the problems. Exploitation of the workforce was common, especially of women and children who were expected to work long hours, in poor conditions, for little pay. A number of Factory Acts were introduced to regulate working standards and employees began to organise themselves into political organisations and trade unions in an attempt to improve their status. Attempts to regulate standards and preserve public order were backed by the formation of a professional police force, or the 'Peelers' as they were known at the time.

The nineteenth century also saw the beginnings of a dilution of the previously strict distinction between social classes. From the middle of the century standards of living and wages rose for the working classes. The middle-class factory owners and industrialists grew in wealth and status, and by contrast the country landowners, who had previously gained their wealth through an agricultural economy, began to struggle. To have a profession was no longer regarded as the lot of the lower classes and those industrialists, scientists, architects and engineers who led the Industrial Revolution came to be regarded as national heroes.

The railways had the effect of making Britain a much smaller place: not only could goods be transported quickly and cheaply, so too could people. Cheap rail travel led to the growth of the tourism industry. This meant that for the first time people began to travel outside the area in which they lived and so led to the creation of holiday resorts. The railways also led to the growth of suburbia as those who could afford it began to move out of the overcrowded cities to the outskirts, from where they could commute to work. Newspapers grew in popularity as people began to take an interest in the world around them, and the fact that news could be disseminated to the far corners of the country in a relatively short period of time meant that its people became increasingly informed on world issues.

The impact that Brunel and his contemporaries had on this rapidly- developing nation was significant. The consequences of their work spread far beyond science, manufacturing and transportation, and helped to create a country that led the world in terms of industry and innovation. This was a time when engineers, designers and scientists were revered and their characters and personalities were almost as well documented as their work. Brunel was famous for his flamboyance of character, the extravagance of his designs and his ruthless determination to achieve his ambitions. There is perhaps no one better placed to portray the true genius of this great man than his own son, Isambard Brunel.

Felicity Ball
1st *February* 2006

# PREFACE

I HAVE not attempted to describe the events of my father's life in chronological order beyond the end of Chapter III, which brings down the narrative to the close of 1835, the year in which the Act was obtained for the Great Western Railway.

Chapter IV contains a general account of my father's railway works, with the exception of the Bridges, which are described in Chapter VII. The history of the Broad Gauge and of the trial of the Atmospheric System on the South Devon Railway is given in Chapters V and VI.

Chapters VIII–XIII contain an account of my father's labours for the advancement of Ocean Steam Navigation. It will be noted that these chapters cover the same period as Chapters IV–VII, namely, from 1835, the year of the commencement of the Great Western Railway and the 'Great Western' Steamship, to 1859, the year of his death, in which the Saltash Bridge and the 'Great Eastern' were both completed.

Chapters VII (on the Bridges) and XIV (on the Docks) have been written by Mr. William Bell, for many years a member of my father's engineering staff; and in regard to Chapter V (on the Broad Gauge), I have to acknowledge assistance rendered me by Mr. William Pole, F.R.S.

For the Note on the Carbonic Acid Gas Engine which follows Chapter I, I am indebted to Mr. William Hawes; and for Chapter VI (on the Atmospheric System) to Mr. Froude, F.R.S.

I have also printed letters, written to me at my request, relating to various incidents in my father's life.

The assistance I received in the preparation of the chapters on Steam Navigation from my friend the late Captain Claxton, R.N., has been referred to in the note to p.183, note 2.

I have throughout availed myself of my brother's professional knowledge.

I have been compelled, in order to bring the work within the compass of a single volume, to omit much that would otherwise have been inserted, and I must therefore be held responsible for the general arrangement of those parts which have been contributed by others, as well as for the chapters which I have written myself.

Lastly, I desire gratefully to thank those friends who, by supplying me with materials and revising the proof sheets, have helped me in my endeavour to make this book, as far as possible, an accurate record of my father's life, written in the spirit of which he would have approved.

I.B.

18 DUKE STREET, WESTMINSTER,

*November,* 1870.

# I

# EARLY LIFE

ISAMBARD Kingdom Brunel was born on the ninth day of April, 1806, at Portsmouth, and was the only son of Sir Marc Isambard Brunel.[1]

Most biographies commence with an account of the parentage of the person whose life is about to be written. If this be permitted in any case, no apology can be needed for prefixing to a Life of Mr. Brunel some particulars of his father's career, since he was indebted to him, not only for the inheritance of many natural gifts, and for a professional education such as few have been able to procure, but also for a bright example of the cultivation of those habits of forethought and perseverance, which alone can ensure the successful accomplishment of great designs.

Sir Marc Isambard Brunel was a native of Hacqueville, a village in Normandy, where his family had been settled for several generations. He was originally intended for the priesthood; but, as he showed no inclination for that calling, and a very decided talent for mechanical pursuits, he was permitted to enter the French Navy; and he served in the West Indies for six years, namely, from 1786 to 1792. On his return home, at the expiration of his term of service, his strong Royalist sympathies made it unsafe for him to remain in France, and with great difficulty he managed to escape to America. He landed at New York in September 1793, and soon obtained employment as a civil engineer. A few years afterwards he was appointed engineer to the State of New York; and, while holding that office, he designed a cannon foundry and other important public works.

In January 1799, when Sir Isambard was in his thirtieth year, he came over to England, and shortly after his arrival married Miss Sophia Kingdom, a lady for whom he had formed an attachment some years before.[2]

The first great work undertaken by him in this country was the machinery for making blocks, which he designed and erected for Government at Portsmouth.

The history of the invention and construction of this system of machinery (for it consisted of *forty-three* separate machines) need not be given at length; but it may be permitted to extract the following passage from Mr. Beamish's 'Life of Sir Isambard Brunel' (pp.97, 99, 2nd edition), in which he points out the benefits which have resulted from its introduction, and the position its inventor is entitled to hold among those who have contributed to the progress of mechanical science:—

Where fifty men were necessary to complete the shells of blocks previous to the erection of Brunel's machinery, four men only are now required, and to prepare the sheaves, six men can now do the work which formerly demanded the labours of sixty. So that ten men, by the aid of this machinery, can accomplish with uniformity, celerity, and ease, what formerly required the uncertain labour of one hundred and ten.

When we call to mind that at the time these works were executed, mechanical engineering was only in its infancy, we are filled with amazement at the sagacity and skill that should have so far anticipated the progress of the age, as to leave scarcely any room, during half a century, for the introduction of any improvement . . .

Beautiful as are the combinations and contrivances in the block machinery, and highly deserving as the inventor may be of credit for originating such labour-saving machines for the production of ships' blocks, there is a far higher claim to the admiration and gratitude of all constructors of machinery, and of all workers in metal. In this block machinery exist the types and examples of all the modern self-acting tools, without the aid of which the various mechanical appliances of the present day could not be produced with the marvellous accuracy which has been attained. It is true that to the trades unions or combinations among the artisans, is in a great measure directly due the introduction of self-acting machines; but the types of all these tools existed in the machines and combinations of Brunel's block machinery. The drilling, the slotting, and the shaping machines, the eccentric chuck, and the slide rest, with the worm wheel motion, are all to be found in his machine.

On the completion of the block machinery Sir Isambard Brunel removed to London, and took a house in Lindsay Row, Chelsea, where he remained until he was obliged to live nearer the works of the Thames Tunnel.

Mr. Brunel's first recollections were of the house at Chelsea; and in 1814, when he was eight years old, he commenced his school life under the Rev. Weeden Butler, who resided in the neighbourhood. Previously to his going to Mr. Butler, he had been taught Euclid by Sir Isambard; and he had also a great talent for drawing, for which he had been remarkable even from four years old. His drawings were beautifully precise and neat, but, when the subject admitted of it, full of vigour and picturesque effect.

After some time he was sent to Dr. Morell's school at Hove, near Brighton. The following extract is taken from one of his letters home in 1820:—

> I have past Sallust some time, but I am sorry to say I did not read all, as
> Dr. Morell wished me to get into another class. I am at present reading

Terence and Horace. I like Horace very much, but not as much as Virgil. As for what I am about, I have been making half a dozen boats lately, till I have worn my hands to pieces. I have also taken a plan of Hove, which is a very amusing job. I should be much obliged to you if you would ask papa (I hope he is quite well and hearty), whether he would lend me his long measure. It is a long eighty-foot tape; he will know what I mean. I will take care of it, for I want to take a more exact plan, though this is pretty exact, I think. I have also been drawing a little. I intend to take a view of *all* (about five) the principal houses in that great town, Hove. I have already taken one or two.

In the intervals of his classical studies he seems to have employed himself, not only in making a survey of Hove in its existing state, but also in a critical examination of the works in progress for its enlargement. It is told of him that one evening he predicted the fall, before the next morning, of some houses which were building opposite the school, and laid a bet on the subject, which his companions readily accepted. He had noticed the bad way in which the work was done, and that the stormy weather, which appeared to be setting in for the night, would probably blow the walls down. In the morning he claimed the wager, for the buildings had fallen in the night.

Except from November 1820 to August 1822, when he was at the Collège Henri Quatre at Paris,[3] Mr. Brunel was so very little absent from home that he became thoroughly acquainted with all his father's undertakings. Among these was the veneering machinery at Battersea, remarkable for the great diameter of the saw, the steadiness of its motion, and the mechanical arrangements for clearing the veneer from the saw; also the works at the Government establishments at Woolwich and Chatham, and the machinery for making shoes. They have been fully described by Mr. Beamish; but the mere mention of their names is enough to show how great were the advantages enjoyed by Mr. Brunel in receiving from his father his early professional education.

From the year 1823 Mr. Brunel was regularly employed in his father's office. It was in the early part of this year that the project of the Thames Tunnel first began to occupy Sir Isambard's attention; but he was also engaged at that time in other works of great importance, among them the suspension bridges for the Ile de Bourbon, and designs for bridges of the same character over the Serpentine, and over the Thames at Kingston.[4] Some account of the Bourbon bridges, and also of experiments with carbonic acid gas, on which Mr. Brunel was engaged, will be found in the notes to this chapter.

The history of the Thames Tunnel will be told, as far as possible, in Sir Isambard Brunel's own words, as given in his journals.[5] Although these extracts do not relate to works for which Mr. Brunel was personally responsible, they have been inserted in the belief that they are valuable, not only as showing the nature and extent of his duties as his father's assistant, but also as displaying, in the most interesting and authentic form, Sir Isambard's character and genius at a time when his son was brought into hourly contact with him, and under circumstances which would cause the influence of his example to make a deep and lasting impression.

Previously to the year 1823 there had been several plans suggested for the construction of a tunnel under the Thames; and it would seem that a great demand was supposed to exist for some such means of communication between the two sides of the river eastward of London Bridge; for after the failure of the operations undertaken by Mr. Vasie in 1805, and Mr. Trevethick in 1807,[6] a high level suspension bridge was proposed, although it was not intended to be used for heavy traffic.[7]

The first reference to the Tunnel in Sir Isambard's journals is dated February 12, 1823. 'Engaged on drawings connected with Tunnel;' and on the 17th and following days of the same month, 'Isambard was engaged on Tunnel.' These entries become more and more frequent in the pages of his diary, until it is evident that Sir Isambard's whole time and thoughts were absorbed in this work.

The spring of 1823 was occupied in preparing drawings and models of his plans, and in enlisting the sympathy and assistance of various influential persons. By the close of the year the designs were matured sufficiently to enable the promoters of the scheme to commence the task of organising a company for carrying it out; and in January 1824 they resolved to call a general meeting of their friends, and invite public subscriptions.

On February 17, Sir Isambard explained his plans at the Institution of Civil Engineers, and on the next day a meeting was held at the City of London Tavern, under the presidency of Mr. William Smith, M.P., more than a hundred persons being present. Resolutions authorising the formation of a company were passed unanimously, and the share list was opened. In the course of an hour one-third of the subscriptions was filled up, namely, 1250 shares; and before the end of the day the number of shares taken was 1381.

Borings were then commenced in order to ascertain the nature of the strata through which the Tunnel would pass. A bed of gravel was found over the clay, which gave Sir Isambard great anxiety. A large pipe or shaft was sunk on the side of the river, and in it the water rose to within three feet of the surface of the ground, and fell about eighteen inches with the tide. 'It is manifest (Sir Isambard writes) from this that unless the Tunnel is enclosed in the stratum of clay, it would be unsafe to drive through the bed of gravel. The Tunnel must, therefore, begin with the substantial clay.'

However, the result of thirty-nine borings in two parallel lines across the river, to the depth of from 23 to 37½ feet, seemed to prove that there was below the gravel a stratum of strong blue clay of sufficient depth to ensure the safety of the Tunnel.[8]

A report to this effect was made to the shareholders at their first general meeting in July, and it was also stated that the works would be completed in three years.

The first operation connected with the works, was the construction of a shaft; and for this purpose land was bought on the Rotherhithe bank, about fifty yards from the river. On March 2, 1825, the ceremony of laying the first stone of the shaft was performed.

> Mr. Smith, our chairman, attended by most of the members of the Court of Directors, and a very numerous cortège of friends invited on the occasion, proceeded from the Tunnel Wharf to the ground, where they were received among the cheers of a great concourse of people. Mr. Smith addressed the assembly in a very eloquent speech suitable to the occasion, and performed the ceremony of laying the first stone. From this day dates the beginning of the work.[9]

The mode in which Sir Isambard decided to construct the shaft was one not uncommonly adopted in the construction of wells; but to apply it to sinking a shaft fifty feet in diameter was a novel and bold undertaking. The brickwork intended to form the lining of the shaft was built on the surface of the ground, and the earth being excavated from within and underneath the structure, it sank gradually down to its final position.

The brickwork was 3 feet thick, bound together by iron and timber ties, and there were built into it 48 perpendicular iron rods, one inch in diameter, fastened to a wooden curb at the bottom, and to another curb at the top of the wall, by nuts and screws.

When the shaft or tower of brickwork was completed up to the top, 42 feet in height, the next step was to remove the blockings on which it rested, and this being done the gravel was excavated and hoisted up, and the shaft descended by its own weight.

The Rotherhithe shaft was only sunk forty feet in this manner; the remaining twenty feet, in order to leave the opening for the Tunnel, was constructed by under-pinning, or underlaying, as it was then termed. The underlaying was commenced in the beginning of June.[10]

By July 4 they had got down to the level of the intended foundation of the shaft, having passed into a stratum of gravel, black pebbles embedded in greenish sand, with little or no water; from which circumstance Sir Isambard was of opinion that it was unconnected with the stratum of gravel above.

*July* 12.—Engaged on a general drawing for the great shield, and in preparing some instructions for moving the same (a very intricate operation!)

*July* 22.—Underlaying is a very laborious mode of proceeding. The sinking of a wall well bound as the first, would evidently be the best and cheapest mode for making another tower of 50 feet diameter.

On the 28th Sir Isambard enters in his journal the following additional observations upon the success of his plans for sinking the shaft:—

Considering the great labour necessary for securing the ground for the underlaying, the waste of planking, and of shores, and the time necessarily taken up in moving about, in securing and in bailing out the water, and the many causes of interruption, and the imperfect way that things are done in underlaying, it is quite conclusive that the original plan of making a shaft, by sinking the structure, is the safest and the most economical. What is done is sound, and when once in place, may be secured with foundations in a very easy manner. The brickwork of the shaft is remarkably hard. Had it been made with brick facings and rubble stone, it would certainly be water-tight, and almost impenetrable by ordinary ways. The *vertical ties* and the *circular wall bands* are not to be dispensed with in a structure destined to be moved as the present has been.

On August 11 the underlaying was completed, and preparations were made for constructing a reservoir in the bottom of the shaft for receiving the permanent pumps. This was finished on October 11, with great difficulty, owing to the nature of the ground, which consisted of loose sand containing a large quantity of water.

*August* 19.—Engaged at home in revising my plans for the manner of carrying on the horizontal excavation, more particularly of penetrating through the shaft. This part of the operation requires *indeed* very great attention, as it presents great difficulties, arising from the wall to be broken through, and chiefly from the angular opening that is to be made at each extremity. Then another consideration is the uniting the brick arches to the brickwork of the shaft.

*September* 16.—Engaged in the early part of the day on revising my plans of future operations in the Tunnel work, and in adapting them to the nature of the ground as it is found at the various depths we have penetrated: namely, to about 73 feet. Went afterwards to Maudslay to request that the great shield may be completed.

*October* 14.—Engaged in the early part of the morning in making some arrangements for the working of the great shield. Too much attention cannot be given to that subject at the early part; for, when once in its place, it would be extremely inconvenient to make any alteration.

*October* 15.—The dome of the reservoir will be covered to-day about noon; the bottom of the shaft will therefore be completed. They are now preparing to apply two frames of the shield. The ground now open in the front is remarkably hard; it consists of pebbles imbedded in a chalky substance, with hard loose stones of the nature of the Kentish rag. Everything is going on well. Devised with Isambard how to make our wells for the descent of the materials, &c.

Thus at last the shaft was completed, and Sir Isambard was able to commence the Tunnel itself, which he ultimately determined to construct in the form of a rectangular mass of brickwork, 37½ feet wide and 22 feet high, pierced by two parallel horseshoe archways, each 14 feet wide and 17 feet high.

Before entering upon the history of this undertaking, some account must be given of the machine which Sir Isambard Brunel devised for effecting its accomplishment.

In order to avoid a quicksand of considerable depth and extent, the Tunnel had to be carried but a short distance below the bed of the river; and, as in all tunnelling through soft soil,[11] the top and sides of the excavation had to be supported until the brickwork was built in; and the front or face had also to be held up as the miners advanced. This support was given by means of a machine called 'the shield,' described on one occasion by Sir Isambard as 'an ambulating cofferdam, travelling horizontally.'[12]

The main body of the shield consisted of twelve independent structures or 'frames' made of cast and wrought iron. They were each 22 feet high, and rather more than 3 feet wide; and, when placed side by side, like books on a shelf, against the face of the excavation, they occupied the whole area of the face, and also the top, bottom, and sides for 9 feet in advance of the brickwork. Each frame stood on two feet resting on the ground, and was divided in its height into three cells by cast-iron floors. In these cells, of which there were thus thirty-six in all, the miners stood, and worked at the ground in front of them.

The duty which the shield had to perform was to support the ground until the brickwork was built within the excavation; but it was essential that this should be done in such a manner as to allow of the mining operations being carried on; and it was also necessary that the machine itself should be capable of being moved forward.

The first point, therefore, which has to be explained in the action of the shield is the manner in which the earth was supported by it.

It has been already stated that each frame rested upon two feet, or large iron plates. These two feet together covered the ground under the frame to which they belonged, and thus the whole of the earth beneath the frames was pressed down by the feet.

The earth above was supported by narrow iron plates, called staves, laid on the heads of the frames parallel to the line of the Tunnel, the ends resting on the completed brickwork behind it. The earth at the sides was kept up by staves resting against the outermost frames.

The arrangement for holding up the earth at the face of the excavation was necessarily of a more complicated character. Each frame supported a series of boards called poling-boards, by means of small screw-jacks or poling-screws, two to each poling-board, which abutted against the frames, and pressed the boards against the earth. The boards were 3 feet long, 6 inches wide, and 3 inches thick, and were arranged horizontally. These poling-boards, more than five hundred in number, covered the whole surface in front of the frames.

To resist the backward thrust of the poling-screws against the frames, each frame was held forward by two large screws, one at the top of the frame, and the other at the bottom, abutting against the brickwork of the Tunnel. The brickwork was completed close up behind the shield as it advanced.

The way in which the earth was excavated, and the shield moved forward, has now to be explained.

The plates or staves which supported the ground at the top and sides of the shield were pushed forward separately by screw-jacks; but in order to advance the poling-boards in front, it was necessary that that portion of the ground against which they pressed should be removed.

The miner, standing in his cell, took down one, or, at the most, two of the poling-boards, commencing at the top of the cell, and having excavated the earth a few inches in advance, replaced the poling-boards against the newly-formed face, pressing them against it with the poling-screws. Thus the excavation was carried on without depriving the ground of the support it received from the shield, except at the point where the miner was actually at work.

The operation of advancing the frames was effected in the following way. When everything was ready for a move, one of the feet which carried the frames on jointed legs was lifted up, and advanced forward a few inches, and then pressed down on to the ground, until in its new position it again bore the weight of the frame. This done, the other foot was lifted, moved forward, and screwed down in the same manner, and then the frame itself was pushed ahead by means of the large abutting screws, which kept it top and bottom from being forced back on the brickwork.

It is, however, evident that these abutting screws would have been unable to push on the frame, while the ground in front was pressing back the poling-boards against it; therefore, during the process of moving a frame, it had to be relieved from the thrust of its poling-screws. Accordingly, when it was desired to advance any one of the frames, the butts of the poling-screws of the tier of boards in front of it were shifted sideways, so as to rest, not against the frame to which they belonged, but against the frames next it on either side. This done, the frame itself was advanced, and was then ready to receive again its own poling-screws, and also those belonging to the adjoining frames, so that they might in their turn be moved forward. It will thus be seen that the whole shield was not moved forward at one time, but that the frames were advanced alternately.

There were many other ingenious arrangements in the design of the shield, which need not be referred to in a description intended only to give such a general idea of the machine as may make the history of its operation intelligible.[13]

When the frames had been completed in Messrs. Maudslay's factory, they were conveyed to Rotherhithe, and lowered down the shaft. An opening had been left at the bottom of the wall, about 37 feet wide by 22 feet high, and against this the shield was erected. It was then ready to commence its progress through the ground below the river.

On November 25, 1825, the shield made its first start. Sir Isambard was unfortunately unable to witness what was in fact the actual commencement of the Tunnel, as he had three days before been seized with a sudden and alarming attack of illness, which kept him at home till December 6. The works were left under the direction of Mr. Brunel.

*December* 8.—The great shield is advancing very slowly, meeting with much interruption by the water, which still runs within the cells, and also by the difficulty of forming abutments for the frames. [*Temporary abutments were necessary until the shield was sufficiently advanced to allow of its being pushed forwards from the brickwork built up behind it.*]

*December* 29.—The frames are very much out of level in the transverse line of the Tunnel. This would be attended with serious inconvenience if I had not provided for the means of recovering any irregularity that might take place, and which, as it appears, cannot perhaps be prevented; but having foreseen this, I have provided the remedy by being able to take down the top of each frame, and to remove the top staves in parts, or the whole, at pleasure—a very important provision it proves to be.

*January* 16, 1826.—Too much precaution cannot be taken, in the management of the frames, to have the leg-screw particularly well secured, as every foot-run of the arch of the Tunnel sustains 82 tons. Each frame carries as follows:—

| The two end frames each | 65 | tons | = | 130 |
|---|---|---|---|---|
| Ten others, each | 52½ | " | = | 525 |
| | | | | 655 tons |

*January* 21.—The ground at the top and sides very good; same in the front. In breaking the ground out of the limits of the shield on the right a great deal of ground fell in. This indicates that, if it was not for the protection of the shield, nothing could be done. This accounts also for the occasional breaking of the ground in making the drift in 1809.

The brickwork of the entrance being carried as far as directed, the body of the Tunnel was begun to-day.

*January* 26.—Isambard went down to Rotherhithe; the water had broken in in great abundance upon the work over Nos. 4 and 5. [*The twelve frames were distinguished by numbers.*] A 4-inch pipe was driven over the shield from inside the shaft, but the water did not follow it, and the stream augmented very rapidly. The frame No. 5 was moved forward, and it checked the water for a moment, but it came again with violence. A heading was immediately ordered by Armstrong [*the resident engineer*] from the east well, in which Isambard concurred.

*February* 3.—Ordered a pit to be opened and made by sinking a curb 8 feet diameter and 18 inches thick, well bound with bolts. [*This pit was a well sunk from the surface to enable the gravel containing water, into which the head of the shield had penetrated, to be removed, and clay substituted.*]

*February* 6.—The shaft begun last night, and was sunk 20 feet to-day, and remarkably true. Had we known the ground as we now know it, we might, by having opened a well contiguous to the great shaft, have sunk the shaft in a week; but for that purpose we must have had two steam engines, one for pumping the water, and the other for taking up the ground.

*February* 10.—Went very early to the Tunnel for the purpose of giving directions to prop up the back of the staves, which, for want of weight at the new shaft, might be overbalanced by the pressure of the ground at the back. I could not rest a moment until it was done, for the consequences might have been fatal, at this moment in particular. What incessant attention and anxiety! To be at the mercy of ignorance and carelessness! No work like this.

*February* 12.—The ground having been opened carefully from under the curb of the pit [*see above on date February* 3], the greater part of the gravel was removed, and stiff clay substituted for it. This was done by driving first some wrought-iron flat bars, which kept the ground up. This shows that the shield is a most powerful protection, and would enable us to penetrate through a bed of gravel. Though the breaking in of the water had somewhat terrified the man in No. 5, he soon returned to his post, and the others have acknowledged their full confidence in the security afforded by the shield. The boring ahead had not yet been attended to: it is owing to the want of this precaution that this accident is chiefly to be ascribed; for had we known as much as we now do, we might have passed through without the pit being opened.

*March* 1.—Water at the back of the frames, but less than before. The men show a great deal of spirit in overcoming the present difficulties. Isambard was very busy yesterday and to-day in the frames, and about the works. He was severely hurt in the leg by a piece of timber falling against it. [*This accident prevented his attending again at the Tunnel until the* 24*th*]

On March 6 the proprietors paid a visit to the Tunnel, and were highly satisfied with what they saw. On that evening Sir Isambard writes:—

> It is of absolute necessity now to provide for everything that is conducive to the more expeditious management of the frames, and to a greater facility in getting up the brickwork. If these two points are realized, then indeed we may soon expect to be moving at a good rate—not less than I have held out, namely, 3 feet per 24 hours.

*March* 11.—Received early in the morning a report from Armstrong stating that the water was completely stopped—that it had been stopped during the night. Aware that we had passed the gravel, it was of course expected that we were under the clay; means were therefore resorted to, to drive clay and oakum at the tail of the top staves, which was productive of a very good effect. The great shield was soon entirely free of water. This shows the efficiency of the shield to oppose difficulties which could not have been overcome without the complete protection it affords, under almost any circumstances. Indeed this has been a tedious operation since January 25, when the water first burst upon No. 5, at the front of the shield. The miners as well as the bricklayers have worked with great spirit and perseverance through the whole, during a period of 44 days. The well that was made at the front of the shaft has been of use in acquainting us with the extent of the open ground we had to pass through. It will be made a useful opening for ventilating the works. By means of this well we have been able to apply the lead pipes with which the water has been diverted: it is not therefore a useless expense. Things were put in better order to prepare for a more expeditious way of working. Directions were given to place the frames in a better condition. Isambard is still too unwell to go to the Tunnel.

*March* 25.—Went to the Tunnel with Isambard. Found that a considerable fall of ground had taken place again at the right side. No one could account satisfactorily for it. I inspected it, and directed that, after making it good, flat bars of iron be driven at the head of the side staves, in order to pin it up, and in order to enable the miners to get at the solid ground. It is very bad and extremely dangerous; the ground is evidently the same as that which, in the report of the first attempt, was found so loose as to have dropped upon the works, leaving a large cavity above, when it is said the man ascended and made good the hole. We should be warned by this, lest we should meet another as fatal as it ultimately was on that occasion. [*This observation refers to the driftway of* 1807.]

*April* 24.—By Armstrong's report the water is entirely out, and the men at work in the morning in removing the dirt, &c. Isambard engaged at the Tunnel, where I am not yet able to attend as often as I could wish. Everything goes on well, much through his exertions.

*May* 11.—One hundred feet will be completed to-night.

*May* 22.—The top plate over the frame No. 1 has been cracked without any particular violence or stress. It appears that it is nothing but the change of temperature that is the cause of that rupture. The accident justifies the opinion I have of cast-iron not being safe upon traction, and the precaution of having had wrought iron bolts at the back of the frames. [*These were vertical rods which took the tensional strain.*] Without these bolts what would have become of the shield, if one casting was to break? The fracture was accompanied with a loud report like that of a gun. Isambard was in the works at the time of breaking: nothing could have prevented it.

*May* 25.—I observed that nothing whatever had been gained to recover the deviation [*the shield had gradually worked 2 feet 3 inches to the westward*], which subjects us to so much inconvenience and loss of time. The only way to bring the shield right is by taking the frames sideways.

*June* 3.—Finding it too laborious and almost fruitless to bring the frames in the right way, I came to the determination of having them brought bodily to the east by cutting the ground on the side. I accordingly gave directions to Armstrong to proceed in making a heading out of No. 12, and by securing the side staves to continue downwards until the ground be clear. The working was accordingly discontinued in front.

*June* 4.—The mode of proceeding by the common way of mining shows the impracticability of carrying a large excavation anywhere, particularly under a considerable body of water. The expense of timbering would be too great, even if it could be sound. The ground above the frames is remarkably good, but under it there is a stratum of silt which breaks and falls in large masses.

*June* 5.—Isambard got into the drift, and gave the line for the better disposition of the staves, which was afterwards done in a proper manner. Isambard's vigilance and constant attendance were of great benefit. He is in every respect a most useful coadjutor in this undertaking.

*June* 10.—The last frame (No. 1) is brought close to the others, and the brickwork brought up to fill the back.

*June* 15.—On inspecting the face of the ground this morning I observed a breach in the front of Nos. 3, 4, and 5, where the ground has given way in the lower cell. This was truly alarming. I ordered iron staves to each floor in order to pin the ground, and thereby to counteract the slipping which would immediately take place.

*June* 19.—The bricklayers left off work, but, on enquiring into the cause, I learned there was no other but to have a libation upon the new arrangement of piece-work.

*June* 29.—Gave positive directions to cut only 4½ to 5 inches thick at a time at the front of the top cells, instead of 9 inches, as they had done for some time.

*July* 3.—The great question is, does the clay undulate at its surface? We should have some reason to apprehend that it does so, because at the beginning we had not proceeded many feet into the clay when we struck again into the gravelly stratum. The surface of the clay must therefore have sunk at that particular spot; which circumstance seems to warn one of the need of great vigilance and great prudence in the progress of the enterprise.

*July* 10.—A cofferdam burst yesterday at the works at Woolwich, having blown up from the foundation. How cautious this should make our men! The cofferdam may be repaired, and very easily too, but an irruption into the Tunnel—what a difference, particularly at this early period!

*August* 10.—Found the lowest cell of No. 1 left by the workmen *without a single poling against the ground.* This is indeed a most unjustifiable neglect.

*August* 12.—At six this morning completed 205 feet.

*August* 21.—This piece-work has not been productive of much effect as to quantity of work. As to quality it is very questionable. A work of this nature should not be hurried in this manner. Fewer hands, enough to produce 9 feet per week, would be far better than the mode now pursued *from necessity,* but not from inclination on my part. Great risks are in our way, and we increase them by the manner the excavation is carried on. The frames are in a very bad condition.

*September* 5.—It is much to be regretted that such a work as the Tunnel should be carried on by the piece. Obliged to drive on, no time is left to make any repair, or to recover any lost advantage. Isambard is most active. Mr. Beamish shows much judgment in his exertions, and zeal in his attendance.[14]

*September* 8.—About 2 P.M. I was informed by Munday that water was running down over No. 9. I went immediately to it. The ground being open, and consequently unsupported, it soon became soft, and settled on the back of the staves, moving down in a stream of diluted silt, which is the most dangerous substance we have to contend with. So oakum was forced through the joints of the staves, and the water was partly checked. Isambard was the whole night, till three, in the frames. At three I relieved him. He went to rest for about a couple of hours; I took some rest on the stage.

*September* 9.—Towards noon the stream changed its character. The clay, being loosened by the water, began to run, but it thickened gradually. It was late in the evening before the loosened clay acquired the consistency of a loose puddling, which covered the staves, and made them a complete shield against further irruption, or rather, oozings of mud. If we consider that at this place we have at the utmost 9 feet between the top staves and the gravel, over which the river flows, it is most satisfactory and most encouraging to have this additional proof of the protection which the shield affords. At nine o'clock at night Isambard sent me word that 'tout va à merveille;' indeed it was so, for it was like a stopper interposed between the river and the top-staves. Instructed as the men were by the first accident, they went on as usual in the irrespective occupations. Pascoe, junr., and Collins were remarkably active and persevering, and some other men equally so; while old Greenwell encouraged them by a speech of his own in high commendation of the security of their situation.

*September* 12.—The water, bringing with it a sort of clay broken in small particles, increased to an alarming degree. In consequence of this continued displacement of the silt and clay, a cavity had been formed above the staves. At about three, when I had gone to the Court of Directors, the ground fell upon the staves with great violence, causing a surge most alarming as to probable consequences. Isambard was at that moment in the upper frames, and he gave directions for increasing the means of security. On my return I found things much worse than I had left them, but every means of security was judiciously applied. During the night in particular things presented a very unfavourable appearance. The men, however, were as calm as if there were no other danger to be dreaded than wet clothes or the splashing of mud. I observed the men in the lower cells were *sound asleep.*

*September* 13.—Every means were resorted to in the course of the night and during the early part of the day to stop the water. The men have shown great zeal and good management in their respective avocations, and above all the utmost confidence. Isambard has not quitted the frames but to lay down now and then on the stage. I have prevailed on him to go to his bed, or rather, used my endeavours to induce him; but he has not since last Friday night (the 8th). Things were rather better at the close of the day.

*September* 14.—Things upon the whole have assumed a more favourable aspect. The situation is nevertheless very critical. Nothing but the utmost precaution in following up what has begun can bring us out of it. This has been a most eventful week!

*September* 18.—Isambard was the greater part of the night in the works, and the benefit of his exertions is indeed most highly felt: no one has stood out like him! Everything is quite safe, the water is kept back, and the work proceeds in a most satisfactory manner.

*October* 22.—It is evident [*from a flow of silt which had taken place on that day*] that with the shield we have passed close under a body of collected water a few inches only above the staves. Isambard is too unwell to stay long in the works.

*October* 24.—The want of the main drain which was originally intended to carry the water to the main reservoir is felt everywhere. This drain is in my original plan, but the committee expressing on several occasions a wish that I should dispense with it, I complied, most reluctantly however, to prove my earnest wish to reduce the expenses. It will not, I apprehend, be found an economy.

*October* 26.—Every step we take shows how much security is derived from the shoes, supporting as they do the shield and the superincumbent weight. They press down in the same proportion the ground on which they bear. They keep it as dense as it originally was, and fit it for the structure that is to come upon it. It is evident, therefore, that what is wanted is that the ground should be kept pressed. It is with this object in view that I have holdfasts and jacks. What incessant vigilance is required, what an incessant call on the resources of the mind, not only to direct, but more particularly to provide for many things that may occur.

*November* 17.—At this date 307 feet 9 inches had been completed.[15]

*December* 8.—The evil [*of not having a proper drain*] is going on with us, and without any remedy except the drain, or a cesspool by way of expedient. How much anxiety must one feel at being so circumstanced! Should any water break in, how should we proceed? This is another source of great solicitude. We have no command of the frames when they rest upon wooden legs, or when the screws are bent; and what is worse is that the men drive on without any consideration or any fear of consequences. This circumstance, and the apprehension of the water breaking in, are matters of the most dreadful anxiety.

*December* 12.—Little do others know of the anxiety and fatigue I have to undergo day and night. Advanced as we are, we have only gained somewhat more experience, but the casualties are just the same. An accident now might be as fatal as it would have been 200 feet back, or as it would be 200 feet forwards. We have not a period when we can think ourselves safe except when we have connected these arches with a shaft on the other side. Loaded as we are with the weight of the river, we have to advance our shield and build our structure under that weight, a weight which varies twice a day, and twice a month to a much greater extent. *The shoes are the great foundation of our security.* When once pressed down with the greatest power that can be applied, they do not give in the least afterwards. They have not yielded even upon loose gravel; we must therefore congratulate ourselves that they have answered so completely. We have now walked our frames upwards of 350 feet; we have had to renew the legs and the heads, but it is not through want of strength so much as from mismanagement.

The first legs were never injured so long as their action was limited to 3 inches, but when it was increased to 6 inches, they immediately gave way one after another, without however any damage to the structure or to the shield. The heads gave way, or began to give way, from the moment the legs did; because, when a leg gave way, it brought upon the contiguous frame an increased weight which broke the heads one after another. That the breaking of so important a part of our shield should not have been attended with any bad consequences is a proof that provision had been made for the casualty. The proof that it had been foreseen and provided for is in the manner these heads were adapted to the frames. By the way they were fixed they would be easily taken down and replaced. Though the heads gave way, the top staves were not materially affected by it, and the service continued until new heads were substituted. Some have fancied that the ground did not bear wholly upon the shield and the arches, but supported itself in parts. Experience proves that the pressure is rather more than that which rests artificially on the frames. The ground is compressed all round by the increased weight of high water: we might therefore conclude that the shield operates as a pillar, that supports beyond the limits of its base or cap. It is a great satisfaction to be able to say that so long as we followed the original plan, nothing gave way except the back screws. These again were damaged by being run out of the sockets. We may therefore ascribe most of the evils and damages to the increased range of action, and still more to the rude implements the men have used, whenever they met with any difficulty in moving the frames. If it is considered that we had no other men to train in the use of this immense machine but excavators and miners, very great allowance must he made for what has occurred.

*December* 20.—An accident of an alarming nature occurred. The poling-screws of Nos. 10 and 12, being on No. 11, Moul, the miner in that frame, removed his butting screw; the consequence was that the frame started back, the polings and poling-screws fell down with a tremendous crash, and the ground followed to a considerable extent. This is the most formidable accident that has yet occurred in the face of the work. The ground was fortunately unusually firm, and no fatal consequences ensued.

*December* 31.—Isambard and nine friends sat down to a dinner *under the Thames!* Now a year is over since we began to make any progress horizontally, for we had only 11 feet of arch when the water broke in on January 24 last. We may therefore say that the whole of what has been made of the Tunnel has been made in that period. It is worthy of remark that until the end of April no fracture whatever, no bending of the legs, had taken place, notwithstanding that we had supported for a period of nearly three months a greater weight than we ever had since. The ground nearly 40 feet high *kept sinking upon us* as we advanced, and yet no stave, no top, no leg gave way. Each leg was capable of carrying nearly 80 tons at the point of fracture, consequently the aggregate strength of 36 legs was equal to 2 tons, which is six times over the greatest effort that could be exerted by the superincumbent weight. The heads, after they had given way, remained in place, some—namely, Nos. 1, 8, 12—for seven months, and the others from four to six months. It cannot be said therefore that there was a want of strength, since the broken heads continued to perform for so many months after being so much damaged; nor is there any defect in the iron. If the frames were, as some have fancied, lanky, which implies weakness in their sides, how could they have supported the alternate stress to which they are put by standing alternately on one leg? Not one single joint has yet started. Every frame has been upwards of 2,000 times in that raking posture, consequently the shield has been upwards of 24,000 times strained under the weight that has broken the heads. One single side has broken, and is now as good as the rest. Is such a machine to be stigmatised as it has been, without looking more minutely into its operations?

*January* 4, 1827[16]—A work that requires such close attention, so much ingenuity, and carried on day and night by the rudest hands possible—what anxiety, what fatigue, both of mind and body! Every morning I say, Another day of danger over!

*January* 12.—It is astonishing how the silt resists the sliding of the top staves. Assured as we were of having stiff clay from 33 to 37 feet, with what confidence we might have looked to making 18 feet per week. There would be no difficulty in having accomplished it. We must not look back, but overcome all difficulties!

*January* 16.—Isambard having been up several successive nights, went to bed at ten, and slept till six the next morning. I am very much concerned at his being so unmindful of his health. He may pay dear for it.

*February* 2.—Work done to this day 405 feet 4 inches.

*February* 3.—I visited the works; and, being in the cabin, I complained of the dust there. *Dust under the Thames!*

*February* 26.—I went to the Tunnel. The arch being well lighted up, and the whole walk completed, a few visitors were admitted. The coup d'œil was splendid. Mrs. Brunel, Emma, Sophia and her three little children were the first. It gave me great pleasure to see the whole of my family in the new scene.

*March* 21.—There being no clay above us, there is much to apprehend from the springs. It would be much better to work slower than we do. It is indeed very hard to be under the necessity of driving. Anxiety increasing daily.

*March* 28.—The top pumps failed; the water rose above the abutting screws. The frames of some of them could not be advanced, nor could the bottom brickwork be laid down—great source of complaint. Isambard called the men in at 10 o'clock; they went on cheerfully. It is surprising that the men are so steady.

*March* 29.—Things are getting worse every day by the influx of water; by which the ground is softened, and the operation rendered extremely complicated and slow. As to the ground, it is evident we are now as Isambard found it by his borings of August last. We have nothing above our heads but clayey silt, and it is of a nature to be detached and run into mud by the action of water.

*April* 3.—The pumping now requires forty hands. There is no exaggeration in saying that the influx of water, and the badness of the ground, cause an extra expense of £150 a week.

*April* 7.—It may now be said that we are contending with the elements above and around, gaining and disputing every inch that we add to the structure.

*April* 9.—Isambard's birthday, he being of age to-day.

*April* 14.—Doing as well as can be expected from the nature of the ground, and the difficulties that increase upon us.

*April* 18.—The faces are found extremely tender; but having proceeded with great caution, no accident occurred. None, I feel confident, would occur if all idea of piece-work were abandoned. It always operates as a stimulant, a very dangerous one. Obliged to drive on, on account of expense, we run imminent risks indeed for it. That a work of this nature, under such circumstances, should be thus carried on, is truly lamentable. It is obvious that the clay we have above our heads has been broken, by the ground beneath it running or breaking in upon us. We shall have to fight it out until we have a stronger or thicker stratum of clay. Sad prospect indeed it is for us!

*April* 20.—The ground at No. 1 broke in again, and occasioned great delay. Some bones and china came down.

*April* 22.—The diving-bell being on the spot, and Isambard having moored it over the shield, he and Gravatt[17] descended at thirty feet water. They found the same substances which had come through the ground into the Tunnel. When Isambard was in the bell, he drove a strong rod into the ground. Nelson, who was in the frame, heard the blows.

*April* 29.—Ground improving as we advance; we are not, however, free from danger: a dreadful alarm took place this morning. While Isambard and Gravatt were at breakfast, the porter came running in, and exclaiming, 'It is all over! The Tunnel has fallen in, and one man only has escaped.' Gravatt was the first to get to the spot, and found all the pumpers upon the floor of the shaft, all stupefied with horror, though every one was there quite safe, and no rush of water was heard. Gravatt and Isambard were soon in the shield, where they observed that a small portion of clay had fallen from the top on the top floor.

*May* 8.—At half-past three in the morning, an irruption took place, bringing down the deposits of the bottom of the river—lumps of clay, stones, bones, wood, nails, &c., &c., with water. The pumpers and men on the stage (Irish) all ran away, some exclaiming, 'The Thames is in! The Thames is in!' Ball and Rogers stood to their post, and soon stopped this most formidable attack.

*May* 10.—Great difficulties present themselves, that oppose our progress; the chief, however, is the lodgment of water above our heads. There it loosens the silt or sand, and runs out, leaving cavities that cause the clay above to break, and run down in lumps and disturbed streams. This is very awful! This opens the way for the river.

*May* 12.—In moving No. 6, they left by some unaccountable neglect the top staves behind, and in that state two top polings were taken down. The ground being very bad, and high water at the same moment, the ground began swelling. Attention was called to several points, and Gravatt continued in No. 6. He drew out at the front of the top staves a shovel, and also a hammer, that had come through the ground above. They are the same which Isambard left at the bottom of the river, when he went down in the diving-bell.

*May* 13.—Notwithstanding every prudence on our part, a disaster may still occur. *May it not be when the arch is full of visitors!* It is too awful to think of it. I have done my part by recommending to the directors to shut the Tunnel. My solicitude is not lessened for that: I have indeed no rest, and I may say have had none for many weeks. So far the shield has triumphed over immense obstacles, *and it will carry the Tunnel through.* During the preceding night the whole of the ground over our heads must have been in movement, and that too at high water. The shield must have therefore supported upwards of *six hundred tons*: it has walked for many weeks with that weight twice a day over its head. What flippancy and inconsistency in some individuals, who, without any knowledge of the subject, without so much as examining the state of the work, will without the least reflection and hesitation obtrude their suggestions upon every case. What shallow conceit for such to pretend they can know better than those that have already the experience that must result from years of deep thought, from days and nights of incessant attention; who have the advantage of the combined talents of several ingenious men, who devote their undivided study, the whole resource of their well-stored minds, to the enterprise; and to add to this, the benefit of the skill of one hundred miners and excavators. Among this class of men, some have been employed in the most perilous enterprises, when each individual must have acted upon his sole judgment, where, in fact, there is no room for an engineer to instruct and direct their efforts. How easy it is to attack everything, to detract from the merits of the best plan. There is always some weak point which may be open to the penetration of the shallowest mind. Then comes the exulting expression, *That I always said would never do,* &c., and all the consequences with it.

*May* 15.—The water increased very much at 9 o'clock. This is *inquiétant*! My apprehensions are not groundless. I apprehend nothing, however, as to the safety of the men, but first the visitors, and next a total invasion by the river. We must be prepared for the worst. I have had no rest for many weeks on this apprehension. Should it occur we must make the best of it, by improving our situation.

*May* 17.—There is no doubt of the ground having improved very materially since last Saturday. Very cheering indeed.

*May* 18.—Visited by Lady Raffles and a numerous party. Having had an intimation by Mr. Beamish of their intended visit, I waited to receive and to accompany them, not only from the interest I felt at being acquainted with Lady Raffles, but also from motives of solicitude, knowing that she intended to visit the frames. Indeed, my apprehensions were increasing daily. I had given some instructions for enquiring where we could obtain clay, that we should have some barges full of clay to be in readiness. I was most anxiously waiting for the removal of the tier of colliers that was over us, being convinced that we should detect some derangement then. I attended Lady Raffles and party to the frames, most uneasy all the while, as if I had a presentiment, not so much of the approaching catastrophe to the extent it has occurred, but of what might result from the misbehaviour of some of the men, as was the case when the Irish labourers ran away from the pumps and the stage. I left the works at half-past five, leaving everything comparatively well: Mr. Beamish continued on duty.[18]

Mr. Gravatt's account is as follows: I was above with I. Brunel looking over some prints, Beamish being on duty. Some men came running up and said to Isambard something I did not hear. He immediately ran towards the works, and down the men's staircase. I ran towards it, but could not get down. I leaped over the fence, and rushed down the visitors' stairs, and met the men coming up, and a lady, who I think was fainting. Met Flyn on the landing-place, who said it was all over. I pushed on, calling him a coward, and got down as far as the visitors' barrier. Saw Mr. Beamish pulled from it. He came on towards the shaft walking. I went up to him to ask him what was the matter. He said it was no use resisting. The miners were all upon the staircase; Brunel and I called to them to come back. Lane[19] was upon the stairs, and he said it was of no use to call the men back. We stayed some time below on the stairs, looking where the water was coming in most magnificently. We could still see the farthest light in the west arch. The water came upon us so slowly that I walked backwards speaking to Brunel several times. Presently I saw the water pouring in from the east to the west arch through the cross arches. I then ran and got up the stairs with Brunel and Beamish, who were then five or six steps up. It was then we heard a tremendous burst.

The cabin had burst and all the lights went out at once. There was a noise at the staircase, and presently the water carried away the lower flight of stairs. Brunel looked towards the men, who were lining the staircase and galleries of the shaft, gazing at the spectacle, and said, 'Carry on, carry on, as fast as you can!' Upon which they ascended pretty fast. I went up to the top and saw the shaft filling. I looked about and saw a man in the water like a rat. He got hold of a bar, but I afterwards saw he was quite spent. I was looking about how to get down, when I saw Brunel descending by a rope to his assistance. I got hold of one of the iron ties, and slid down into the water hand over hand with a small rope, and tried to make it fast round his middle, whilst Brunel was doing the same. Having done it he called out, 'Haul up.' The man was hauled up. I swam about to see where to land. The shaft was full of casks. Brunel had been swimming too.

The first alarm, as I heard it, was as follows: Goodwin, in No, 11, said to Roger, in the next box, 'Roger, come, help me.' Roger said, 'I can't, I have my second poling down, and my face will run in.' In a little time Goodwin said, 'You must come,' which Mr. Beamish directed him to do. Roger turned round and saw Goodwin through a sheet of water. Corps, a bricklayer, went to help Goodwin: he was knocked down. Roger made his way alone, calling to Mr. Beamish, 'Come away, sir, 'tis no use to stay.' Roger saw Corps fairly washed out of his box like a lump of clay.

Sir Isambard's journal continues:—

*May* 19.—Relieved as I have found myself, though by a terrible catastrophe, of the worst state of anxiety, that which I have been in for several weeks past, I had a most comfortable night. Isambard and Gravatt descended with the diving-bell, and stood upon the tails of Nos. 10, 11, and 12.

*May* 20.—Having descended into the hole and probed the ground, I felt that the staves were in their places, and that the brickwork was quite sound. It is evident that the great hole has been a dredging spot. A large mass of bags full of clay, and united together with ropes, was let down. The Rotherhithe curate, in his sermon to-day, adverting to the accident, said it was a fatal accident, that it was but just judgment upon the presumptuous aspirations of mortal men, &c.! The poor man!

*May* 23.—Went with the diving-bell to examine the ground and the bags, which do apparently well, but it is working rather in the dark. It cannot, however, fail of making a much better stratum than that we had before. The plan is therefore good.

On the 30th a raft was sunk over the shield, and the water in the shaft was brought so low that the last flight of steps was visible. However, on the next day the river broke in again; and as it was found that the raft was open at the west side, it was raised and towed on shore.

*June* 5.—There is much danger in getting out of the diving-bell, the bags are so loose in some places. One might sink and be swallowed, which had very nearly happened to-day. Isambard and Pinckney being down, the latter lost his hold. The footboard being accidentally carried away, he could not have recovered himself had not Isambard stretched out his leg to his assistance.

*June* 17.—Visited by Charles Bonaparte. Isambard took him into the arch with the yawl. Isambard fell overboard.[20]

On June 19, a general meeting of the proprietors was held, to consider the position of the company. Sir Isambard addressed the meeting, and also presented a long report, in which he entered very fully into the circumstances of the recent accident and the causes which led to it. He then described the means he had taken to restore the works by sinking bags of clay and gravel. He adds: 'I have already succeeded in closing the hole through which the water first penetrated, and feel confident that the second opening which afterwards appeared is also stopped, but a short time is necessary to elapse for the new ground over the shield to settle and consolidate. It has already supported a head of water of thirty-five feet.'

*June* 25.—At 7 P.M. made preparations to re-enter the shield. Isambard, mustering the men who had been the last to quit the frames, told them they would be the first to take possession of them again—a precedence due, as he said, to them. Rogers, Ball, Goodwin, Corps, and Compton, were accordingly ordered to trim themselves for the expedition, provided with a phosphorus box, and dressed in light clothes, to be fit for a swim.

At about ten o'clock, Isambard and Mr. Beamish, accompanied by Ball and Woodward (miners), went down with the punt, and got to the large stage, the head of the crane just emerging. It was found impossible to get into the frames, as a mound of clay and silt closed the entrance. The centering was in place and quite sound, and of course the brickwork. Finding that they could not get nearer, they gave three cheers, which were rapturously answered by the men at the mouth of the Tunnel. Having placed candles upon the ground that closed the entrance, and upon the head of the crane, they returned. Isambard, having promised that the men who had left the frames last should be the first to re-enter, returned with them. This is a great day for our history!

*June* 27.—Mr. Beamish was able to get to the frames, which he found firm and undisturbed.

A small tarpaulin was now spread over the frames, and operations commenced for cleaning them. This was a most difficult and dangerous work, especially as the water was still so high that the frames could only be approached by boats. The men, even the best hands, were at first greatly alarmed at the danger they were in; but the example set by Mr. Brunel and Mr. Beamish produced, as Sir Isambard notes, the best effect, and they soon became reconciled to their situation.

*July* 7.—Very uncomfortable in the frames; the candles cannot burn, the ventilation cannot act. Isambard went several times to-day down in the diving-bell. On one occasion the chain slipped through the stoppers, but most providentially it jammed itself tight before being altogether run out. *The consequence might indeed have been fatal.* Can there be a more anxious situation than that which I am constantly in? Not one moment of rest either of mind or body. Mr. Beamish always ready. Poor Isambard always at his post too, alternately below, or in the barges, and in the diving-bell.

On July 11, Sir Isambard thought that matters had so far advanced that a large tarpaulin, which it was proposed to sink over the frames, 'would have its full effect.' It was accordingly sunk on the following day, under the superintendence of Mr. Brunel. Sir Isambard adds to his account of the operation—'This reflects great credit on Isambard, and the apparent facility with which it was effected evinces his presence of mind, for a single *faux pas* would have spoilt the whole.'[21]

*July* 21—During the early part of the night an alarm was given, by Fitzgerald calling for clay wedges, and exclaiming that the whole of the faces were coming in altogether. Rogers collected a quantity of wedges to go to the frames, but no boat was to be seen. He called to the men in the frames, but received no answer.

Taking the small boat in the east arch, he reached the frames, but found nobody, nor any appearance of derangement in the ground. Conjecturing they might be drowned, he explored further, and saw the four men stretched on the small stage, not drowned, but sound asleep!

*July* 26.—Water nearly out of the arches. For the first time we could walk to the frames—a most gratifying circumstance indeed! *Two months and eight days.*

*September* 30.—How slow our progress must appear to others; but it is not so, if it is considered how much we have had to do in righting the frames and in repairing them; what with timbering, shoring, shipping and refitting—all these operations being in confined situations, the water bursting in occasionally, and the ground running in: in short, it is truly terrific to be in the midst of this scene. If to this we add the actual danger, magnified by the re-echoing of the pumps, and sometimes (still more awful warning!) the report of large pieces of cast iron breaking, it is in no way an exaggeration to say that such has been the state of things. Nevertheless, my confidence in the shield is not only undiminished—it is, on the contrary, tried with its full effect, and it is manifest now that it will soon replace us in good ground, and in a safe situation. No top staves have given way. That is our real protection.

*October* 17.—At 2.15 A.M. Kemble, having first called upon Gravatt, came to Isambard in a hurry, and, quite stupefied with fright, told him that the water was in. Says Isambard—'I could not believe him. He said it was up the shaft when he came. This being like positive, I ran without a coat as fast as possible, giving a double knock at Gravatt's door in my way. I saw the men on the top, and heard them calling earnestly to those whom they fancied had not had time to escape. Nay, Miles had already, in his zeal for the aid of others, thrown a long rope, and was swinging it about, calling to the unfortunate sufferers to lay hold of it, encouraging and cheering those who might not find it, to swim to one of the landings. I immediately, I should say instantly, flew down the stairs. The shaft was completely dark. I expected at every step to splash into the water. Before I was aware of the distance I had run, I reached the frames in the east arch, and met there Pamphillon, who told me that nothing was the matter, but a small run in No. 1 top, where I found Huggins and the *corps d'élite.* They were not even aware that any one had left the frames. The cause of the panic was one of the labourers; hearing the man in No. 1 call for Ball, he ran away, jumping off the stage, crying, "*Run, run, murder, murder; put the lights out.*" His fellow-labourers followed like sheep, making the same vociferations.'

*November* 10.—Isambard gave his entertainment to nearly forty persons, who sat at table in the Tunnel. Nothing could exceed the effect for brilliancy. About 120 men partook of a dinner in the adjoining arch.

As the year drew to a close, the difficulty of working the silt increased, and with this difficulty increased also the expense of maintaining the staff of men required. On December 18, Mr. Brunel, writing for his father, who was absent from town for a few days, thus describes the nature of the soil through which they were then passing.

> The state of the ground over Nos. 1, 2, and 3 top has caused considerable delay, particularly this week, although not such as to give any cause of anxiety as to our future rate of progress, or to have any serious effect except the increased expense incidental to this delay. My father desired me to describe to the Board the causes of these difficulties. There is a considerable spring at this point, and a corresponding soft part in the bed of the river, which seems to indicate the rising of the spring. The ground in the neighbourhood is affected by this spring in rather a peculiar manner: at the half-flood tide the pressure is greatest: dry hard clay oozes with great force through openings hardly observable, the silt and water running by starts. At high-water the pressure and quantity of water begin to diminish and on the ebb-tide the ground is hard and dry, and can be worked with ease. On the flood-tide there are as many as twelve and fifteen of the best hands, besides myself (or one of my assistants) and the foreman, engaged entirely at one face.

On January 1, 1828, Sir Isambard returned to London; and on the 12th, when about 600 feet of the Tunnel had been completed, a second irruption occurred, which put a stop to the works for seven years.

The particulars of this accident are thus described by Mr. Brunel, in a letter to the Directors of the Company:—

> I had been in the frames (shield) with the workmen throughout the whole night, having taken my station there at ten o'clock. During the workings through the night, no symptoms of insecurity appeared. At six o'clock this morning (the usual time for shifting the men) a fresh set or shift of the men came on to work. We began to work the ground at the west top corner of the frame: the tide had just then begun to flow; and finding the ground tolerably quiet, we proceeded by beginning at the top, and had worked about a foot downwards, when on exposing the next six inches, the ground swelled suddenly, and a large quantity burst through the opening thus made. This was followed instantly by a large body of water. The rush was so violent as to force the man on the spot, where the burst took place, out of the frame (or cell) on to the timber stage behind the frames. I was in the frame with the man, but upon the rush of the water I went into the next box (or cell), in order to command a better view of the irruption, and seeing that there was no possibility of then opposing the water, I ordered all the men in the frames to retire. All were retiring, except the three men who were with me, and they retreated with me. I did not leave the stage until those three were down the ladder of the frames, when they and I proceeded about twenty feet along the west arch of the

Tunnel. At this moment the agitation of the air, by the rush of water, was such as to extinguish all the lights, and the water had gained the height of our waists. I was at that moment giving directions to the three men, in what manner they ought to proceed in the dark to effect their escape, when they and I were knocked down, and covered with a part of the timber stage. I struggled under water for some time, and at length extricated myself from the stage, and by swimming and being forced by the water, I gained the eastern arch where I got a better footing, and was enabled by laying hold of the railway rope, to pause a little, in the hope of encouraging the men who had been knocked down at the same time with myself. This I endeavoured to do by calling to them. Before I reached the shaft the water had risen so rapidly that I was out of my depth, and therefore swam to the visitors' stairs, the stairs for the workmen being occupied by those who had so far escaped. My knee was so injured by the timber stage that I could scarcely swim, or get up the stairs, but the rush of the water carried me up the shaft. The three men who had been knocked down with me were unable to extricate themselves, and I am grieved to say, they are lost; and I believe also two old men, and one young man, in other parts of the work.

This statement Sir Isambard embodied in a report to the Directors of January 28, which was circulated among the proprietors.

As soon as the first excitement caused by the irruption had ceased, Mr. Brunel directed the diving-bell to be prepared in order to ascertain the state of the shield and the extent of the disturbance of the bed of the river caused by the rush of water into the Tunnel.

He was, however, so seriously injured that he could not actively superintend the preparations, but his orders were given with his usual clearness, calmness, and decision; and as soon as the barge containing the diving-bell was properly moored over the Tunnel, he was carried out and laid upon a mattress on the deck of the barge, that he might direct what was to be done.

As evening came on he became so much worse that he was taken into the cabin; but everything which took place was reported to him.

At length, the bell being ready, it was lowered early on the Sunday morning, but the chain not being long enough, proceedings were delayed until a longer chain could be obtained.

As, however, a chain of the right size and length could not be obtained, the strongest cable which could be procured in the neighbourhood was substituted for the chain. A controversy then arose between the assistant engineers and the foremen as to the sufficiency of the strength of the cable; and it was agreed to consult and to abide by the opinion of Mr. Brunel, who was then lying in great pain in the cabin.

No answer could be obtained from him for some minutes, and then he only said, 'Don't go down.' This not being satisfactory to the advocates of the sufficiency of the cable, it was agreed to lower the bell empty, which was done, and it was brought up safely; but just as it was swung over the barge, the rope broke and the bell fell on to the stage.

The next day Mr. Brunel was taken home, when it was found that, besides the injury to his knee which he received while endeavouring to save the lives of the three men who were with him,[22] he had received serious internal injuries, which kept him under medical treatment for several months.

When he was able to return to Rotherhithe all hope of continuing the works was for the time abandoned. When they were resumed, in 1835, he was entirely engrossed in the independent pursuit of his profession; and, with the exception of a few occasions when he acted for his father, he had no further connection with the Tunnel.

It is not, therefore, necessary to continue the narrative in detail; but a brief summary of the subsequent history of the enterprise may be interesting to those who are unacquainted with it.

The Tunnel was cleared of water, and efforts were made, unfortunately without success, to raise funds for the completion of the undertaking. Great enthusiasm was exhibited by the general public and by many eminent persons, including the Duke of Wellington; but the money was not forthcoming, and nothing was left but to brick in the shield, and wait for more favourable times.

It was not till the beginning of 1835 that the Company was able, by the aid of a loan from Government, to recommence the works. The old shield was removed and a new one substituted, in which considerable improvements were introduced. Slings connecting the frames were added, which enabled each frame to support its neighbours when necessary, and important alterations were also made in the arrangements for keeping the frames at the right distance from one another, and for giving greater facility of adjustment to the various parts.

Before the Wapping side was reached there were three more irruptions of the river, namely, August 23, November 3, 1837, and March 21, 1838; but in October 1840 the shaft on the Wapping shore was commenced. It differed from the Rotherhithe shaft, in being sunk the whole depth without underpinning, and was made of a slightly conical form, to reduce the friction in sinking, and had a larger quantity of iron hoops introduced into the brickwork, in order to increase its strength. When this structure had been sunk to the required depth (70 feet), the excavation of the Tunnel was resumed, and at last the shield was brought up to the brickwork of the shaft. The operation of making the junction between the Tunnel and the shaft was one of much difficulty, but it was at length satisfactorily accomplished, and the Tunnel was opened to the public on March 25, 1843— eighteen years and twenty-three days after the commencement of the work.[23]

Sir Isambard Brunel, whose health had for some time been failing, now retired altogether from his professional labours. After passing a few years in peaceful and happy seclusion; surrounded by those he loved, and watched over by their affectionate care, he died on December 12, 1849, in his 81st year, having been spared to carry to completion his greatest work, and to see his son following in his footsteps with a success which must have exceeded his most sanguine expectations.

The education Mr. Brunel received from his father was well calculated to form the foundation of his future career. During the later and more arduous part of the contest, which was ended by the irruption of January 1828, he held both the nominal and actual post of Resident Engineer of the Tunnel; but from the commencement of the works, when he was only nineteen years old, he had been, as stated by Sir Isambard, 'a most valuable coadjutor in the undertaking.' While placed in this responsible position he acquired habits of endurance and of self-reliance, and learnt to act with promptitude and decision in the application of those measures which experience had shown to be effective in each particular class of emergency. But beyond all other advantages, he had before him the example of his father's character, in which a rare degree of gentleness and modesty of disposition was joined to unflinching energy, and a determination to overcome all difficulties.

## NOTE A
### *The Bourbon Suspension Bridges*[24]

The suspension bridges designed by Sir Isambard Brunel for crossing rivers in the Ile de Bourbon were two in number. One of them had two spans of 122 feet each in the clear, and 131 feet 9 inches between the points of suspension of the chains. The second had but one span of the same dimensions as those of the larger bridge. In the design of these bridges one of the most important points to be attended to, was to render them secure against hurricanes, which are both frequent and severe in the Ile de Bourbon.

In the larger bridge there was a pier of masonry, built in the middle of the river up to the level of the roadway of the bridge. The suspension chains of the bridge were in three groups, 9 feet 8 inches apart, so as to leave room for two roadways, each about 8 feet 9 inches wide. Each of these groups of chains consisted of two chains side by side. Each chain was made with long links like those of the chain cables used for moorings.

These links, which were made of iron 1.36 inch in diameter, were 4 feet 8 inches long, inside measure, and were each connected together by two short coupling links, 8¾ inches long, inside measure, of iron 1.36 inch by 1 inch, and two pins, each 2 inches in diameter.

The two chains of each group were placed side by side, with the links upright; one of the pins at each joint was made long enough to serve for both chains, and, in the middle of its length between the two chains, was passed through an eye at the upper end of one of the suspending rods of the bridge. Thus to every joint in each group of the main chains, or at intervals of about 5 feet, there was a suspending rod. These rods were 1¼ inch in diameter.

The pins of the joints of the main chains had half heads at each end of them. They could thus be easily inserted in erecting the bridge, but once in place were quite secure. At every fourth joint in the main chains one of the joint pins was made in two halves, with wedges inserted between them for adjusting the length of the main chains.

Thus there were six chains, and as the links of these had each two parts of iron 1.36 inch in diameter, the total sectional area of the six chains was 17.4 inches.

Each group of the main chains was supported at a height of 25 feet 6 inches above the roadway at the centre pier, and at a height of 5 feet 3 inches at each of the side piers, the lowest portion of the curve of the chain being about 1 foot below the points of suspension of the side piers.

The upright standards, carrying the chains both at the centre pier and at the side piers, consisted for each group of chains of a triangular framework of cast iron, strengthened by long bolts of wrought iron. There were thus three of these triangular frames parallel to each other at each of the piers, and those at the centre pier were braced together over the carriage road. The main chains were not bolted to the standards, but were slung from them by a vertical suspension link, which thus allowed them to move a little lengthways. This link, in fact, performed the function of the rollers now generally put under the saddles of suspension bridges.

The ends of the main chains were held by back stays, formed of bars 3 inches broad by 1¼ inch thick, and 10 feet long, with joints made with short links, and 2⅜ inch pins. The ends of those back stays, were secured to holding down plates 3 feet in diameter, sunk deep in the ground and well loaded.

As there was a vertical suspension rod at each joint of the main chains, there was a suspension rod hanging from each of the three groups of chains at about every five feet of the length of the bridge. To each set of these rods was attached a cross girder of cast iron of a T section, with a large rounded bead at the lower edge of the upright web; and connecting these under each of the main chains was a longitudinal timber beam about 8 inches square.

The cast-iron cross girders carried longitudinal teak planking, the planks on which the carriage wheels ran being 12 inches wide and 4 inches thick, protected at the top by wrought-iron plates running longitudinally. The horse-path was protected by iron plates arranged crosswise.

Under each span of the bridge were four chains curved upwards and also sideways. These chains were fastened at their ends into the piers, and were connected to the roadway by ties drawn up tight and attached to the main longitudinal bearers of the platform; the object being to stiffen the platform.

These under tie chains were made each of a set of rods 1¼ inch in diameter with eyes at their ends, the ends being connected by short joint links and 1¼ inch pins; and to these joint links were attached the tie rods which connect these inverted chains with the platform of the bridge, and so prevented its being lifted or blown sideways by the force of the wind.

In the smaller bridge, which, as has been said, consisted of one span of 131 feet 9 inches between the points of suspension, these points were 15 feet 5 inches above the roadway, and the lowest part of the chain was 9 feet 7 inches below the points of suspension. The details of this bridge were similar to those of the larger one.


## NOTE B
### *Experiments with Carbonic Acid Gas*

In 1823 Mr. Faraday made the important discovery that under certain conditions of temperature and pressure many gases could be liquefied, and that these liquids exerted great expansive force by slight additions of temperature, returning quickly with regularity and certainty to their original state upon the application of cold.

The discovery of this new force appeared of such importance, that Mr. Faraday lost no time in publishing it to the world; and Sir Isambard Brunel very soon afterwards commenced a series of experiments to determine the value of the liquid gas as a mechanical agent.

The first experiments were made at Chelsea; but the prosecution of them was soon transferred to the care of Mr. Brunel at Rotherhithe, where he devoted all his spare time to the construction of his father's proposed 'Differential Power Engine.'

That the progress of this discovery, and of the experiments made with a view to the application of the liquid gas, as a motive power, may be understood, it is necessary to state that in March 1823 Mr. Faraday communicated to the Royal Society the results of his first experiments on the liquefaction of gases.

The fluid was then produced by the decomposition of the hydrate of chlorine by heat in a closed tube, the amount of gas evolved being so great as to produce a pressure in the tube sufficient to condense the gas into a fluid of the same volume.

This interesting experiment was followed by others with that rapidity and success so remarkable in everything undertaken at that time in the laboratory of the Royal Institution; and within a month another paper was read before the Royal Society, in which the degrees of pressure and temperature at which several gases could be liquefied were recorded, and the means employed to produce and liquefy each gas accurately described.

On April 17 a third paper was communicated by Mr. Faraday, 'On the Application of Liquids produced by the Condensation of Gases as Mechanical Agents.'

The question is thus stated: 'The ratio of the elastic force dependent upon pressure is to be combined with that of the expansive force dependent on temperature; and the development of latent heat on compression and the necessity of its reabsorption in expansion must awaken doubts as to the economical results to be obtained by employing the steam of water under very great pressures and very elevated temperatures.

'No such doubt can arise respecting liquids, which require for their existence even a compression equal to thirty or forty atmospheres, and where slight elevations of temperature are sufficient to produce an immense elastic force, and where the principal question arising is whether the effort of mechanical motion is to be most easily produced by an increase or diminution of heat by artificial means.'

Difficulties were suggested by Mr. Faraday as to the possibility of obtaining sufficient strength in the apparatus, but the small difference of temperature required to produce an elastic force of many atmospheres, he considered would render the risk of explosion small.

To construct the machinery whereby this new force could be practically applied as a substitute for steam, occupied the time of Sir Isambard Brunel and his son at intervals for several years; for although Mr. Brunel was satisfied at an early period of the enquiry that the liquefied gases could only be advantageously employed where the cost of motive force was secondary to economy of space and to the avoidance of the cumbrous apparatus required for the use of steam, still he was so impressed with the importance of the subject, if the difficulties he foresaw in its application could be overcome, that he continued his experiments for a long period with unflagging energy and perseverance.

The facts relating to the liquefaction of the gases, their elastic force when liquefied under different temperatures, the rapidity with which they could be alternately expanded and condensed, and the best mode of producing each gas, were determined by Mr. Faraday; and as Mr. Brunel was at that time attending the morning chemical lectures at the Royal Institution, he was in constant communication with him, and thoroughly conversant with his experiments.

After Mr. Brunel had made a few preliminary experiments, Sir Isambard determined to employ liquefied carbonic acid gas for the motive power of the proposed new engine, the facility and cheapness of its production, its great expansive force, and its neutral character distinguishing it from any other gas; but it was long before vessels were constructed, in which gas could be produced in sufficient quantity and purity to exert the force required to liquefy it in its own volume, for it was soon found to be impossible to obtain the required pressure with pumps.

Carbonate of ammonia and sulphuric acid were the elements used, and the generator was so arranged that it could be charged, emptied of atmospheric air, and the joints made perfect, before the commencement of the formation of the gas which was to be liquefied.

To the generator was attached a receiver, which could be surrounded with a freezing mixture, so that the temperature of the gas in the cylinder might be below that in the generator.

The gradual formation of the liquid, the development of its elastic force, and the regularity and rapidity with which it increased or diminished by each degree of heat or cold, were carefully watched through a glass gauge, and the receiver when filled with liquid could be disconnected from the generator.

The mechanical difficulties as they arose, one after the other, in the construction and arrangement of the various parts of the generator and receiver were at length overcome; and the receiver was not only filled with liquid gas, but found to be capable of retaining it, whether exerting an elastic force of 30 atmospheres at ordinary temperatures, or of 100 atmospheres when subjected to a slight degree of heat.

The receiver being satisfactorily completed, the next object of attention was the design and construction of a working cylinder capable of resisting at least 1,400 lbs. pressure on the square inch; a task which was one of great anxiety, as any weakness might have caused a serious accident.

It was only after the trial of every known method of making joints to resist high pressures had failed, that an arrangement was devised, requiring the most perfect workmanship, by which packing of any kind was dispensed with, and the cylinder fitted for use.

With the improved tools of the present day it is not easy to realise the difficulties, delays, and disappointments which forty-five years ago occurred from the failure, first of one part of a joint, and then of another; but the construction of vessels capable of producing and also of retaining the gas in its liquid state, with the means of alternately expanding and condensing it from thirty or forty to eighty or one hundred atmospheres, having been accomplished, the object of the expenditure of so much labour and inventive power appeared to be within reach.

The construction of the machinery to utilise the elastic force contained in the cylinder was now proceeded with. Day by day new difficulties arose, and each as it was successfully met seemed but to leave another of greater importance to be surmounted.

It is not necessary in this Note to describe the various arrangements which were devised for transferring the great elastic force in the cylinder of small diameter to a piston in another cylinder of much larger dimensions; it is sufficient to say, that after the devotion of much valuable time extending over several years, and a very large expenditure of money, and after carefully considering the cost of the liquid carbonic acid gas, the difficulty of preventing waste, and the necessarily very expensive character of the machinery, Mr. Brunel was satisfied 'that no sufficient advantage in the sense of economy of fuel can be obtained by the application of liquefied carbonic acid gas as a motive power'; but so thoroughly did he exhaust the subject before he committed himself to this opinion, that no one has since renewed the enquiry or attempted to make a machine to be moved by the elastic force of liquefied gases, the construction of which, it was well known, had baffled the inventive genius of Sir Isambard Brunel and his son.

1.  To avoid confusion, Sir Isambard Brunel has been called throughout by that designation, the one by which he is generally known: he was knighted on March 24, 1841.

    His Life has been written by Mr. Richard Beamish, F.R.S. (London, 1862.)

2.  Lady Brunel survived her husband five years. Of their children, three lived to maturity, one son, Isambard Kingdom, and two daughters, Sophia, wife of the late Sir Benjamin Hawes, K.C.B., Under Secretary of State for War, and Emma, wife of the Rev. George Harrison, Rector of Sutcombe.

3.  He was sent to Paris to recover his knowledge of French, which had got rather rusty at school, and also to study mathematics. He retained through life a great admiration of the method of teaching this subject which was adopted in France.

    In addition to the time spent in the study of mathematics and languages, Mr. Brunel occupied himself on his holidays in examining the various engineering works going on in Paris, and he used to send his father drawings and descriptions of them.

4.  Sir Isambard was also consulted upon a proposed suspension bridge over the Tamar at Saltash, where Mr. Brunel subsequently built the Royal Albert Bridge.

5.  This history has been written by Mr. Beamish in his *Life of Sir Isambard Brunel*, pp.202–304, and also, up to the year 1828, in the very valuable work by Mr. Henry Law, C.E., entitled 'A Memoir of the several Operations, and the Construction, of the Thames Tunnel,' and published by the late Mr. Weale in his *Quarterly Papers on Engineering*.

6.  For an account of these earlier attempts see Law, pp.3-7

7.  This expectation does not seem to have been realised, as there was never any considerable traffic through the Thames Tunnel. Perhaps, however, it would have been otherwise had the large descents for carriages and horses been constructed.

8.  The results obtained by these borings were no doubt fallacious, but not to the extent which has sometimes been imagined. At a meeting of the Institution of Civil Engineers, in November 1849, Dean Buckland called attention to 'the evils arising from the ignorance of the engineers who reported to Sir Isambard Brunel, previous to the commencement of the Thames Tunnel, that the whole of the bottom of the river at that spot was London clay.' Whereupon Mr. Brunel rose and said, that he 'agreed that knowledge of every kind was most desirable, and that it would be well if engineers were generally much better informed on many subjects which would be useful, and more particularly on matters connected with geology; at the same time he could not admit that they were deficient in that knowledge of the surface of the earth which was necessary for the purpose of guiding them in their work. It might be true that many members of the profession were, like himself, not perfectly well acquainted with the minute geological characteristics of the soils they had to deal with, but he thought the education and the practical experience of the profession generally rendered them well acquainted with those features and characteristics which were necessary for their guidance in the design or execution of work. He must also say a few words in defence of those persons (now nearly all dead) who

made the borings in the Thames, and were stated to have made so fallacious a report previous to the commencement of the Tunnel. Now, although that statement had by constant repetition become a sort of historical fact, it was really only one of those popular fallacies which obtained too ready credence in the world. The position of the Tunnel was not determined by any report, or by the result of any borings, but with a view to establishing a communication between particular localities for encouraging the traffic which was anticipated from the facility of access to the docks, and for other local reasons, such as the general direction of the roads and streets on both shores. After the position was settled, and not until then, borings were made to ascertain what soils might be expected in that part of the river. It must be remembered that these borings were made full twenty-five years ago, when boring in the bed of a river through a depth of water of nearly thirty feet was not an ordinary occurrence. The tool then generally employed was the worm, and tubes were not even used in such cases. The borings showed the existence in that spot of something which, in the of ordinary acceptation of the term, might have been inadvertently called London clay, but he had no recollection of its geological designation having ever been thought of. It was reported and shown to be a very fair clay for working in . . . The errors which were made in giving the results of the borings did not, in fact, arise from ignorance, but from mechanical defects in the tools, for it was subsequently discovered that the worm frequently carried a portion of the upper tenacious clay through the softer strata beneath, and brought it up again. The tenacious clay might have been called London clay, but no value was attached to that particular designation; they cared little in engineering for its denomination, provided it was of a good tenacious quality. This mistake in terms (supposing it to have occurred) could not have had any influence on after proceedings; for, before the Tunnel was far advanced, he conducted with great care a series of borings extending across the Thames, and, as he used improved tools and worked through tubes, the holes were kept so dry that a candle was frequently lowered down to the bottom in order to see the amount of infiltration. By this means he was enabled to construct a correct section of the bed of the Thames at that spot, showing every layer of shells and gravel as well as every variation of the surface of the silt, &c. He entered more at length into these details than might perhaps appear necessary, because he felt it was incumbent upon those who had the conduct of works to show that they did not proceed so ignorantly or so recklessly as had been assumed, in the design or execution of large undertakings.'

9.   The paragraphs in small type, without any reference, are from Sir Isambard's journals.

10.  The shaft subsequently made on the Wapping shore was sunk to its full depth without any under-pinning.

11.  Professor Rankine, in his work on *Civil Engineering*, p.599, describes the Thames Tunnel works under the significant heading 'Tunnelling in Mud'.

12.  *Proceedings Inst. C.E. i.* 34. The circumstances which led Sir Isambard to conceive the idea of a shield, and the earlier designs he made for it, are described, with illustrations, by Mr. Law, pp.7–10.

13. Mr. Law's memoir contains a detailed description of every part of the shield, illustrated by careful drawings.

14. Mr. Beamish had joined the works on August 7.

15. On November 20 Mr. Brunel mentions in his diary that he had 'passed seven days out of the last ten in the Tunnel. For nine days on an average 20⅓ hours per day in the Tunnel and 3⅔ to sleep.'

16. On the previous day, Mr. Brunel had been formally appointed resident engineer.

17. Mr. Gravatt had been appointed an assistant engineer six months before.

18. Sir Isambard's journal of this eventful night consists—as he was not himself present—of Mr. Beamish's journal, with a few words in warm commendation of that Gentleman's 'judgment, coolness, and courage,' followed by observations upon the stability of the shield. He then gives a statement made by Mr. Gravatt, and taken down in shorthand. No extracts are given in the text from Mr. Beamish's narrative, as he has already inserted it in a condensed form in his *Life of Sir Isambard Brunel,* pp.244–248.

19. Mr. Michael Lane, at this time foreman bricklayer, became one of Mr. Brunel's most valued assistants, and was employed by him on the Monkwearmouth Docks and the Great Western Railway. After filling various posts in the service of that company, he was in 1860 appointed their principal engineer; an office which he held till his death, in February, 1868.

20. On this occasion an amusing incident occurred. Mr. Brunel was exceedingly unwilling to permit his visitors to make this expedition into the arch; but on the assurance that they could all swim perfectly well, he consented to take them, with the understanding that, if he jumped overboard, they were immediately to follow his example, and swim after him to the shaft. While they were in the arch Mr. Brunel (as Sir Isambard mentions) fell overboard. As soon as he recovered himself, and turned to swim back to the boat, he remembered that he had unwittingly given to his companions the signal to jump out into the water. He was much amused, on looking up, to see that they were not swimming after him, but were still sitting in the boat clinging to the gunwale, with faces expressive of blank despair.

21. Mr. Brunel's comment in his diary is as follows:—'Without ascribing any particular merit to myself, I cannot help observing, for my future guidance, that being alone, and giving few but clear orders, and those always to the men who were to execute them, I succeeded in an operation not altogether mean, and which a very trifling want of precaution or order might have caused to be a total failure.'

22. On January 15, 1828, the Directors of the Thames Tunnel Company passed the following resolution, which they ordered to be advertised in the *Times, New Times, Herald, Ledger and Courier*:—'That this court, having heard with great admiration of the intrepid courage and presence of mind displayed by Mr. Isambard Brunel, the company's resident engineer, when the Thames broke into the Tunnel on the morning of the 12th instant, are desirous to give their public testimony to his calm and energetic endeavours, and to that generous principle which induced him to put his own life in more imminent hazard to save the lives of the men under his immediate care.'

23. The Thames Tunnel was not successful as a commercial undertaking; but it has always been considered, especially by foreigners, one of the most interesting sights in London, and has been visited by several millions of persons. In 1865 it was purchased by the East London Railway Company, and trains now (March, 1870) run through it. The possibility of using the Tunnel as a railway had been considered in Mr. Brunel's lifetime, and the idea was approved of by him.

24. This description is based on the translation given by Mr. Drewry (*Suspension Bridges.* London 1832, p.75), from the *Memoire sur les Ponts Suspendus,* by M. Navier (Paris, 1823, p.49). M. Navier saw the bridges when they were erected at Sheffield in May 1823.

# II

# THE CLIFTON SUSPENSION BRIDGE

AFTER Mr. Brunel had recovered from his accident in the Thames Tunnel, he went for a trip to Plymouth, where he examined with great interest the Breakwater and other engineering works in the neighbourhood. He notes in his diary that he went to Saltash, and that he thought the river there 'much too wide to be worth having a bridge.' This remark was no doubt made in consequence of his father having some years before been consulted as to the construction of a suspension bridge at this place, which Mr. Brunel himself eighteen years afterwards, selected for the crossing of the Tamar by the Cornwall Railway, and built there the largest and most remarkable of his bridges.

For the remainder of the year 1828, and during the greater part of 1829, Mr. Brunel kept himself fully employed in scientific researches, and in intercourse with Mr. Babbage, Mr. Faraday, and other friends; but he was without any regular occupation, until, in the autumn of 1829, he heard that designs were required for a suspension bridge over the Avon at Bristol, and he determined to compete.

This project originated in a bequest made in 1753, by Alderman William Vick, of the sum of £1,000 to be placed in the hands of the Society of Merchant Venturers of Bristol, with directions that it should accumulate at compound interest until it reached £10,000, when it was to be expended in the erection of a stone bridge over the river Avon, from Clifton Down to Leigh Down. Alderman Vick stated that he had heard and believed that the building of such a bridge was practicable, and might be completed for less than £10,000.

The legacy was duly paid to the Society of Merchant Venturers, and invested by them. The interest accumulated; and in 1829, when the fund amounted to nearly £8,000, a committee was appointed to consider in what way it would be possible to carry out Alderman Vick's intentions.

An estimate for a stone bridge was procured, but as it gave the cost at £90,000, it was evident that this scheme must be abandoned.

The committee then advertised for designs for a suspension bridge. Mr. Brunel, on hearing through a friend of the proposed competition, went to Bristol; and, after examining the locality, he selected four different points within the limits prescribed by the instructions of the committee, and made a separate design for each of them. His plans were sent in on the day appointed, Nov. 19, 1829, with a long statement, from which the following description of them is taken.

The first design was for a bridge of 760 feet span between the points of suspension, the length of the suspended floor being 720 feet. In order to obtain a height of 215 feet above high-water mark (which was the least that the levels allowed of), towers 70 feet high would have had to be built on the cliffs to carry the chains. The total length of chain, including the land-ties, was about 1,620 feet. Mr. Brunel did not approve of this design, as the situation was not favourable to architectural effect, a point to which the committee attached great importance; but he suggested it from its being somewhat more economical in construction than his other plans.

In another design, the situation being some way further down the river than that of the design last mentioned, towers would also have been necessary. The distance between the points of suspension was 1,180 feet, with a suspended floor of over 900 feet. It is probable that Mr. Brunel only proposed this plan because the site came within the limits of deviation, as he does not say anything in favour of it in his report.

The two remaining plans are the most interesting of the series, as there can be no doubt that, if Mr. Brunel had had his own way, he would have adopted one of them for execution; and it appears from a little sketch on the top of one of his earliest letters from Bristol, that his first idea for the bridge was that which is carried out in these two designs. The site selected was one where the rocks rise perpendicularly for a considerable height above the proposed level of the bridge, and therefore piers and land-ties were dispensed with, the chains being hung directly from the rock. No masonry was required except for architectural effect.[1]

The principal difference between these two designs is that in the second a short tunnel is avoided at one end. The style of architecture selected for the tunnel-front and the face of the rock, as shown on the drawings sent in to the committee, is Norman. There are also extant many beautiful sketches made by Mr. Brunel for different parts of the design.[2]

In determining upon the mode of construction, which was the same in the four designs, Mr. Brunel acted upon the principle which guided him in all his subsequent undertakings which was, as he states in his report, 'to make use of all that has been found good in similar works, and to avail himself of the experience gained in them, and to combine with all their advantages the precautions which time and experience had pointed out.'

He dismissed in a few words the plan of breaking the span into two or three lengths. This was in his opinion unnecessary, and he computed that the cost of a pier built up from the water's edge to sufficient height above the bridge to carry the chains, would be at least £10,000. For this reason he recommended the adoption of spans, the smallest of which far exceeded any up to that time constructed.

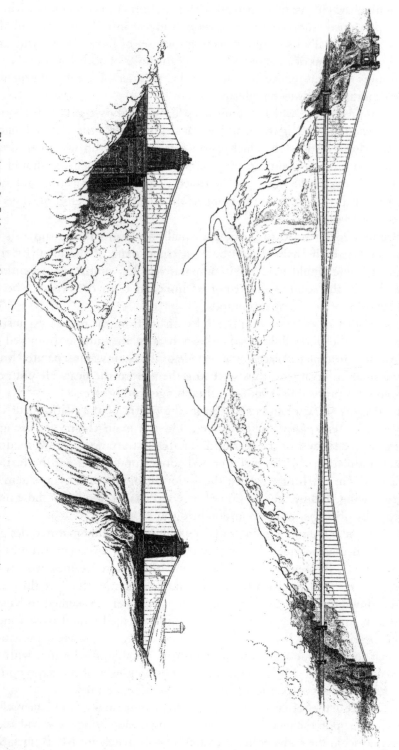

CLIFTON SUSPENSION BRIDGE

Elevation of Drawing N° 3 of Mr Brunel's Designs in the first competition. AD 1829.

Elevation of the Bridge according to the Design on which the works were commenced. AD 1836.

In designing the chains, he dispensed with the short connecting links, which had been previously adopted in suspension bridges, introducing instead the method now universally used, of connecting each set of links directly with the adjoining one by means of a pin passed through the holes of both. The number of joints and pins was thus reduced one half, and a considerable saving of expense, as well as diminution of weight, effected.

Another improvement, which diminished still further the weight of the chains, was making the links in lengths of 16 feet, or nearly double that of the longest links at the Menai bridge. The chief reason for this alteration was to ensure a near approximation to equality in the strains on the different links, should all the distances between the holes not be exactly equal. This improvement was afterwards carried still further in the Hungerford Suspension Bridge, the links of which were 24 feet long.[3]

Mr. Brunel also intended to introduce equalising beams in the supports of the floor, so that each chain should bear an equal share of the load. By this arrangement, there would have been comparatively few points of suspension, and 'the view of the scenery would not be impeded from the observer being surrounded by a forest of suspension rods.'

The disturbance of the strains on the links arising from the greater expansion of the metal of the outer links by the direct heat of the sun, he proposed to obviate by sheet-iron plates placed on each side of the chains, but separated from them by a small interval, and thus screening them from the heat. He did not, however, use this protecting covering at the Hungerford bridge.

All the designs show a camber or rise in the centre of the platform of the bridge, to the extent of two or three feet; and the main chains are brought down almost to the level of the platform. To this last arrangement, as tending to prevent undulation, Mr. Brunel attached some importance; and he further intended to stiffen the bridge against the action of high winds by a system of transverse bracing, and by the addition of inverted chains, similar to those used with success by his father in the Bourbon bridges.[4]

Such, then, were the main features of the bold and carefully matured designs placed by Mr. Brunel before the committee. Out of twenty-two plans submitted, only those of Mr. Brunel and four other competitors were deemed worthy of consideration. He and his friends were naturally much gratified at this, and were full of hope for his ultimate victory. But now, when he seemed to have a fair chance of success in a contest which he justly deemed would have a most important bearing upon his future professional career, an obstacle presented itself, which for the time seemed almost insurmountable; for he met with an unexpected opponent in Mr. Telford, the foremost engineer of the day, and the designer of the famous suspension bridge over the Menai Straits.

The committee of the Society of Merchants had, not unnaturally, found themselves unable to decide upon the merits of designs for a suspension bridge, and had asked Mr. Telford to act as their adviser in the matter. Unfortunately for Mr. Brunel, Mr. Telford was of opinion that the maximum span admissible was that of the Menai bridge, *i.e.* under 600 feet, and that Mr. Brunel's proposed bridge, though very pretty and ingenious, would most certainly tumble down in a high wind.

This decision was, of course, fatal to the success of any design which substituted one large span for two or more smaller ones, and dispensed with pillars. Mr. Brunel therefore obtained permission to withdraw his plans from the competition.

Mr. Telford then reported to the committee that none of the remaining designs were suitable for adoption without the introduction of such material alterations as would in fact, constitute a new design. Whereupon the committee took the only course which, under the circumstances, was open to them, and requested Mr. Telford to prepare a design himself.

Mr. Brunel was not a little disappointed at the turn matters had taken; but, having, as he said, 'smoked away his anger,' he took leave of his friends at Bristol, and went for a visit to some of the principal manufacturing towns in the north.

Meanwhile Mr. Telford prepared his design, and it was exhibited in Bristol in January 1830. It consisted of a suspension bridge of three spans (the centre span 360 feet, and the side ones 180 feet each), the chains being supported at the intermediate points by tall stone piers rising from the river's banks at just sufficient distance apart to avoid interfering with the roadways on either side of the stream. The style of architecture was a florid Gothic; and, in order to display the peculiar features of that style, the faces of the piers were covered with elaborate panelling, and the chains ornamented with fret-work.

This design was received with a flourish of trumpets; numerous engravings were published, exhibiting the bridge from various points of view, and 'thousands of copies were disposed of;' but, after a time, it would appear that the captivating effect of the Gothic belfries wore off, and that the more the citizens of Bristol looked at Mr. Telford's plan, the less they were satisfied with it; for, although it was deposited in the Private Bill Office, on application being made for an Act of Parliament, the trustees who were appointed under the Act determined to invite a second competition.

On this occasion, Mr. Telford appeared as a competitor and not as a referee, that office being filled by Mr. Davies Gilbert, sometime President of the Royal Society.

The site of the bridge was fixed, being that selected by Mr. Telford; but the trustees expressly left it to the judgment of the competitors to decide whether there should be intermediate piers or one unbroken span.

Of the thirteen designs sent in, five, including those submitted by Mr. Telford and Mr. Brunel, were reserved for further examination. On March 17, 1831, Mr. Davies Gilbert (who had been assisted by Mr. Seward) made his report. Mr. Telford's design was put aside, 'on account of the inadequacy of the funds requisite for meeting the cost of such high and massive towers as were essential to the plan which that distinguished individual had proposed.'

Mr. Brunel's design was placed second.[5] Although Mr. Gilbert reported that it presented every desirable strength and security, he saw objections to many of the details, and therefore did not recommend it for adoption. However, on the following day, March 18, he stated to the trustees that he had seen Mr. Brunel, and that it gave him much pleasure to state that the explanations made by Mr. Brunel had materially altered his views as to the details of the plans, which he (Mr. Gilbert) was now satisfied were quite equal to those which he had placed first, and that, considering the superiority of Mr. Brunel's design in the essential particular of strength, he should judge it preferable to any of the others.

Thereupon the trustees, 'having considered Mr. Davies Gilbert's report, and referred to all the plans, including Mr. Telford's, unanimously gave the preference to Mr. Brunel's,' and appointed him their engineer.

Subscriptions came in but slowly, and it was not till 1836 that the works were commenced.

The first stone of the abutment on the Leigh woods or Somersetshire side of the river was laid on August 27 by the Marquis of Northampton, President of the British Association, which was then holding its meeting in Bristol.[6]

The span of the bridge is greater than that of Mr. Brunel's design for the second competition, but much less than the spans of the earlier designs, to which he had given the preference.[7] On this point, as well as on the question of site, he had to conform to the wishes of the trustees.[8] The span approved of by them necessitated the building of a very large abutment on the Leigh woods side, the height of which, from the surface of the rock to the level of the roadway, is 110 feet. Above the roadway, the tower to carry the chains is built to a height of 86 feet. On the Clifton side, the base of the tower is formed by one of the boldest of the range of St. Vincent's rocks, which here rise almost perpendicularly to a height of 230 feet above high water, and consequently a very small abutment was required. The tower on this side is 3 feet higher than that on the Leigh woods side, and the roadway has a general inclination of about 1 in 233. Mr. Brunel thought that if the roadway were level, it would have the appearance of falling towards Clifton, owing to the ground there being precipitous, while on the Leigh woods side it is sloping.

He intended, in the construction of the bridge, to have followed out the ideas embodied in his report of 1829, and would have preferred to have had only one chain on each side of the bridge, and that much stronger than was usually adopted; but, in deference to public opinion, he put two chains, though he doubted if they would expand equally. 'A rigid platform would in some degree prevent the unequal distribution of load thus caused, but he endeavoured to lessen the effect of unequal expansion by arranging a stirrup at the top of each suspending rod, so as to hold equally at all times on both chains, and thus to cause each to sustain its proportion of the load.'

The road platform was to have had beneath it 'a complete system of triangular bracing, which would render it very stiff.'

In order to lessen the action of wind on the bridge, he brought down the main chains in the centre nearly to the level of the platform, and intended to apply the system of brace chains at a small angle to check vibration. There were, moreover, to be two curved chains lying horizontally, and attached underneath the platform, so as to resist the lateral action of the wind.[9]

He here introduced movable saddles to carry the chains on the top of the towers, with rollers running on perfectly flat and horizontal roller beds.[10] By this arrangement no pressure except a vertical one could come on the towers.

He also devised means, by levers and hydraulic presses, for relieving the rollers and roller beds from pressure, in the event of their requiring renewal.

Mr. Brunel ultimately determined to adopt the Egyptian style of architecture. His brother-in-law, Mr. John Callcott Horsley, R.A., gives the following account of the proposed designs for the towers:—

> His conception of the towers or gateways at either end of the bridge was peculiarly grand and effective, as may be seen from his sketches still existing. They were to be purely Egyptian; and, in his design, he had caught the true spirit of the great remains at Philæ and Thebes. He intended to case the towers with cast iron, and, as in perfect accordance with the Egyptian character of his design, to decorate them with a series of figure subjects, illustrating the whole work of constructing the bridge, with the manufacture of the materials—beginning with quarrying the iron ore, and making the iron, and ending with a design representing the last piece of construction necessary for the bridge itself. The subjects would have been arranged in tiers (divided by simple lines) from top to bottom of the towers, and in the exact proportion of those found upon Egyptian buildings. He made very clever sketches for some of these proposed figure subjects, just to show what he intended by them. I remember a group of men carrying one of the links of the chainwork, which was excellent in character. He proposed that I should design the figure subjects, and he asked me to go down with him to Merthyr Tydvil, and make sketches of the iron processes. We accomplished our journey, and all the requisite drawings for the intended designs were made.

The works were commenced with the Leigh abutment, which was completed in 1840, great delay having been caused by the failure of the contractors. This misfortune led to a large excess of expenditure over the original estimates. In 1843 the whole of the funds raised (amounting to £45,000) were exhausted, and there still remained to be executed the ornamental additions to the piers (the cost of which was estimated at about £4,000), half of the iron work, the suspension of the chains and rods, the construction of the flooring, and the completion of the approaches, &c., the estimate for the execution of which was £30,000.

Unfortunately, all efforts to raise further subscriptions were unsuccessful; and in July 1853, when the time limited for the completion of the bridge had expired, the works were closed in, and the undertaking abandoned.[11]

Several proposals for completing the bridge were made in Mr. Brunel's lifetime, and he took every opportunity of furthering this object, which he had very much at heart. It was not, however, till about a year after his death that the superstructure of the bridge was actually commenced.

A company was formed in 1860 by some of the principal members of the Institution of Civil Engineers, 'who had an interest in the work as completing a monument to their late friend Brunel, and at the same time removing a slur from the engineering talent of the country.'[12] Mr. John Hawkshaw, F.R.S., and Mr. W.H. Barlow, F.R.S., were appointed the engineers, and Mr. Brunel's old friend Captain Christopher Claxton, R.N., the secretary. The works were carried on with vigour; and the bridge was opened with much ceremony on December 8, 1864.

The chains were brought from the Hungerford Suspension Bridge, then in process of demolition. A description of the Hungerford bridge will be found in the note to this chapter.[13]

Although the Clifton bridge was not completed by Mr. Brunel, his connection with it forms a very important passage in the history of his life. Doubtless, if he had never heard of the proposed competition in 1829, or if he had been one of the disappointed competitors, he would have found some other opportunity of making a name in his profession; but, as a matter of fact, the Clifton bridge competition did give him the opportunity he desired, and all his subsequent success was traced by him to this victory, which he fought hard for, and gained only by persevering struggles. He never forgot the debt he owed to Bristol, and to the friends who helped him there; and he would have greatly rejoiced to see the completion of his earliest and favourite work.


NOTE
*The Hungerford Suspension Bridge*

The suspension bridge which spanned the Thames at Charing Cross, on the site of the present railway bridge, was designed and constructed by Mr. Brunel between the years 1841 and 1845. It consisted of a centre span of 676 feet, and two side spans of 343 feet each. Being intended for foot passengers only, its width was 14 feet. The versed sine, or deflection of the middle of the catenary, was 50 feet. The two river piers, which still exist up to the level of the railway, and form piers of the present bridge, were of brickwork, with large footings at the bottom, so as to distribute the pressure over a considerable area. The whole structure was made hollow and as light as possible. From the level of the footway the piers were carried up as ornamental campanile towers, the weight of the chains being taken by four solid pillars of brickwork, 7 feet 3 inches square, forming the angles. Mr. Brunel introduced here many of the arrangements he had designed for the Clifton bridge. In order that the pressure from the chains might be always vertical on the piers, the saddles rested on rollers working in oil, on the level surface of a large cast-iron bed-plate. By this arrangement it was rendered possible for the chains of the land spans to leave the tower at a greater inclination than those of the

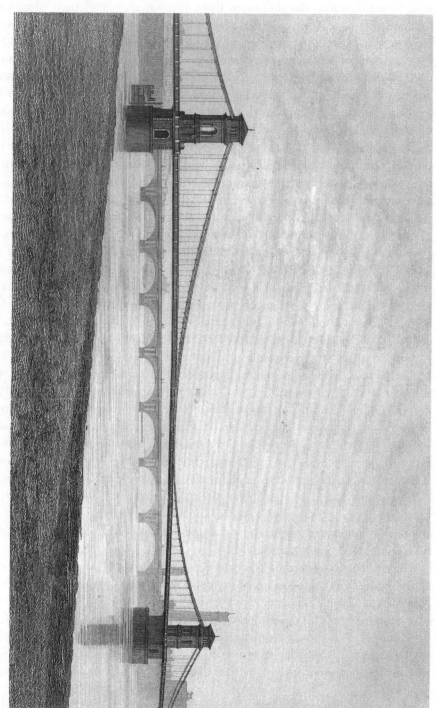

HUNGERFORD SUSPENSION BRIDGE

middle span, so that the chains were made shorter, and as they were at a lower level where they met the abutment, there was less change in their direction at that point, and consequently less thrust on the brickwork. Freedom of horizontal motion was also secured, so that, in the case of unequal loading of the spans, the chains might accommodate themselves to the strains, and move horizontally until equilibrium was restored. At each of the land abutments the chains passed down over a fixed saddle, at an inclination, to anchorages placed at the bottom of the abutment. The brickwork under the fixed saddle was so disposed as to resist directly the thrust resulting from the change of direction between the main chains and the anchor chains. To resist any movement of the abutments, the piles on which they rested were driven obliquely, with their heads inclined from the river. These piles were very numerous, the abutments spreading out so as to cover a large area at the foundations. Nearly all the spaces between the longitudinal, cross, and outside walls were filled with concrete, in order that the abutments might be as massive as possible. The details of the brickwork in the piers and abutments showed Mr. Brunel's skill in the economical employment of this material. The chains were constructed so that the sectional area was proportional to the strain; the total area at the centre was 296 square inches, while near the piers it was 312 square inches. There were four chains, two on each side of the bridge, placed one above the other, and consisting each alternately of ten and eleven links. The links were 24 feet long and 7 inches in depth, the thickness varying so as to give the requisite sectional area.

The relative diameter of pin, and proper form of the ends of the link, were subjects of much consideration, and many experiments were made in order to determine these points. The fact that two specimens of iron, apparently identical in every respect, sometimes exhibit considerable difference in their breaking weights, shows that an average of a great number of experiments is required in order to test satisfactorily any proposed refinements of construction. Mr. Brunel, however, convinced himself by experiment that he had practically arrived at such a form of link and diameter of pin that the chain would have no tendency to break at one point rather than another. The links were forged with shoulders near the eyes, in order that by means of clamps the pin could be taken out and the links disengaged, if necessary.

The efficient action of the rollers was demonstrated shortly after the completion of the bridge. On the occasion of the opening of the Corn Exchange by Prince Albert, one of the land spans was crowded with people, while the centre span was nearly empty. In consequence of this the land chains became depressed considerably below their normal position; and the saddles on the top of the tower nearest to the loaded span moved horizontally on the rollers to the extent of 3 inches; and, when the crowd had dispersed, they returned to their original position.

Many years after the completion of the bridge a proposal was made to widen it for carriage traffic; but this was not carried out, and eventually the superstructure was removed, to make way for the bridge of the Charing Cross Railway. As the

Hungerford Suspension Bridge has ceased to exist, an engraving has been given of it (p.55), in order that some record of its appearance may remain.

1.  The dimensions of these designs were as follows:—

    (a.) Length of floor                                890 feet
         Distance between points of suspension          980  "
         Length of chain                              1,300  "
         With a capacity to bear excessive load of      650 tons

    (b.) Length of floor                                916 feet
         Distance between points of suspension        1,160  "
         Length of chain                              1,468  "
         With capacity to bear excessive load of        650 tons

2.  On p.47 is given (fig. 1) a facsimile on a smaller scale of the drawing sent in by Mr. Brunel for the last-mentioned (b) of these two designs.

3.  See p.54

4.  See above, p.37.

5.  The dimensions proposed in this design were as follows:—

        Distance between points of suspension    600  feet
        Versed sine                                60   "
        Width of roadway                           32   "

6.  A few days before this ceremony, an iron bar, 1½ inch diameter, and about 1,000 feet in length, was hung across the valley from Clifton Rocks to Leigh Down, to facilitate the works. It was traversed by a basket pulled by ropes. The first few journeys of this machine were somewhat perilous. It was intended that Mr. and Mrs. Brunel should be the first passengers; but, when all was ready, one of Mr. Brunel's assistants started on a clandestine trial trip, and owing to a bend in the bar, the basket stuck half way, and the mast of a passing steamer caught in the rope. The rope was however cut, and he was drawn back. When the apparatus had been put to rights, on another occasion, when Mr. Brunel was in the basket, it got jammed, and he had to climb up the connecting link and get upon the bar, before he could release the basket.

7.  Span                          702 feet 3 inches
    Versed sine                    70   "
    Roadway above high-water      248   "

8.  The lower drawing in the illustration on p.49 shows an elevation of the bridge according to the designs on which it was commenced.

9.  See Mr. Brunel's remarks:—*Proceedings Inst. C.E.* for 1841, pp.78, 79

10. Rollers on an arched surface had been used previously in several bridges.

11. The chains were used in the construction of the Saltash bridge.

12. Speech of the Chairman, the late Captain Mark Huish, at the first general meeting, August 2, 1861.

13. Some re-arrangement of Mr. Brunel's design was rendered necessary in order to adapt the Hungerford bridge chains to the Clifton bridge, and there are three chains instead of two, as in Mr. Brunel's design. The platform is stiffened by wrought-iron girders instead of by timber trussing, and the whole bridge is stiffened transversely by the wrought-iron girders at the sides, which are connected throughout by diagonal bracing. The clear width of the bridge is 30 feet, 5 feet less than originally intended. It should be added, that no attempt has been made to complete the towers according to Mr. Brunel's architectural designs.

# III

# EARLY HISTORY OF THE GREAT WESTERN RAILWAY

BEFORE entering upon the history of the Great Western and the other railways of which Mr. Brunel was the engineer, it may be useful to give a brief sketch of the development of the railway system, previous to the period when he first became engaged in works of this description.

The first railway in England designed for the conveyance of general merchandise and passengers, was the Stockton and Darlington. An Act of Parliament authorising the construction of this line was passed in 1821.

In 1823, a further Act was obtained, in which a clause was inserted, at the request of Mr. George Stephenson, then the engineer of the company, taking power to work the railway by locomotive engines, and to employ them for the haulage of passengers. This railway, which consisted of a single line with four sidings in the mile, was opened for traffic on September 27, 1825. Its success led at once to the promotion of similar works in other parts of the country.

Next in order must be noticed the celebrated railway between Liverpool and Manchester. A project for constructing a line of railway between these important towns was discussed as early as the year 1822; but a company for carrying it out was not formed till two years later. In 1825, the directors applied to Parliament for an Act; and after a long contest before a committee of the House of Commons, the preamble approving of the construction of the railway was carried by a majority of one. The Bill was, however, withdrawn, as the first two clauses empowering the company to make the line, and to acquire land for that purpose, were lost.[1] In the following year the Act was obtained, and the works were commenced under the direction of Mr. George Stephenson. The line was opened for traffic on September 15, 1830.

In 1824, Mr. George Stephenson wrote a report on a proposed line connecting Liverpool and Birmingham. Surveys were made, and plans deposited; but the Bill was thrown out on standing orders. A similar fate attended the introduction of a Bill in 1826. In 1830, a new line was surveyed by Mr. Joseph Locke and Mr. Rastrick, under the direction of Mr. George Stephenson. The Act was obtained in 1833, and the railway, which was called the Grand Junction, and is now a part of the London and North-Western system, was constructed by Mr. Locke.[2]

In 1830, surveys were commenced by Mr. Robert Stephenson for a line between London and Birmingham, and a Bill was introduced into Parliament in 1832. The Liverpool and Manchester Railway had now been opened for some time, and the promoters of the Birmingham line had the advantage of being able to give in evidence the results of the working of the earlier undertaking. Those results, it is said, were such as to startle most of those who heard them. It was shown that a speed had been attained double that of the fastest stage-coach, that the cost of travelling had been diminished by one half, and that out of 700,000 persons carried since the opening of the railway, only one had met with a fatal accident. The amount of travelling between Liverpool and Manchester had increased four-fold, and the value of the shares of the railway had risen one hundred per cent. Similar evidence was given as to the results of the working of the Stockton and Darlington Railway, and the promoters endeavoured to prove that advantages at least as great would arise from the construction of a railway between Birmingham and London. They were successful in the House of Commons; but, they failed to convince the Upper House that the benefits which such a railway would confer on the country traversed by it were sufficient to entitle its promoters to receive for it the sanction of the legislature. The Bill was again introduced in the following session (1833); and, strange to relate, it passed both Houses almost without opposition.[3]

Meanwhile, the principal merchants of Bristol, who had in 1825 made an attempt to get up a railway company, were urged forward, both by the inadequacy of their communications with the metropolis, and by the success of the Liverpool and Manchester Railway, to make another effort. In the autumn of 1832 a committee was formed of members of the corporation, and other public bodies of the city of Bristol, to carry out the project of a railway to London.

The first step taken by the committee was the appointment of an engineer to make the preliminary surveys, and to prepare an estimate of the cost of the undertaking.

Among the candidates for the post was Mr. Brunel. He was well known in Bristol as the engineer of the Clifton Suspension Bridge, and of the works for the improvement of the Floating Harbour. He had made many friends among the leading citizens, and they used their best exertions to procure his election; but there were several other candidates in the field who had great local interest, and the contest was a close one.

While the issue was yet undecided, an unexpected difficulty arose. Some members of the committee resolved to select their engineer by means of a competition among the candidates, as to which of them would provide the lowest estimate. Upon this being announced, Mr. Brunel declared that he must withdraw his name, as he could not consent to become a party to so objectionable a proceeding. 'You are holding out,' he wrote to the committee, 'a premium to the man who will make you the most flattering promises. It is quite obvious that the man who has either least reputation at stake, or who has most to gain by temporary success, and least to lose by the consequences of disappointment, must be the winner in such a race.' Happily, this plan was abandoned; Mr. Brunel obtained a majority of votes, and was appointed engineer on March 7, 1833.

He commenced the survey without delay; and in addition to his strictly professional duties, he assisted in forming a committee in London, and took a leading part in the consultations which were held upon various important matters connected with the general interests of the undertaking.

A hasty survey of the country between London and Bristol occupied him till the middle of June; and as soon as it was completed, and the course of the line settled on, preparations were made for placing the scheme before the public.

The first public meeting was held on July 30, 1833. Mr. Brunel thus refers to it in his diary:—'Got through it very tolerably, which I consider great things. I hate public meetings: it's playing with a tiger and all you can hope is, that you may not get scratched, or worse.' The result, however, seems to have been successful, and in a month's time a company was formally constituted, and the Parliamentary survey commenced.

Mr. Brunel organised a staff of assistants, at that time rather a difficult task, and set them to work on various parts of the line. His own duty of superintendence severely taxed his great powers of work. He spent several weeks travelling from place to place by night, and riding about the country by day, directing his assistants, and endeavouring, very frequently without success, to conciliate the landowners on whose property he proposed to trespass.

His diary of this date shows that when he halted at an inn for the night, but little time was spent in rest, and that often he sat up writing letters and reports until it was almost time for his horse to come round to take him on the day's work. 'Between ourselves,' he wrote to Mr. Hammond, his assistant, 'it is harder work than I like. I am rarely much under twenty hours a day at it.'

A great portion of this labour was for the time thrown away, for as November 30 drew near, it became evident that subscriptions were coming in to the extent which would enable the directors to lodge a Bill for the whole line in the session of 1834.

The directors therefore determined to apply to Parliament for powers to make a railway from London to Reading, and from Bath to Bristol, 'as a means of facilitating the ultimate establishment of a railway between London and Bristol;' postponing till a future session their application for an Act to enable them to complete the undertaking by making the line from Reading to Bath.

The Bill was introduced into the House of Commons, and on March 10, Lord Granville Somerset moved that it be read a second time. This motion was seconded by the Earl of Kerry, and supported by several influential members, amongst whom were Mr. Labouchere (the late Lord Taunton) and Mr. Daniel O'Connell.

The second reading was carried by a majority of ninety in a House of 274 members.

The Bill was then referred to a committee which met on April 16, Lord Granville Somerset being in the chair. Evidence was called to prove the advantages of the railway to the agricultural and trading community of the country through which it would pass, even if only the two proposed divisions of the line were constructed.

The traffic in merchandise between Bristol and London was at this time principally carried on by means of water carriage, consisting, first of the river Avon navigation from Bristol to Bath, next of the Kennet and Avon Canal from Bath to Reading, and lastly of the river Thames from Reading to London. The evidence went to show that the distance between London and Reading, which by railway would be thirty-six miles, amounted by the river to nearly eighty; that the delays and impediments arising from drought, flood, and frosts on the rivers, were such as sometimes to detain barges for several weeks; and that so great were the consequent uncertainties and inconveniences of this navigation, that goods which came as far as Reading by the canal, were frequently sent thence to London by road, although at a great increase of expense. Even under the most advantageous circumstances, goods could not be conveyed from Reading to London in less than three days, or in less than a day by the river Avon from Bath to Bristol. It was therefore contended, that to form a railway which should supersede, or at all events come in aid of, the worst portions of the navigation between London and Bristol, would be an important public benefit.[4] The various advantages of the measure were most fully discussed in an investigation which lasted during fifty-seven days. Against the Bill was arrayed every class of opponent that a private Bill could possibly encounter. Those interested in the canals, rivers, and stage-coaches, opposed it from the fear of competition; the inhabitants of Windsor opposed it, because the railway did not run so near to the town as they wished; the corporation of Maidenhead opposed it, because they thought that all the traffic which paid toll on their bridge over the Thames would be diverted to the railway; landowners and farmers near town opposed it, because they feared it would bring produce to London from a distance, as cheap as that supplied by themselves.

There was another very formidable class of opponents to the Bill, consisting of landed proprietors and owners of houses in the immediate neighbourhood of London.

Many engineers were called by these several opponents to show that a more advantageous line of railway might have been selected; but, upon sifting the merits of the various new lines proposed, it became apparent that the one chosen by Mr. Brunel was the best. Indeed, although some trifling deviations of his line were suggested, the opposing engineers admitted that in all essential features the railway had been most skilfully laid out. It was generally agreed that the line through the valley of the Thames, and thence in a direction north of the Marlborough Downs, was the only proper course for a railway between Bristol and London, as the levels were much better, and communication could be made with much greater ease with the northern and South Wales districts, than if the route to the south of the Marlborough Downs had been selected.

The plans proposed for entering London raised great opposition. In this respect public feeling has greatly changed, for now no railway is thought complete which has not a terminus in the heart of London; and it is considered an advantage for houses to be within easy reach of a railway station; but in 1834 such a neighbour was looked upon with horror and dismay—a nuisance to be, if possible, absolutely prohibited.

When Mr. Brunel commenced the survey for the London terminus, he had some idea of bringing the railway in on the south side of the Thames; but this was abandoned, as it was found to involve very heavy works, and the line proposed in the first Bill was made to terminate on the north side of the river at Vauxhall Bridge. It was to have been carried on a viaduct 24 feet high, with a parapet 6 feet 6 inches high, to prevent the passengers looking into the windows of the neighbouring houses.

The owners of the land through which this part of the line would pass were influential members of the Upper House, and therefore the directors thought it useless to brave their opposition; accordingly, on the thirteenth day of the hearing, they abandoned the last two miles of the viaduct, and proposed to stop at the 'Hoop and Toy,' a public-house near the site of the South Kensington Station of the Metropolitan Railway.

But although the opposition of some of the landowners was conciliated by this concession, that portion of the line through Brompton which had not been abandoned was attacked with unabated energy. The residents in Brompton opposed the Bill from the apprehension that the railway would interfere with their quiet and seclusion; Brompton being at that time considered, at any rate by one of the counsel for the opposition, 'the most famous of any place in the neighbourhood of London for the salubrity of its air, and calculated for retired residences.' They could not, indeed, be blamed for indulging in these apprehensions, if they really believed in their counsel's statement that 'streams of fire would proceed from the locomotive engines.'

Others objected to the viaduct itself as being an undertaking of so colossal a nature as hardly to be practicable; and the supposed increase of traffic and consequent obstruction in Piccadilly and other leading thoroughfares brought down upon the promoters the opposition of the Commissioners of Metropolitan Roads.

All these objections were made the ground of much argument in committee, and doubtless had great influence over the minds of those who voted against the Bill.

The engineering evidence occupied, as might be expected, the greater part of the forty-two days during which witnesses were examined before the committee, and of these forty-two days no less than eleven were taken up by the cross-examination of Mr. Brunel. So protracted a cross-examination has probably never been heard in any court or committee-room. One of those present thus describes it:—

> The committee-room was crowded with landowners and others interested in the success or defeat of the Bill, and eager to hear Brunel's evidence. His knowledge of the country surveyed by him was marvellously great, and the explanations he gave of his plans, and the answers he returned to questions suggested by Dr. Lardner, showed a profound acquaintance with the principles of mechanics. He was rapid in thought, clear in his language, and never said too much, or lost his presence of mind. I do not remember ever having enjoyed so great an intellectual treat as that of listening to Brunel's examination, and I was told at the time that George Stephenson and many others were much struck by the ability and knowledge shown by him.

In his evidence, Mr. George Stephenson stated that he did not know any existing line so good as that proposed by Mr. Brunel. 'I can imagine (he said) a better line, but I do not know of one so good.'[5]

At length, on the fifty-fourth day of the sittings of the committee, Mr. Harrison, K.C., rose to reply on behalf of the promoters, and on the conclusion of his address the Bill was passed.

In the House of Lords the second reading was moved by Lord Wharncliffe. It was opposed, and on a division being taken, the motion was lost by a majority of seventeen (30 content and 47 non-content). The Bill was therefore thrown out.

The directors, undaunted by their defeat lost no time in making preparations for bringing a Bill before Parliament in the session of 1835, with such improvements as the experience of the past campaign suggested to them. Taking into consideration the various grounds on which opposition had been raised to the plans they had proposed for entering London through the Brompton district, they opened negotiations with the London and Birmingham Railway Company, and arrangements were concluded by which the traffic of the Great Western Railway was to be carried upon the London and Birmingham line for the first four miles out of London, the junction being made a little to the west of the Kensal Green Cemetery.

They had also during the autumn raised money enough to enable them to apply to Parliament for powers to construct the whole of the line from London to Bristol. They thus escaped all the sarcastic observations which had been made upon the scheme of 1834, of which it had been said, that it would be a head and a tail without a body, and neither 'Great' nor 'Western,' nor even a 'railway' at all, but 'a gross deception, a trick, and a fraud upon the public, in name, in title, and in substance!'

On March 9, the earliest day allowed by the standing orders, the Bill was read a second time and committed. A division being taken on the motion for committal, there appeared in favour of the motion 160, and against it none but the tellers.

Shortly after its first meeting, the committee, of which Mr. Charles Russell, then member for Reading, was chairman, came to the resolution that, inasmuch as the evidence given in the previous year as to the public advantages of a Bristol railway had been referred to them by order of the House, they needed no further evidence on that subject. Counsel were therefore directed to confine their case as much as possible to the merits of the line proposed.

Evidence was called by the opponents chiefly with a view to show the advantages of a proposed line from Basing to Bath, and the inexpediency of granting an entirely new line of 115 miles in length to the Great Western Railway Company, which involved the construction of a 'monstrous and extraordinary,' 'most dangerous and impracticable tunnel' at Box, and this, when 44 miles of railway in a western direction—viz. as far as Basingstoke, had already been sanctioned by the legislature in the Southampton Railway Act, passed in the previous session. The promoters of the Bill contended that the levels of the Basing and Bath line were not so good as those proposed for their own, and that the Great Western Railway would approach almost every town of importance situated on the proposed Basing and Bath line, by means of short branches; whilst at the same time it presented the great advantage of being capable of easy extension to Gloucester and Wales, and to Oxford, an object wholly unattainable by the other line. In reply to these assertions, the opponents maintained that although the levels of the Basing and Bath Railway presented greater inclinations than those of the Great Western, yet that they were so balanced as that the rises and falls compensated one for another, so as to render the line practically level. The enunciation of this theory called forth a remark by the chairman that according to this principle the Highlands of Scotland would be as good as any other place for the construction of a railway.

The preamble was voted proved, and the Bill passed the House of Commons without further opposition, and on May 27 was read a first time in the Lords. On June 10, the second reading was carried after a sharp debate, the numbers being 46 contents, and 34 non-contents.

Lord Wharncliffe was chairman of the committee.[6] The proceedings began by an opposition on the standing orders, which, after much skirmishing, were voted to have been complied with. The promoters, however, judged from the nature of the first day's proceedings, that they had to expect a contest of no inconsiderable duration; and the result proved their anticipations to have been correct. For forty days the battle was fought with a degree of earnestness and vigour on both sides, almost unequalled in any similar proceedings.

The committee soon came to the same decision as the House of Commons, that, with regard to the advisability of a Bristol railway, they were satisfied, and needed no further evidence. The case became then one of mere comparison between the relative merits of the two lines proposed.

The case in support of the Bill occupied eighteen days, and was closed with a speech by the Hon. John Talbot.

Mr. Serjeant Merewether, whom the opponents had chosen as their leader in the House of Lords, was then heard on their behalf; and occupied no less than four days in the delivery of his speech, in which certainly no argument that ingenuity could devise was omitted to strengthen his case. There was hardly any conceivable injury which, according to the learned serjeant's notions, the Great Western Railway would not inflict. It was said that the Thames would be choked up for want of traffic, the drainage of the country destroyed, and Windsor Castle left unsupplied with water. As for Eton College it would be absolutely and entirely ruined: London would pour forth the most abandoned of its inhabitants to come down by the railway and pollute the minds of the scholars, whilst the boys themselves would take advantage of the short interval of their play hours to run up to town, mix in all the dissipation of London life, and return before their absence could be discovered. Moreover, while the beauty of the country and the retirement of private dwellings would be destroyed, the interests of the public would be far more effectually served by the adoption of the Basing and Bath line, and a line from the London and Birmingham Railway to Gloucester. This was in fact the point at issue, and on this the result of the contest depended. The promoters of the Bill had called, in support of their line, in addition to Mr. Brunel, who being engineer to the company might be considered an interested witness, Mr. Locke, Mr. Palmer, Mr. Price, Mr. George Stephenson, and Mr. Vignoles. They expressed their unqualified approbation of the line chosen by Mr. Brunel, and of the estimates he had prepared.

The preamble was proved, and after an unsuccessful opposition the Bill was read a third time, on August 27. The Royal Assent was given on the last day of that month.[7]

During this contest Mr. Brunel made among his fellow-labourers many deep and lasting friendships. One of the most intimate of these friends, Mr. St. George Burke, Q.C., has, in compliance with a request made to him, furnished the following reminiscences of his intercourse with Mr. Brunel during the progress of the Bill through Parliament.

March 9, 1869

My dear Isambard,—You wish me to supply you with reminiscences of my old associations with your father, in order that, in your biography of him, you may present a true picture of those features of his character which so endeared him to his most intimate friends.

For many years it was my good fortune to enjoy his friendship, and many of the pleasantest hours of my life were due to it.

For a period of nearly three years, viz. during the contest for the Great Western Railway Bill, I think that seldom a day passed without our meeting, whether for purposes of business or pleasure, both of which his buoyant spirits enabled him to combine in a manner which I have seldom seen equalled.

It would be wearisome to detail the many incidents which occurred illustrative of the singularly facile manner in which, in the midst of the heaviest and most responsible labours, he could enter into the most boyish pranks and fun, without in the least distracting his attention from the matter of business in which he was engaged; but all who knew him as I did could bear testimony to this characteristic of his disposition.

I believe that a more joyous nature, combined with the highest intellectual faculties, was never created, and I love to think of him in the character of the ever gay and kind-hearted friend of my early years, rather than in the more serious professional aspect under which your pages will, no doubt, rightly depict him.

In 1833 your father and I occupied chambers facing each other in Parliament Street, and as my duties involved the superintendence, as Parliamentary agent, of the compliance with all the Standing Orders of Parliament, and very frequent interviews and negotiations with the land-owners on the line, we were of necessity constantly thrown together. To facilitate our intercourse, it occurred to your father to carry a string across Parliament Street, from his chambers to mine, to be there connected with a bell, by which he could either call me to the window to receive his telegraphic signals, or, more frequently, to wake me up in the morning when we had occasion to go into the country together, which, it is needless to observe, was of frequent occurrence; and great was the astonishment of the neighbours at this device, the object of which they were unable to comprehend.

I believe that at that time he scarcely ever went to bed, though I never remember to have seen him tired or out of spirits. He was a very constant smoker, and would take his nap in an arm-chair, very frequently with a cigar in his mouth; and if we were to start out of town at five or six o'clock in the morning, it was his frequent practice to rouse me out of bed about three, by means of the bell, when I would invariably find him up and dressed, and in great glee at the fun of having curtailed my slumbers by two or three hours more than necessary.

No one would have supposed that during the night he had been poring over plans and estimates, and engrossed in serious labours, which to most men would have proved destructive of their energies during the following day; but I never saw him otherwise than full of gaiety, and apparently as ready for work as though he had been sleeping through the night.

In those days we had not the advantage of railways, and were obliged to adopt the slower, though perhaps not less agreeable, mode of travelling with post-horses. Your father had a britzska, so arranged as to carry his plans and engineering instruments, besides some creature comforts, never forgetting the inevitable cigar-case among them; and we would start by daybreak, or sometimes earlier, our country excursions, which still live in my remembrance as some of the pleasantest I have ever enjoyed; though I think I may safely say that, pleasurable as they were, we never lost sight of the business in which we were engaged, and for which our excursions were undertaken.

I have never known a man who, possessing courage which to many would appear almost like rashness, was less disposed to trust to chance or to throw away any opportunity of attaining his object than was your father. I doubt not that this quality will be fully exemplified in the details which you will have received of his engineering experiments; but I speak of him also in the character of a diplomatist, in which he was as wary and cautious as any man I ever knew.

We canvassed many landowners together, and I had plenty of opportunities of judging of his skill and caution in our discussions with them, though we had many a good laugh afterwards at the arguments which had been addressed to us as to the inutility and impolicy of the scheme in which we were engaged, and the utter ruin it would be sure to entail on its promoters, as well as on the country affected by it.

I frequently accompanied him to the west of England, and into Gloucestershire and South Wales, when public meetings were held in support of the measures in which he was engaged, and I had occasion to observe the enormous popularity which he everywhere enjoyed. The moment he rose to address a meeting he was received with loud cheers, and he never failed to elicit applause at the end of his address, which was distinguished as much by simplicity of language and in modesty of pretension as by accurate knowledge of his subject.

Yours very truly,

St. George Burke.

Isambard Brunel, Esq.

The following is an extract from Mr. Brunel's diary, written at the end of the year 1835:—

53 Parliament Street, December 26.
What a blank in my journal [*the last entry is dated January 1834*], and during the most eventful part of my life. When last I wrote in this book I was just emerging from obscurity. I had been toiling most unprofitably at numerous things, unprofitably, at least, at the moment. The railway was certainly being thought of; but still very uncertain. What a change. The railway now is in progress. I am thus engineer to the finest work in England. A handsome salary, on excellent terms with my directors, and all going smoothly. But what a fight we have had, and how near defeat, and what a ruinous defeat it would have been. It is like looking back upon a fearful pass; but we have succeeded.

And it is not this alone, but everything I have been engaged in has been successful. Clifton bridge—my first child, my darling, is actually going on: recommenced work last Monday—glorious!! [*Here follows a list of the undertakings on which he was then engaged.*] I think this forms a pretty list of real sound professional work, unsought for on my part, that is, given to me fairly by the respective parties—all, except the Wear Docks, resulting from the Clifton bridge, which I fought hard for, and gained only by persevering struggles . . . And this at the age of twenty-nine. I really can hardly believe it, when I think of it. I am just leaving 53 Parliament Street, where I may say I have made my fortune, or, rather, the foundation of it, and I have taken 18 Duke Street.

1. A graphic account of this famous parliamentary contest will be found in the third volume of Mr. Smiles' *Lives of the Engineers,* chapter xi.
2. See Mr. Smiles' *Life of George Stephenson,* p.325
3. See Mr. Smiles' *Lives of the Engineers,* vol. iii. chap.xv.
4. By means of the railway (it was said) goods would be conveyed with ease from London to Reading in *three or four hours,* and from Bath to Bristol in *one hour.*
5. During Mr. Stephenson's cross-examination, several questions were put to him as to the dangerous consequences which might be expected to result from travelling through a tunnel a thousand yards long. At length he lost all patience at the ignorance displayed by the questions put to him by counsel, and the following passage of arms took place:—

    *Mr. Stephenson.* I wish you had a little engineering knowledge—you would not talk to me so.'

    *Counsel.* I feel the disadvantage.

    *Mr. Stephenson.* I am sure you must.

    In other parts of the engineering evidence there are some statements which read strangely enough at the present day, as for example the following: 'The noise of two trains passing in a tunnel would shake the nerves of this assembly. I do not know such a noise. No passenger would be induced to go twice.'
6. At this time the Lords' committees were open to all peers who chose to sit on them, and it was not considered indecorous for peers who had not attended any of the previous sittings to vote on the division.
7. The Great Western Railway was constructed with but few deviations from the line sanctioned in 1835. The only alteration of any importance was at the London end, where, by an Act passed in 1836, the line was taken to Paddington, instead of joining the London and Birmingham Railway near Kensal Green. This change of plan was rendered necessary by reason of a difficulty having arisen between the two companies as to the terms of their agreement, and not, as has been often stated, in consequence of the adoption of the broad gauge on the Great Western line.

# IV

# RAILWAY WORKS

IN the extract from Mr. Brunel's diary given at the close of the last chapter he refers to the successful issue of the contest for the Great Western Railway Act as a very important event in his life.

As the result proved, he did not take too hopeful a view of his future prospects; for from that time to his death he was fully employed as the engineer of railways which, in number and importance, were not inferior to those of any of his contemporaries. Of the main lines he constructed, one extends uninterruptedly from London to the Land's End, and another to the extremity of South Wales, at Milford Haven, 285 miles from Paddington.

It would be impossible to describe in detail all the engineering works which are to be found on Mr. Brunel's railways, the aggregate length of which is upwards of 1,200 miles; but in this chapter it is proposed to give a brief sketch of the lines he constructed, omitting all that can be more properly inserted in the three chapters which follow, relating to the broad gauge, to the Atmospheric System, and to the bridges and viaducts.

The Great Western Railway was opened to Maidenhead, a distance of nearly twenty-three miles, in June 1838, and to Twyford, eight miles farther on, in July 1839. A description of the Wharncliffe Viaduct at Hanwell and of the Maidenhead bridge will be found at p.133. The line from Twyford to Reading was opened in March 1840, and from Reading to Chippenham by May 1841. Meanwhile the portion from Bristol to Bath had been opened in August 1840. The last division, namely, that from Chippenham to Bath, containing the Box Tunnel, was opened on June 30, 1841; and the railway was completed throughout its whole length.

A considerable part of the history of the Great Western Railway is connected with the adoption on it of the broad, or 7-foot gauge, and will be dealt with in the next chapter, in which is also given some account of the longitudinal system of permanent way.

The bridges are described in Chapter VII; but some of the other works may be noticed here.

In laying out the line, Mr. Brunel endeavoured to make it as straight and as level as possible throughout, and to concentrate those changes of level, which could not be avoided, into short inclines, to be worked, if necessary, by auxiliary power.

Accordingly the line is thus divided:—

|  | Miles | Yards |
|---|---|---|
| Level, or with an inclination not exceeding 4 feet in the mile | 67 | 88 |
| Above 4 feet, and not exceeding 8 feet in the mile | 47 | 110 |
| Steep inclines | 3 | 1210 |

The steep inclines are two in number, of a gradient of 1 in 100, or about 53 feet in a mile, and descend towards the Bristol end of the line.

The Wootton Basset incline, 85½ miles from London, is 1 mile 550 yards long.

The second incline is at Box, 99 miles from London, and is 2 miles 660 yards in length. An assistant engine is still occasionally used to work heavy trains at this point.

On this incline the line passes through the Box Tunnel. This tunnel is the first out of London, and could only have been dispensed with by taking a circuitous route several miles longer than that adopted by Mr. Brunel.

The tunnel is 1⅞ mile in length, and is ventilated by six shafts. They are 30 feet in diameter, and from 70 to 300 feet deep.

The Box Tunnel had been the subject of much criticism before the works were commenced; and during its actual construction it did not escape the unfavourable notice of those who were ignorant of the difficulties which presented themselves, and the means which had been taken to overcome them. Indeed, for some time after the opening of the line, there were travellers who used to avoid the terrors of the tunnel by posting along the turnpike road in that part of their journey.

Mr. Brunel never troubled himself about the ordinary gossip which is always circulated concerning any remarkable work; but matters assumed a different aspect when, a year after the completion of the tunnel, doubts were expressed as to its safety by an eminent geologist, at a meeting of the Institution of Civil Engineers. Mr. Brunel, who was on terms of friendly intercourse with the speaker, addressed to him the following letter:—

June 21, 1842.

I assure you, my dear Sir, that when my attention had been drawn to the statements reported to have been made by you on this subject, I refrained as far as possible from expressing any opinion. I thought it my duty to read the notes taken, but I never said that I thought your statements were correct. Indeed, I had hoped to have avoided the necessity of making any observations upon these statements; but as a letter to Mr. Saunders on such a subject is almost the same thing as if it were addressed to myself, and as it was shown to me, it would not be candid towards you if I now refrained from saying, that the opinions you are reported to have expressed with respect to the Box Tunnel are by no means considered correct, either by myself or by those others who, from being intimately acquainted with the rock as it was really found, and the works as they were really executed, are capable of judging.

In the notes shown to me the observations alluding to this work in particular, as an illustration of the views you were explaining, appear to have been curtailed, and the allusions rendered somewhat less direct; but still the inference unavoidably to be drawn from them is, that the back joints, as we call them, and other defects which exist originally, or which show themselves after a time, in this rock, are not well known, and tolerably well understood and guarded against, by practical engineers, and even by our workmen. In this opinion I assure you you are mistaken. Ignorant as I may probably be myself of the science of geology, I cannot have been engaged for several years in making very extensive excavations, probably the largest hitherto made, in this particular rock, having also the opportunity of examining very old and large quarries in the same rock and close to the line, having among my assistants men not meanly acquainted with this particular branch of geology, and surrounded by workmen of considerable experience, I cannot have gone through such a study without acquiring a very intimate and practical knowledge of the structure and peculiarities of the particular mass of rock which is now in question; and I will say frankly what I feel upon this point, which is, that I ought now to possess a more thorough and practical knowledge of this particular rock and its defects, and the best mode of remedying them, than even you yourself, with your immeasurably greater scientific knowledge of rocks generally.

The opinion you are said to have expressed of there being great danger of some serious accident occurring in the tunnel is, I am firmly convinced, erroneous; at all events the reason given convinces me that you have not become acquainted with the means which have been taken by me to examine and to ascertain the security of every part of the rock, to remove or to support with masonry any part not so ascertained to be secure, or with the precautions taken to prevent any such accidents as those you have imagined. And notwithstanding the heavy responsibility which rests upon me, from all which you gentlemen of science are, happily for yourselves, so free, I feel that as regards the works of the Box Tunnel everything necessary has been done to render them secure, and that the doubts and fears you have so easily raised, but which it might be more difficult again to set at rest, are entirely unfounded.

In conclusion, I must observe that no man can be more sensible than I am of the great advantage it would be to me as a civil engineer to be better acquainted with geology, as well as with many other branches of science, that I have endeavoured to inform myself on the subject, and that I have not altogether thrown away the many opportunities afforded me in my professional pursuits; but that if from a conviction that you possessed information far more extensive than mine, if from doubts of the sufficiency of my abilities or the means I was likely to bring to bear upon the subject, if from a fear of such consequences as you now anticipate, you had kindly, on any one of the many occasions when I have had the pleasure of meeting you, intimated that you had any suggestions to make to me, I should have been anxious to have availed myself of your assistance. But after the lapse of years, the first intimation I have of such doubts is the very public expression of a very strong opinion, which, if weight be attached to it, must tend to alarm the public unnecessarily, and to injure the value of the property of individuals who have embarked several millions in that property.

Between Chippenham and Bristol the nature of the building stone enabled Mr. Brunel, at moderate cost, to make the bridges, tunnel fronts, and stations ornamental features in the picturesque scenery through which the railway passes.

He took great pleasure in finishing minutely the various designs, and making them correct in their proportions and details. One tunnel front, near Bristol, may be singled out for especial mention. During its construction a part of the ground behind slipped away, and it became unnecessary to complete the top of one of the side walls. It was therefore left unfinished, and was planted with ivy so as to present the appearance of a ruined gateway.

The roofs of the Bath and Bristol stations are of large span, and are handsome architectural structures. They are each in the form of a Tudor arch; the Bristol roof is 72 feet span, and the Bath roof 50 feet span. The framing is an example of a peculiar form of construction, somewhat analogous to that adopted in the large ship building sheds in the dockyards. There are no cross tie-rods, but each principal of the roof is formed of two frameworks, like cranes, meeting in the middle of the roof; the weight being carried on columns near the edge of the platform, and the tail ends of the frames held down by the side walls. As the two frames do not press against each other at their meeting point at the ridge of the roof, there is no outward thrust. The side walls being on a viaduct could not without difficulty have been made to resist a horizontal thrust.

At the Bristol station Mr. Brunel introduced hydraulic machinery for working lifts. By these the waggons were lowered to and raised up from the goods shed, which was placed at the level of the ground, about 12 feet below the railway.

Although the works already described were completed in 1841, the permanent terminus at Paddington was not commenced till the year 1849. It was completed in 1854. Previously to that time a temporary station had been used, the booking offices being under the arches of the Bishop's Road bridge.

As the level of the railway was lower than that of the surrounding land, no exterior architectural effect could be produced; but Mr. Brunel took this opportunity to carry out his views as to the proper structural use of metal in works of this description.

In the design of the ornamental details, he obtained the assistance of Mr. (now Sir Matthew) Digby Wyatt.

The interior of the principal part of the station is 700 feet long and 238 feet wide, divided in its width by two rows of columns into three spans of 68, 102, and 68 feet, and is crossed at two points by transepts 50 feet wide, which give space for large traversing frames. The roof is very light, consisting of wrought-iron arched ribs, covered partly with corrugated iron and partly with the Paxton glass roofing, which Mr. Brunel here adopted to a considerable extent. The columns which carry the roof are very strongly bolted down to large masses of concrete, to enable them to resist sideways pressure.

This station may be considered to hold its own in comparison with the gigantic structures which have since been built, as well as with older stations. The appearance of size it presents is due far more to the proportions of the design than to actual largeness of dimension. The spans of the roof give a very convenient subdivision for a large terminal station, dispensing with numerous supporting columns and at the same time avoiding heavy and expensive trusses. The graceful forms of the Paddington station, the absence of incongruous ornament and useless buildings, may be appealed to as a striking instance of Mr. Brunel's taste in architecture and of his practice of combining beauty of design with economy of construction.

The goods station was erected at about the same time as the passenger station, and is remarkable for the completeness of its arrangements, and for the great use made of hydraulic machinery. This is also applied in the passenger station.[1]

In connection with the Paddington station mention may be made of the Great Western Hotel, which was built at the extremity of the land belonging to the railway company.

When, in 1854, no tenant could be found for it, a few of the shareholders of the Great Western Railway, being unwilling that the building should remain empty and be a loss to the proprietors, formed themselves into a company to lease and work the hotel. Mr. Brunel became a Director, and shortly afterwards (in December 1855) the chairman. He occupied this post till his death, by which time the hotel had become very prosperous. He found attendance at the meetings of the Directors and the supervision of the management of the hotel a very agreeable relaxation from the more important duties which took him to Paddington.

The branches and extensions of the Great Western Railway, as far as their history affected the general interests of the company, are referred to in the chapter on the broad gauge. Branches were opened to Oxford in 1844, to Windsor in 1849, to Wycombe in 1854, to Uxbridge in 1856, to Henley in 1857, and to Brentford in July 1859.

The Bristol and Exeter Railway is a continuation of the Great Western Railway, and was opened to Exeter in 1844. The two portions of it, from Bristol to Taunton, and from Taunton to Exeter, are in marked contrast to each other. The former part of the line is almost level, and has very easy curves. Between Taunton and Exeter it passes over the high ground on the borders of Devonshire, with the Whitehall Tunnel at the summit, ⅝ mile in length. On this part of the line there are long gradients of from 1 in 80 to 1 in 120. Mr. Brunel resigned the position of engineer in 1846, in consequence of differences having arisen between the Bristol and Exeter and the Great Western Companies, which, in Mr. Brunel's opinion, made it impossible for him to continue engineer to both railways.

The South Devon Railway, and the adoption on it of the Atmospheric System, are described in Chapter VI. In connection with this line is the important Torquay branch, and the railway in continuation of it to Dartmouth. This was completed as far as Paignton during Mr. Brunel's lifetime.

The South Devon and Tavistock Railway branches off from the South Devon Railway near Plymouth, and has several large viaducts.

On the Cornwall Railway from Plymouth to Truro, and the West Cornwall Railway from Truro to Penzance, the most remarkable works are the viaducts, and the Royal Albert Bridge.

In the case of the Cornwall Railway, it became necessary to reduce the capital expenditure, even at the cost of increasing the charges for maintenance. With this object the line was re-examined and modifications introduced, principally by an increase in the extent of viaduct. These lines pass through a very difficult country; involving the adoption of steep gradients and sharp curves. Mr. Brunel, in a memorandum written in 1845, after having explained his reasons for considering that the prejudicial effects of gradients and curves were commonly overrated, gives the following opinion in reference to the proposed Cornwall Railway:—

> I must not be understood to argue against the advantage of straight lines or large and easy curves, but I wish to show that where small curves are unavoidable, they can in practice be so constructed as not to be very prejudicial; and I consider that the character of the country in Cornwall is such that no railway can be constructed at any moderate expense without either sacrificing all consideration for the interests of localities and the position of the population to the mere choice of levels, or without steep gradients and sharp curves.

The principal lines branching off from the Great Western, and since incorporated with it, are, on the south, the Berks and Hants, from Reading to Basingstoke and Hungerford; and the Wilts and Somerset, to Weymouth and Salisbury. On the north-west is the Cheltenham and Great Western Union Railway, from Swindon to Cheltenham and Gloucester. This line passes through the Cotswold Hills at Sapperton by a tunnel 1⅜ mile in length. The Gloucester and Dean Forest Railway runs from Gloucester to Grange Court, and thence to Ross and Hereford.

The South Wales Railway, which extends from Grange Court to Milford Haven, contains a tunnel at Swansea ⅓ mile long, and some of Mr. Brunel's most important works, including the Chepstow bridge and several other bridges of considerable size, and the viaducts at Landore and Newport. There are also on this line four opening bridges across navigable channels. The works at the termination of the line at Neyland, in Milford Haven, are described in Chapter XIV. Mr. Brunel considered that Milford Haven, with its excellent harbour, which can be entered at all times of tide by the largest vessels, would probably become a great port for ocean steamers, and especially for the 'Great Eastern' and ships of her class.

There are also in South Wales the following railways: the Taff Vale, the Vale of Neath, the Llynvi Valley, and the South Wales Mineral. The Taff Vale, a line from Cardiff to Merthyr, was opened on the narrow gauge in 1841.[2] On this railway is the lofty masonry viaduct at Quaker's Yard.

On the Vale of Neath Railway, from Neath to Aberdare and Merthyr, there is a tunnel, near Merthyr, 1¼ mile long, and 650 feet below the summit of the hill.

Full advantage is taken on this railway of the facilities which the broad gauge offers for heavy traffic. The line has long steep gradients, and the locomotives used on it, of the class known as tank engines, are of great power. One of these gradients is 4½ mile long, with an inclination of 1 in 50. Large quantities of coal are brought down by this railway to the Swansea and Briton Ferry Docks. The coal of South Wales is of a friable nature, and, in order to avoid the breakage consequent on the ordinary mode of shipping coal, by tipping it down a shoot, Mr. Brunel introduced on a large scale the use of trucks carrying four iron boxes, each box about 4 feet 8 inches cube, and containing two and a half tons of coal. At the docks machinery is provide by which each box is lowered down into the hold of the ship, and the under side being allowed to open, the coal is deposited at once on the bottom of the vessel.

The Llynvi Valley Railway is a short line, leading from the South Wales Railway at Bridgend into the coal and iron districts.

The South Wales Mineral Railway is another line of the same class. It passes through a very heavy country, and has on it a self-acting incline of 1 in 9, ¾ mile long, worked by a rope, and a tunnel ⅝ mile long, and 470 feet below the surface.

In connection with the South Wales district is the Bristol and South Wales Union Railway, a line running from Bristol to the banks of the Severn, across which the traffic is carried by a steamer to a short branch from the South Wales Railway on the other side. This railway had been for a long time contemplated, and Mr. Brunel devoted much time to a careful investigation of the Severn in order to determine the most suitable point for the crossing. He decided that the best place would be at what is known as the New Passage. The arrangements had to be made in accordance with the requirements of the Admiralty. Trains run to the end of timber piers extending into deep water, and there are staircases and lifts leading to pontoons, alongside which a steamer can come at all times of tide. The tide at this part of the Severn rises 46 feet.

The three railways last mentioned were not completed during Mr. Brunel's lifetime.

The Bristol and Gloucester Railway, on which is the tunnel at Wickwar, ¾ mile long, was opened in 1844, and passed into the hands of the Midland Company in 1846.

The northern extensions of the Great Western Railway are the Oxford and Rugby, constructed as far as Fenny Compton; the Birmingham and Oxford Junction; and the Oxford, Worcester, and Wolverhampton Railways.

Mr. Brunel ceased to be engineer of the last-mentioned company in 1851, and the works were completed by Mr. Fowler. The Oxford, Worcester, and Wolverhampton line has since, under the title of the West Midland, become a part of the Great Western Railway.

In Ireland Mr. Brunel was engineer of the line from Dublin to Wicklow, round Bray Head, and of a line from Cork to Youghal.

He laid out two railways in Italy, the line from Florence to Pistoja, and that across the Apennines from Genoa to Novi and Alessandria, in the direction of Turin and Milan. He acted as engineer during the construction of the former line; but the works of the latter were carried out by the Sardinian Government.

One of the last of Mr. Brunel's important railways was the Eastern Bengal Railway, a line of about 100 miles in length, in a north-easterly direction from Calcutta. He took a great interest in the work and devoted much time to the special arrangements and designs, and to the best way of crossing the Ganges and its branches in the future extension of the railway; but no part of it was opened during his lifetime.

It was impossible for Mr. Brunel to look after all his works to the same extent as he had done in the case of the Great Western Railway and he was compelled to spend a very considerable amount of time in attendance on Committees of the Houses of Parliament on behalf of the railways to which he was engineer. Mr. Brunel frequently regretted this, and considered it a great evil that engineers were prevented by their duties during the session from attending properly to the construction of their works. He endeavoured as far as possible to superintend the execution of his different undertakings. He availed himself of every opportunity of examining them, and was acquainted throughout with all the designs which were prepared. He would take advantage of two or three free days to go down to the distant works in South Wales or in Cornwall, looking after details, such as the pickling tanks for timber, and the masonry of the viaducts.

He was fortunate in the selection of the members of his staff, and in his organisation of it. He had a rare power of utilising the capabilities of his different assistants; and although he had to deal with a great variety of men, he managed that they should work in harmony with him. From the complete personal supervision Mr. Brunel sought to maintain over all his works, his assistants had not perhaps so many opportunities of independent action as they might otherwise have obtained, but they had, on the other hand, the advantage of constant personal communication with their chief.

After the time when Mr. Brunel's works became so numerous, and his time so much occupied, that he could not exercise in person that general supervision which he conceived to be necessary, he was ably assisted by Mr. Robert Pearson Brereton, who, on the death of Mr. Hammond in 1847, became the chief of his engineering staff.

Mr. Brunel rarely made any changes in the *personnel* of his office. Mr. Brereton had become one of his assistants in 1836; and his secretary, the late Mr. Joseph Bennett, came to him in the same year.

The Great Western Railway retained its early place in his affections, and among his most valued friends were members of the Board of Directors, Mr. Saunders the Secretary, and other officers of the company. When in the last year of his life he was obliged to go to Egypt for his health, it was a matter of deep anxiety to him lest his absence from England should cause any alteration in his relations with the Company and it was a source of great pleasure to him that no such consequences followed.

Mr. Brunel's position as confidential adviser of so large a number of railway companies gave him frequent opportunities of acting as mediator between contending parties; and his decisions were always received with respect, for he was known to be scrupulously just.

Besides the more friendly task of reconciling opponents he had a large practice as a referee under Acts of Parliament and orders of the superior courts; and displayed in these matters great judicial abilities.

In all the causes and parliamentary contests affecting the various companies of which he was engineer, Mr. Brunel was a very important member of the preliminary consultations, and during the proceedings counsel relied with confidence on his suggestions.

One of the most arduous parts of his duty, as engineer of the Great Western Railway Company, was connected with the conduct of the great cases of Ranger and MacIntosh. To the former of these reference is made in the letter printed below, at p.350. The MacIntosh case, which was commenced shortly after the opening of the line, was not concluded before Mr. Brunel's death. He was compelled to devote a considerable portion of his time to it, even after his return home in the evening, during the launch of the 'Great Eastern.'

Mr. Brunel had a very high reputation as a witness. Mr. St. George Burke, Q.C., has communicated a memorandum on this subject:—

As a witness he could always be relied on as a perfect master of the case he had to support, and he had the rare quality of confining his answers to a simple reply to the questions put to him, without appearing as an advocate. He was, however, extremely particular as to the questions which should be put to him in his examination in chief, and was therefore never satisfied to entrust the preparation of his proof to the solicitor, without revising it himself.

In his cross-examinations he was generally a match for the most skilful counsel, and by the adroitness of his answers would often do as much to advance his case as by his examination in chief.

He was almost as much of a diplomatist as an engineer, and knew perfectly well how to handle a case in the witness-box so as to leave no loophole for his opponents to take advantage of. At the same time he was a perfectly honest witness, and while he avoided saying more than was necessary for the advancement of the cause in which he was engaged, he would have scorned to say or imply anything by his evidence inconsistent with strict truth.

Although he had attained to great celebrity as a witness, the committee room being crowded to hear him, he always declined to engage in the very lucrative work of a professional witness. He made a rule never to appear except on behalf of undertakings of which he was the engineer, or with which his own companies were interested. To help a friend, he occasionally but very, rarely broke through this resolve; but, whether he appeared in support of his own plans or those of others, there were few, if any, professional men whose evidence carried more weight than his did before Parliamentary Committees.

The following memorandum from Mr. George T. Clark, of Dowlais, formerly one of Mr. Brunel's assistants, contains his recollections of Mr. Brunel during the construction of the Great Western Railway:—

I made your father's acquaintance, rather characteristically, in an unfinished tunnel of the Coal-pit Heath Railway; and when the shaft in which we were suspended cracked and seemed about to give way, I well remember the coolness with which he insisted upon completing the observations he came to make. Shortly afterwards I became, at his request, his assistant; and during the parliamentary struggle of 1835, and the subsequent organisation of the staff and commencement of the works of the Great Western, I saw him for many hours daily, both in his office and in the field, travelled much with him, and joined him in the very moderate recreation he allowed himself.

These two years, and the preceding year, 1834, were, I apprehend, the turning points of his life. His vigour, both of body and mind, were in their perfection. His powers were continually called forth by the obstacles he had to overcome; and the result of his examinations in the committee rooms placed him, in the opinion of the members of the legislature, and of his own profession, in the very first rank of that profession, both for talents and knowledge.

I wish I could convey to you even a tolerable idea of your father as he was in those years, during which I knew him intimately, and saw him often under circumstances of great difficulty.

He was then a young man, but in the school of the Thames Tunnel he had acquired a close acquaintance with all kinds of masons' and carpenters' work, the strength and cost of materials, bridge building, and constructions under water, and with the working of the steam engine as it then stood. It happened not unfrequently that it was desirable to accept the tender of some contractor for railway work whose prices upon certain items were too high, and then it became the engineer's business to go into the details and convince the contractor of his error. On such occasions Brunel would go step by step through the stages of the work, and it was curious to see the surprise of the practical man as he found himself corrected in his own special business by the engineer. Thus, I remember his proving to an eminent brickmaker who had tendered for the Chippenham contract that the bricks could be made much cheaper than he supposed. He knew accurately how much coal would burn so many bricks, what it would cost, what number of bricks could be turned out, what would be the cost of housing the men, what the cartage, and how many men it would require to complete the work in the specified time. The contractor was astonished; asked if Mr. Brunel had ever been in the brick trade, and finally took and made money by the contract at the proposed figure.

In the case of the Maidenhead bridge, the contractor being alarmed at learning that the arch was the flattest known in brick, Brunel pointed out to him that the weight which he feared would crush the bricks, would be less than in a wall which he, the contractor, had recently built, and he convinced him by geometry, made easy by diagrams, that the bridge must stand. Knowledge of detail Brunel shared with the carpenter, builder, or contractor for earthwork, and he was their superior in the accuracy and rapidity with which he combined his knowledge, and arrived at correct conclusions as to the cost of the work and the time it would take to execute it.

In talking to landowners and others whose opposition it was important to overcome, I have often been struck by your father's great powers of negotiation. The most absurd objections—and there were many such—were listened to with good humour, and he spared no pains in explaining the real facts, so that it sometimes happened that he converted opponents into supporters of the railway. In the course he took there was much skilful diplomacy, but there was no dishonesty, no humbug. He was very frank and perfectly sincere. His object was to impart his own convictions, and in that he often succeeded.

I never met his equal for sustained power of work. After a hard day spent in preparing and delivering evidence, and after a hasty dinner, he would attend consultations till a late hour; and then, secure against interruption, sit down to his papers, and draw specifications, write letters or reports, or make calculations all through the night. If at all pressed for time he slept in his armchair for two or three hours, and at early dawn he was ready for the work of the day. When he travelled he usually started about four or five in the morning, so as to reach his ground by daylight. His travelling carriage, in which he often slept, was built from his own design, and was a marvel of skill and comfort. This power of work was no doubt aided by the abstemiousness of his habits and by his light and joyous temperament. One luxury, tobacco, he indulged in to excess, and probably to his injury. At all times, even in bed, a cigar was in his mouth; and wherever he was engaged, there, near at hand, was the enormous leather cigar-case so well known to his friends, and out of which he was quite as ready to supply their wants as his own.

His light and joyous disposition was very attractive. At no time was he stern, but when travelling or off work he was like a boy set free. There was no fun for which he was not ready. On the old Bath road, on a Wiltshire chalk hill-side, is cut a large horse, the pride of the district and only inferior in reputation to that of the famous Berkshire vale. The people of the district, afraid to lose their coach traffic, were violently opposed to the Great Western Railway Bill. Talking over this one evening, some one suggested turning the horse into a locomotive. Brunel was much amused at the idea, and at once sketched off the horse from memory, roughly calculated its area, and arranged a plan for converting it into an engine. Ten picked men were to go down in two chaises, and by moonlight to peg and line out the new figure, and then cut away the turf, and with it cover up as much of the horse as might be left. From the tube was to issue a towering column of steam, and below was to be inserted in bold characters the offensive letters G.W.R. It was, of course, not intended to carry this joke into execution, but Brunel often alluded to it, and laughed over the sensation it would have created.

He possessed a very fine temper, and was always ready to check differences between those about him, and to put a pleasant construction upon any apparent neglect or offence. His servants loved him, and he never forgot those who had stood by his father and himself in the old Tunnel days of trouble and anxiety.

No doubt the exertions of those three years, though they laid the foundation, or rather built the fabric, of his reputation, also undermined his constitution, and eventually shortened his life. Everything for which he was responsible he insisted upon doing for himself. I doubt whether he ever signed a professional report that was not entirely of his own composition; and every structure upon the Great Western, from the smallest culvert up to the Brent viaduct and Maidenhead bridge, was entirely, in all its details, from his own designs.

In the press of work and the altered circumstances under which he superintended the construction of his later railways, many changes inevitably followed. The open britzska gave place to a close travelling carriage, which in its turn became useless; and no time was left for fun or practical jokes; but the same energy of mind and the same kindliness of heart remained uninfluenced by increasing occupations or advancing years.

1.   Sir William Armstrong's hydraulic machinery at Paddington is described by him in a communication printed in the *Report of the British Association* for 1854, p.418: 'I have also applied it [water pressure machinery] extensively to railway purposes chiefly under the direction of Mr. Brunel, who has found a multitude of cases involving lifting or traction power in which it may be available. Most of these applications are well exemplified at the new station of the Great Western Railway Company in London, where the loading and unloading of trucks, the hoisting into warehouses, the lifting of loaded trucks from one level to another, the moving of turn-tables, and the hauling of trucks and traversing machines are all performed, or about to be so, by means of hydraulic pressure supplied by one central steam engine with connected accumulators.'

2.   See p.102, note 2.

# V

# THE BROAD GAUGE

THE railways designed by Mr. Brunel were, with a few exceptions, distinguished from those in all other parts of England by a peculiarity in the width between the two rails forming each line of way, or in what is called the *gauge*. In most railways, the distance between the internal edges of the rails is 4 feet 8½ inches, being what is termed the *narrow gauge*; on Mr. Brunel's railways, it was seven feet, or what is termed the *broad gauge*.

The gauge of the earlier railways, which were but a modification of the old wooden tramway, was made that of the tram plates which they superseded; and this had been originally fixed to suit the distance between the wheels of the country carts in the north of England.

When Mr. George Stephenson introduced the locomotive engine, the gauge of the lines in the Northumberland district had been already fixed. In laying out the Stockton and Darlington line (1821–1825) he saw no reason to depart from the gauge he had previously adopted; and, indeed, some of the waggons to be used on this line were brought from the Northumberland collieries. In this way the first important railway in England was made with the gauge of 4 feet 8½ inches; not from a deliberate choice of this width on the ground of any peculiar advantages, but from the mere fact of its already being established elsewhere.

In the construction of the Manchester and Liverpool Railway, in 1826, the same gauge was adopted as on the Stockton and Darlington; this course was also followed by the Grand Junction and the London and Birmingham Railways, and thus the 4 feet 8½ inches gauge became established in that part of the country.

Long experience appears to have determined the general type of wheeled vehicles: the wheels being of somewhat large size, and the body placed between them, so as to come down close upon the axle-tree.

This type, which gives obvious advantages in a mechanical point of view, appears to have been adhered to in all railway vehicles used before the opening of the Liverpool and Manchester Railway; these, however, were chiefly coal-waggons. But on the Liverpool and Manchester Railway it was soon perceived that the great increase of carrying power which the railway afforded must be met by a corresponding increase of space in the rolling stock, as it was necessary to accommodate light bulky goods and passenger traffic. The available width between the wheels was limited to about 4 feet 6 inches, and to carry in this width any large amount of cotton goods, or of passengers, would have required a train of an inordinate length. To meet this difficulty a new form of vehicle was designed; the wheels were made small, and the body was raised and widened out, projecting on either side over the tops of the wheels.

The earliest description of this form of waggon is contained in the second edition of Wood's 'Practical Treatise on Railroads,' published in 1832, about two years after the opening of the Liverpool and Manchester Railway. In Plate III, Mr. Wood shows a truck with a raised platform overhanging the wheels, and adapted for carrying loose boxes of coals; adding, in the description:

> Although the drawing shows only the form of boxes used for the conveyance of coals, yet it will readily occur that the form can be varied to suit the carriage of any kind of articles; the framework or body of the carriage being raised above the wheels, the breadth can be extended to any width which the distance between the railways [i.e. between the up and down lines of road] will admit (p.75).

Such was the state of matters when, in the year 1833, Mr. Brunel was appointed Engineer of the Great Western Railway. With the view of leaving the question of gauge open for future consideration, he procured the omission in the Great Western Act of a clause defining it. He came to the conclusion that it would be desirable to adopt a wider gauge, and he recommended this measure to the Directors in a report dated October 1835.[1]

In October 1836 a Royal Commission, consisting of Mr. Drummond, Under-Secretary for Ireland, Mr. R. Griffith, Colonel (now Field-Marshal) Sir John Burgoyne, R.E., and Professor Barlow, of Woolwich, was appointed to report on the establishment of railways in Ireland. They considered carefully the question of gauge, and their arguments in favour of an increase in the gauge were afterwards stated by Mr. Brunel to be identical with his own.

They drew attention to the advantage of large wheels, the use of which would be facilitated by a wider gauge; and they thought it a matter of importance to be able to place the bodies of the carriages between the wheels, instead of over them.

It was the width of the carriages, and not the distance between the rails, that determined the general dimensions, and therefore the cost, of the works of a railway. Mr. Brunel saw many advantages to be gained by an increase in the gauge, even while retaining the existing dimensions of carriages; and he thought it unwise at the commencement of a work of such magnitude as the Great Western Railway to retain a limit the inconvenience of which had already become apparent.

He says, in his evidence before the Gauge Commission:

> Looking to the speed which I contemplated would be adopted on railways,
> and the masses to be moved, it seemed to me that the whole machine was
> too small for the work to be done, and that it required that the parts
> should be on a scale more commensurate with the mass and the velocity
> to be attained. (Q. 3924.)

The width between the rails being the fundamental dimension of 'the whole machine,' on which its entire development must depend, Mr. Brunel proposed to begin by the enlargement of this dimension, and recommended that on the Great Western Railway the gauge should be seven feet. He considered that the whole of the part of the railway and of its rolling stock would be susceptible of continual, though gradual improvement, and that it was highly advisable to remove, in the outset, a great obstacle in the way of this progress.

He did not in the first instance propose any important change in the details as consequent on the wider gauge; and in regard to one of the principal points, the diameter of the wheels, he said:—

> I am not by any means prepared at present to recommend any particular
> size of wheel, or even any great increase of the present dimensions. I
> believe they will be materially increased; but my great object would be
> in every possible way to render each part capable of improvement, and
> to remove what appears an obstacle to any great progress in such a very
> important point as the diameter of the wheels, upon which the resistance,
> which governs the cost of transport and the speed that may be obtained,
> so materially depends. (Report in Appendix I p.388-9)

Mr. Brunel also looked forward to the advantages which a wider gauge would give for the construction of the locomotive engines. Difficulties had been already experienced from the limited width between the wheels, which cramped the machinery, rendering it difficult of access for repairs; it also limited the size of the boiler and fire-box on which the power depended. For this reason Mr. Brunel considered that a wider gauge would present great advantages, as it would allow the locomotives to be constructed of greater power, and with their machinery arranged in a more advantageous manner. He also thought that the greater width of base for the carriages would give increased steadiness and smoothness of motion, with greater safety, particularly at high speeds, and that there would be the advantage of being able to use larger wheels for the carriages. Moreover, he had in view the possibility which the broad gauge would give of adopting wheels of a still larger diameter without raising the centre of gravity, the body of the carriage being placed between them, as in the original type of common road vehicles.

The broad gauge was also considered by Mr. Brunel in prominent connection with the peculiarly favourable circumstances of the Great Western line, in regard to its gradients and curves. He thought that 'it would not have been embracing all the benefits derivable from the gradients of the Great Western Railway, unless a more extended gauge was adopted.' In the first place, it was evident that a diminution of the frictional resistance would present the greatest advantage where the gradients were flat. In regard to curves, the wider gauge was at that time considered by him to be more advantageously applied where the curves were of large radius than where they were sharp.[2] On the Great Western Railway both gradients and curves were remarkably good; with the exception of two inclines of 1 in 100, on which auxiliary power was proposed to be used, there was no gradient between London and Bristol steeper than 1 in 660, the greater part of the line being nearly level, and except between Bath and Bristol there was no curve sharper than about one mile radius.

For these reasons Mr. Brunel thought that unusually high speed might easily be attained for passengers, and great tractive power for goods. He said:—

> I shall not attempt to argue with those who consider any increase of speed unnecessary. The public will always prefer that conveyance which is the most perfect, and speed, within reasonable limits, is a material ingredient in perfection in travelling. (Report in Appendix I p.389)

In deciding that the distance between the rails should be seven feet, Mr. Brunel seems to have been guided by the principle that the wheels should be put sufficiently far apart to admit of an ordinary carriage body being placed between them.[3]

Mr. Brunel did not anticipate that the difference between the gauge he proposed and that of other railways would lead to any important inconvenience. The views he held on this subject were expressed fully by him in a report of December 13, 1838. He says, speaking of the difficulty of communication between the Great Western and other railways:—

> This is undoubtedly an inconvenience; it amounts to a prohibition to almost any railway running northwards from London, as they must all more or less depend for their supply upon other lines or districts where railways already exist, and with which they must hope to be connected. In such cases there is no alternative.
>
> The Great Western Railway, however, broke ground in an entirely new district, in which railways were unknown. At present it commands this district, and has already sent forth branches which embrace nearly all that can belong to it; and it will be the fault of the company if it does not effectually and permanently secure to itself the whole trade of this portion of England, with that of South Wales and the south of Ireland; not by a forced monopoly, which could never long resist the wants of the public, but by such attention to these wants as shall render any competition unnecessary and hopeless. Such is the position of the Great Western

Railway. It could have no connection with any other of the main lines, and the principal branches likely to be made were well considered, and almost formed part of the original plan; nor can these be dependent upon any other existing lines for the traffic which they will bring to the main trunk.

Mr. Brunel was not singular in holding the opinion that it would be desirable to allot a given district to one railway, which might conveniently serve it by means of a trunk line and branches of a special gauge. In the Eastern Counties line, designed in 1836, by Mr. Braithwaite, and opened in 1839, a gauge of 5 feet was adopted; and Mr. Robert Stephenson, as engineer of another line, the Northern and Eastern Railway, branching out from the Eastern Counties to the northward, adopted the same gauge. It was not till the Northern and Eastern line was extended, some years afterwards, that the rails of the whole system were altered to the narrow gauge.

Mr. Brunel's recommendation was adopted by the Directors of the Great Western Railway; and, in their report of August 25, 1836, after observing that the generally level character of the line would greatly facilitate the attainment of a higher speed of travelling, they pointed out the advantages of the broad gauge, and stated that engines had been ordered specially adapted to the nature of the line, which would be capable of attaining with facility a rate of from thirty-five to forty miles per hour.

The line was opened between Paddington and Maidenhead on June 4, 1838, and the performance of the engines was considered satisfactory, trains of eighty tons and upwards being drawn at speeds of from thirty-eight to forty miles per hour.

Notwithstanding these favourable results, the change in regard to the gauge did not pass unquestioned. Attacks were made on it in various quarters, and considerable excitement was caused among the shareholders and the public.

It was asserted that the width of 4 feet 8½ inches was exactly the proper width for all railways, and that a deviation from it was tantamount to the abandonment of an established principle which experience had proved to be correct. It was further alleged that the cost of all the works connected with the formation of the line would be greatly increased; that the carriages must be stronger and heavier, that they would not run round the curves, and would be liable to run off the rails, and particularly that the increased length of the axles would render them liable to be broken. These were not advanced as difficulties which, existing in all railways, might be somewhat increased by the increase of gauge, but they were assumed to be peculiar to the broad gauge, and fatal to it. Some urgent representations appear to have been made to the Directors; for in their report of August 15, 1838, they state, that as the gauge and the permanent way, which had also been the subject of adverse criticism, had been sources of some anxiety to them, they had applied to three of the most eminent authorities on the construction and working of railways—Mr. James Walker, President of the Institution of Civil

Engineers, Mr. Robert Stephenson, and Mr. Nicholas Wood, of Newcastle-on-Tyne—to undertake a thorough inspection of the line, to investigate the working of it, and to give their opinion on the plan adopted.

Mr. Walker and Mr. Stephenson declined the task, on the ground that they did not wish to become embroiled in professional controversy, but Mr. Wood undertook it; and a similar commission was afterwards given to Mr. Hawkshaw.

In order to put the shareholders fully in possession of all the information in their power, the Directors published a very complete statement by Mr. Brunel on the arrangements adopted by him. It will be seen that in this report, which is given in Appendix I he states his original arguments, and answers the objections brought against his plans; and he contends that the result of experience establishes their success. In regard to the gauge, he says:—

> Everything that has occurred in the practical working of the line confirms
> me in my conviction that we have secured a most valuable power to the
> Great Western Railway, and that it would be folly to abandon it.

But the two engineers, to whom the consideration of this matter had been referred, differed materially in opinion from Mr. Brunel. The nature of their investigations and reports, and of Mr. Brunel's replies, is stated in the extracts given below from the report of the Directors in January 1839.

In addition to the question of gauge, another important matter referred to the consideration of Mr. Wood and Mr. Hawkshaw was the construction of the permanent way.[4] On the Great Western a construction had been introduced by Mr. Brunel differing materially from that ordinarily used; and as defects had shown themselves after the opening of the railway, some anxiety was felt in reference to it by many of the shareholders.

The subject of the permanent way adopted on the Great Western Railway does not necessarily belong to the gauge question, and would, perhaps, have been more properly considered in the chapter on Mr. Brunel's railway works; but, as a matter of fact, the controversy concerning it became so interwoven with that of the broad gauge, that in a historical account it would be difficult to separate them.

It appeared to Mr. Brunel that, with a view of applying the engine power to the greatest advantage, particularly in attaining high speed, more attention ought to be paid to the construction of the permanent way. He says, in a report dated February 1837:—

> It appears to be frequently forgotten that although lofty embankments
> and deep cuttings, bridges, viaducts, and tunnels are all necessary for
> forming the level surface upon which the rails are to be laid, yet they are
> but the means for obtaining that end; and the ultimate object for which
> these great works are constructed, and for which the enormous expenses

consequent upon them are incurred, consists merely of four level parallel lines, not above two inches wide, of a hard and smooth surface; and upon the degree of hardness, smoothness, and parallelism (which last has hitherto been very much neglected) of these four lines depend the speed and cost of transport, and in fact the whole result aimed at . . .

In forming all my plans I have looked to the perfection of the surface on which the carriages are to run, as the great and ultimate desideratum; and in the detail of construction of this last operation, without which all the previous labour is comparatively wasted, I have always contemplated introducing all the perfection of materials and workmanship of which it is capable.

With a view to improvement on this point Mr. Brunel considered it would be advantageous if two important changes were made. He proposed, in the first place, to abolish the use of stone blocks for the rails to rest on, and to substitute timber; and, secondly, to apply the support uniformly and continuously along the whole length of the rails, instead of only at intervals.

The first of these changes, namely, the substitution of timber for stone, was not wholly new, for transverse wood sleepers were often used in exceptional situations; but it was the general opinion that stone, where it could be applied, formed the best support for the rails,[5] and the exclusive employment of timber was considered a great innovation.

The other principle, that of 'continuous bearing,' was similar to that of the old wooden and stone tramways; and, even as applied to iron rails it had been extensively used before, as Mr. Brunel mentions in his report of August 1838 (see Appendix I p.391)

Mr. Brunel designed for this continuous bearing a peculiar shape of rail, which, from the form of its section, acquired the name of the 'bridge rail.' The rail was bolted down to the longitudinal timbers, and the timbers of the two rails were connected together at intervals by cross-pieces, called transoms, bolted to them; these served to keep the two rails at a proper distance apart. The longitudinal timbers lay on gravel or other 'ballast,' which had been found to form the best foundation, as being firm and solid, easy of adjustment, and allowing free drainage.[6]

Mr. Brunel, however, thought there would be difficulty in giving the longitudinal baulks a sufficiently solid bearing on the gravel below them.

A similar difficulty had already been experienced with the heavy stone blocks used on other railways. As a remedy for this, Mr. Stephenson caused each block to be lifted and dropped several times on its place, so as to consolidate the ballast below.

The same thing could not be done with a long wooden baulk, and Mr. Brunel therefore contrived another mode of overcoming the difficulty. Piles were driven into the ground between the rails, and their heads bolted to the cross-transoms, the object being to hold the timber framework firmly down.

The gravel was then rammed hard under the longitudinal baulks, to give the consolidation desired. The result, however, of this mode of construction was far from successful, and the state of the road, when run over by the trains, was in many places very defective.[7]

In the course of his enquiry Mr. Wood tried a large number of experiments on the Great Western and other lines. He was of opinion that stone blocks afforded a permanently firmer base, and so caused less resistance to the train, but that there was less noise with continuous timber bearings, and that they gave a smoother and a more perfect road for high rates of speed. He thought, however, that the piles were objectionable, and that the weight of the trains would in the course of time sufficiently consolidate the foundation. Mr. Brunel accepted Mr. Wood's conclusions and abandoned the piling, adopting at the same time larger timbers and heavier rails.

The experience of the permanent way, as thus altered, fully justified the favourable anticipations Mr. Brunel had formed of the continuous timber bearing.[8]

After the reports of Mr. Wood and Mr. Hawkshaw, with Mr. Brunel's replies to them, had been circulated among the shareholders, a special general meeting was called in London, to receive and consider these documents. It was convened for December 20, 1838, but was adjourned till January 9, 1839. This meeting was of great importance not only to the Company, but to Mr. Brunel personally, as on the resolutions to be passed depended whether or not his plans should be proceeded with.

He had, however, the warm support of the Directors, as will appear from the following extracts from their report:—

It may be here concisely stated, that Mr. Wood deduces from experiments upon the performance of engines on the Great Western and other lines, that although a higher rate of speed has been attained on the former, it would appear only to have been accomplished by the increased power of the engines, with a much greater consumption of coke when calculated per ton per mile. He ascribes this result principally to the resistance presented by the atmosphere to the motion of railway trains, especially at high rates of speed. His remarks on that subject are qualified, however, by the expression of a doubt as to the value to be assigned to the single set of experiments on each of two inclined planes, which are quoted as the authority for the degree of atmospheric resistance supposed to have been discovered.

The reduction of friction by the employment of wheels of increased diameter, and the benefit of lowering the carriages between the wheels, are affirmed by Mr. Wood as incontrovertible. The increased stability, and consequent increased steadiness of motion to carriages on the wider base, are also admitted by him.

The various propositions of doubtful advantage from the wide gauge, as well as of alleged objection to it, appear to have been thoroughly considered in the report in question. The experiments on the consumption of coke at high velocities were unfavourable, and, in connection with the theory of atmospheric resistance, appear to have influenced the mind of Mr. Wood to consider that a seven-feet gauge was beyond the width which he would deem the best. At the same time, upon a review of all the circumstances, and considering that there are counteracting advantages, incidental to an increased width of gauge, he does not think that the result of his enquiries would justify a change in the dimensions adopted on this line, and he recommends the present width should be retained.

The advice thus given by Mr. Wood, upon mature reflection, being directly at variance with the conclusion at which Mr. Hawkshaw had previously arrived upon an investigation similarly delegated to him, it became the duty of the Directors to consider most attentively the train of reasoning and argument which led the latter to urge such an opposite course. Naturally expecting from that circumstance to find in his report a clear and definite statement of the positive loss or disadvantages accruing from the increased width of gauge, the Directors could not fail to remark with some surprise that he enforces his recommendation, not upon any ascertained injury or failure in the plan, but almost exclusively upon the presumption that all railways, however disconnected or locally situated, should be constructed of one uniform width. While he appears to think that it might be an improvement to have an addition of a few inches, five or six at the most, he still questions the expediency of any variation from the 4 feet 8½ inches gauge. Mr. Hawkshaw, in his report, also considers any additional expense upon the gauge, as well as upon the improvement of gradients, to be undesirable, and assumes it at a scale of augmentation far beyond the real difference of cost. His estimates on that head are impeached in the engineer's observations, and no doubt exists in the minds of the Directors, that the subject, reduced to a mere question of figures, in its present position, would undeniably show a pecuniary loss to be borne by the Company by any such change of system as he advocates, even if it were on other grounds deemed advisable. The objection that the wide gauge might prevent a junction with other lines seems both to Mr. Wood and the Directors to have but little weight, as applied to the Great Western Railway. Already has the same width been contemplated and provided for in the extension lines through Gloucestershire to Cheltenham and from Bristol to Exeter. Any local branches hereafter to be made would undoubtedly follow the same course, and the proprietors, therefore, may be satisfied that no apprehension need be entertained by them on that head.

The advantage of following Mr. Wood's advice, in not making any alteration in the width of way, has been since most forcibly shown by more recent experiments, which have entirely changed the results upon which the chief objections to the gauge were founded. The performance of the engines, shown by Mr. Wood's experiments in September, gave such a disproportionate result in their power upon the attainment of high velocities, as to render it all but impossible that the effect could be entirely produced by the action of the atmosphere on the trains. All doubts were shortly removed by its being ascertained that a different cause (a mere mechanical defect in the engine itself) had been in operation. If

Mr. Wood had witnessed these recent performances of the engines, he must unquestionably have changed his opinions as to the means and practicability of carrying full average loads at a high speed, without the great increased expense of fuel. The Directors have satisfied themselves of this very important fact, by personally attending an experiment (accompanied by several gentlemen, among whom was a very eminent practical mechanic), on which occasion the 'North Star' took a train of carriages, calculated for 166 passengers, and loaded to 43 tons, to and from Maidenhead, at a mean average speed of thirty-eight miles per hour, the maximum being forty-five miles per hour, consuming only 0.95, or less than 1 lb. of coke per net ton per mile, instead of 2.76, say 2¾ lbs., as previously shown. This was accomplished by a mere altered proportion in the blast pipe of the engine, in the manner explained by Mr. Brunel, being a simple adaptation of size in one of the parts, which admits a more free escape of steam from the cylinder, after it has exerted its force on the piston, still preserving sufficient draft in the fire.[9]

It must be almost needless to point out to those who have perused the reports, how importantly this change bears upon the subject in almost every relation of the enquiry. It negatives the assumption that the velocity can only be attained by a ruinous loss of power. It establishes beyond doubt that the consumption of fuel as now ascertained, in proportion to the load, is only one-third of that which from the former experiments had been the basis of Mr. Wood's arguments. An analysis in the report of the performance of the Great Western engines, with heavy loads varying from 80 tons to 166 tons, shows in every respect a peculiarly satisfactory result at a small cost of fuel, and warrants the expectation of very great benefit to the Company from the economical transport of goods on the line. That the expenses of locomotive repairs, especially on that heavy class of repair which arise from lateral strains on the wheels and framing of the engines, have been materially less than on other lines is ascertained by very detailed accounts, accurately made and submitted to the Board by the superintendent of that department. The experience of some months has now enabled the Directors to witness the progressive improvement in the practical working of the railway. A higher rate of speed has been generally maintained than on other lines, and at the same time, with that increased speed, great steadiness of motion has been found in the carriages, with consequent comfort to the passengers. If speed, security, and comfort, were three great desiderata in the original institution of railway travelling, the Directors feel sure that the public will appreciate and profit by any improvements in those qualities, the Company deriving ample remuneration in the shape of increased traffic. A saving of time upon a long journey, with increased comfort, will necessarily attract to one line in preference to another many travellers from beyond the ordinary distance of local connection, and will thus secure a valuable collateral trade which would not otherwise belong to it. It has also a decided tendency to avert competition, which may with much reason be regarded as the chief peril to which railway property is subjected.

The Directors, upon a deliberate reconsideration of all the circumstances affecting the permanent welfare of the undertaking, divesting the question of all personal partialities or obstinate adherence to a system, unanimously acquiesce in the abandonment of the piles, in the substitution of a greater scantling of timber, and of a heavier rail, retaining the width of gauge with the continuous timber bearings, as the most conducive to the general interests of the Company.

The views of the Directors were approved of by the majority of the shareholders (the numbers being 7,792 for, and 6,145 against); and the construction of the line was proceeded with according to Mr. Brunel's plans.

By June 30, 1841, the whole length of the Great Western Railway was opened from London to Bristol. Some of the Directors' reports mention the fact that the speed uniformly maintained by the engines much exceeded the ordinary rate of railway travelling, and allude to the 'general testimony borne to the smoothness and comfort of the line and carriages.'

As has been before mentioned, extensions and branches on the same gauge, to all of which Mr. Brunel was engineer, were projected and ultimately carried out, in accordance with the original scheme of the undertaking, to Exeter, Plymouth, and Cornwall, and to Gloucester, Hereford, and South Wales, as well as to Oxford, Windsor, and other towns in the immediate neighbourhood of the line.

About 1844, the attention of the Company began to be directed to projects involving extensions of a much more serious character, and which were destined to have a powerful influence on the position of the gauge question. During the railway mania, the Great Western Company found it impossible to stand aloof from the contests which were going on around them, and thought it necessary, in order to protect their own interests, to extend their lines beyond the district to which they had originally intended to confine themselves.

At the general meeting in August, an extension from Oxford to Rugby was determined on, as 'of the greatest importance to the Great Western line.' About the same time a broad-gauge line was promoted from Oxford to Worcester, and thence by Kidderminster and Dudley to Wolverhampton, in order to open an immediate communication with the Staffordshire and Worcestershire districts. There were also rival projects on the narrow gauge, promoted by the London and Birmingham Company; and the competing plans were referred, as was the custom at that time, for the examination of the Railway Department of the Board of Trade.

In regard to the communication from north to south, through Oxford, the question was, where the break of gauge should be.[10] The Board of Trade saw nothing in the relative merits of the gauges to determine this question, and from commercial considerations, they recommended that the change of gauge should be made at Oxford. On this and other grounds they considered that the narrow gauge schemes to the north of Oxford were preferable to those of the Great Western Railway.

The rival schemes then went before Parliament, and after a protracted enquiry, obstinately fought between the parties, the decision was given in favour of the Great Western lines, contrary to the recommendation of the Board of Trade. It was, however, stated by the chairman of the Commons Committee that the decision had been founded on the local and general merits of the respective lines, without any reference to the comparative merits of the two gauges. On this account some peculiar provisions were made in the Acts; for though the lines were sanctioned on the broad gauge, the proprietors were bound also to lay down narrow gauge rails upon them, if required to do so by the Board of Trade. At the same time the House of Commons, on the motion of Mr. Cobden, passed a Resolution praying her Majesty to refer the gauge question to a Royal Commission.

A Commission was issued in July; the Commissioners being three in number— Sir J.M. Frederic Smith, R.E.; Mr. G.B. Airy, Astronomer Royal; and Professor Barlow, of the Royal Military Academy, Woolwich. They took a large amount of evidence, both oral and documentary, and made some examinations of the working of the two gauges. Their report was presented to Parliament early in the session of 1846.

Of forty-eight witnesses, thirty-five were advocates of the narrow gauge; and against these were arrayed but four champions of the broad gauge, all officers of the Great Western Railway:—Mr. Charles Alexander Saunders, the secretary; Mr. Seymour Clarke, the traffic superintendent; Mr. (now Sir Daniel) Gooch, the locomotive superintendent; and Mr. Brunel.

The report was of considerable length, and in it the Commissioners addressed themselves to three heads of enquiry, viz.:—

> 1. Whether the break of gauge was an inconvenience of so much importance as to demand the interference of the legislature.
> 2. What means could be adopted for obviating or mitigating such inconvenience.
> 3. Considerations on the general policy of establishing a uniformity of gauge throughout the country.

The general conclusions arrived at on these points were thus summed up by the Commissioners:—

> 1. That, as regards the safety, accommodation and convenience of the passengers, no decided preference is due to either gauge, but that on the broad gauge the motion is generally more easy at high velocities.
> 2. That, in respect of speed, we consider that the advantages are with the broad gauge; but we think the public safety would be endangered in employing the greater capabilities of the broad gauge much beyond their present use, except on roads more consolidated, and more substantially and perfectly formed, than those of the existing lines.

3. That, in the commercial case of the transport of goods, we believe the narrow gauge to possess the greater convenience, and to be the more suited to the general traffic of the country.

4. That the broad gauge involves the greater outlay, and that we have not been able to discover, either in the maintenance of way, in the cost of locomotive power, or in the other annual expenses, any adequate reduction to compensate for the additional first cost.

Therefore, esteeming the importance of the highest speed on express trains for the accommodation of a comparatively small number of persons, however desirable that may be to them, as of far less moment than affording increased convenience to the general commercial traffic of the country, we are inclined to consider the narrow gauge as that which should be preferred for general convenience, and therefore, if it were imperative to produce uniformity, we should recommend that uniformity to be produced by an alteration of the broad to the narrow gauge.

Guided by the foregoing considerations, the Commissioners recommended that 4 feet 8 ½ inches should be fixed by law as the standard gauge of the country; and that as to the existing broad gauge lines, either they should be altered to the narrow gauge, or some course adopted which would admit of narrow gauge carriages passing along them.[11]

This adverse report was a great surprise to the supporters of the broad gauge system, as rumours had led them to hope for a different result. Immediately after its appearance, several documents were published, containing powerful and severe strictures on the proceedings and opinions of the Commissioners. The most important of these was written by Mr. Saunders, Mr. Daniel Gooch, and Mr. Brunel. It occupied fifty closely printed folio pages, and was entitled, 'Observations on the Report of the Gauge Commissioners, presented to Parliament.' To this, 'Supplemental Observations' were added, after the publication of the Evidence and the Appendix to the Report.

In the conclusion of the 'Observations' the writers gave a summary of the points they considered to have been proved in the controversy, namely—

That the question of 'break of gauge' originated as a cloak to a monopoly.

That even if the gauge were uniform, through trains would be impracticable.

That the transfer would be of little inconvenience.

That any advantage of small waggons was applicable to the broad gauge, but that the advantage of large waggons was not applicable to the narrow.

That the competition between the two systems was advantageous.

That the final recommendations of the Commissioners were at variance with their separate conclusions.

That it would be unjust to refuse to allow the broad gauge to be laid down on lines for which it was already sanctioned by Parliament.

> That the enquiry before the Commissioners was not properly conducted, and that consequently no legislation ought to be founded on it.
>
> That the data published by the Commissioners were often wrong, and in some cases led to the reverse of their conclusions.
>
> That greater economy was proved on the broad gauge.
>
> That the broad gauge was superior in the points of safety, speed, and conveyance of troops.
>
> That the experiments made in the presence of the Commissioners had demonstrated beyond all controversy the complete success of the broad-gauge system.

For these and other reasons, a strong protest was made against any legislative interference with the broad-gauge system.

A reply was published to these arguments; and during the controversy a large number of pamphlets, articles, and other publications appeared on both sides.

Mr. Brunel's views on the whole question, about this time, are concisely expressed in the following letter, written to a friend in France, who asked for information on the subject of the broad gauge:—

August 4, 1845.

I am just off for Italy, but write a few hasty lines in reply to Mons. ——'s queries, and which you must scold him for not addressing direct to me. Nobody can answer such questions but myself; and I am compelled to be very brief.

In answer to the *first*, I send a drawing.

*Secondly.* I see no reason why the ordinary construction of rails, chairs, and sleepers should not be equally applicable to the wide gauge as to the narrow. I have used them occasionally. I should think 75 lbs. per yard heavy enough for any purposes.

*Thirdly.* Within all ordinary limits, certainly in curves of more than 250 metres [12½ chains] radius, the gauge does not affect the question of curves. The effect of a curve of larger radius than this appears, both from much observation as from theory, to arise merely from two causes, the one centrifugal force, which is easily neutralised, and is independent of gauge; the other from the axles not being able to travel in the direction of the radius, and consequently the wheels not running in a tangent to the curve. This also is unaffected by the width of gauge. Practically I believe the conditions are not altered.

*Fourthly.* The expenses of construction are not dependent on the breadth of gauge unless the total width allowed for the loads or carriages is thereby or for other reasons increased, which is not a necessary consequence of a seven-feet gauge.

The wide gauge could be laid upon the London and Birmingham Railway without altering any of the works, but in constructing the Great Western Railway I thought it desirable to provide for carrying larger bodies, and I placed the centres of the two railways 13 feet apart, instead of 11 feet, and therefore my railway became *four* feet wider in total width.

The increased cost of this, including the cost of land, will vary from £300 to £500 per mile.

*Fifthly.* The increase of width will not increase the weight of an engine (of the same power) 500 lbs., but I avail myself of the larger width to get more powerful engines, and they weigh, with water in the boiler, 18 to 21 tons. I send a drawing of one; the stroke is 18 inches.

*Sixthly.* The passenger carriages are all on six wheels, and excessively strong; at present the framework of carriages and the whole of the waggons are made of iron. The first-class carriages weigh, with wheels, &c., 7 tons 16 cwt. (17,472 lbs.), and carry 32 passengers. Second-class about the same weight, and hold 72.

*Seventhly and Eighthly.* The comparison being on different railways under different managements and totally different circumstances, no strictly correct comparative results can be given; and of course the most opposite opinions are entertained and expressed. I believe we travel much quicker at the same cost and with more ease, and certainly the wear and tear of engines and carriages is *very much less* with us than with the other lines; but for the reasons above stated it cannot be made matter of exact proof, but remains matter of opinion.

The report of the Gauge Commission, on being presented to Parliament, was referred by the House of Commons to the Board of Trade, who reported on it in June 1846. They did not, however, concur with the Commissioners to the full extent of their recommendations; for, while admitting the break of gauge to be an evil, they could not, having regard to the circumstances under which the broad gauge companies had been established, and the interest they had acquired, recommend either that the broad gauge should be reduced to narrow, or that rails should be laid down for narrow gauge traffic over all their lines. Such measures would involve great expense, and they were unable to suggest any equitable mode of meeting it.

This conclusion necessarily affected the opinion of the Board of Trade in regard to the several lines under construction with the Great Western Railway, which the Board recommended should be all made on the broad gauge.

In regard to the broad-gauge lines sanctioned by Parliament from Oxford to Rugby, and from Oxford to Worcester and Wolverhampton, the Board determined to exercise their powers in requiring the narrow gauge to be laid down, in addition to the broad.

The House of Commons adopted the recommendations of the Board of Trade, and passed a series of resolutions in conformity thereto, and 'An Act for regulating the Gauge of Railways' received the Royal Assent on August 18, 1846.

It was enacted that it should not be lawful to construct any new passenger railway on any other gauge than 4 feet 8½ inches in England, and 5 feet 3 inches in Ireland.

Exceptions, however, were made in favour of certain lines in the west of England and South Wales.

The provisions relating to the gauge in the Acts for the Oxford and Rugby, and the Oxford, Worcester, and Wolverhampton Railways, were left in force.

The Act also generally excepted 'any railway constructed or to be constructed under the provisions of any present or future Act containing any special enactment defining the gauge or gauges of such railway or any part thereof.'

This Act, while it professed to establish the narrow gauge as the standard throughout the kingdom, did so only nominally; in reality, by the words 'present or future,' in the passage above quoted, it left the question of the gauge of any new railway open for the consideration of the committee on the particular bill; and it only obliged the promoters of the undertaking to adopt the narrow gauge when no case could be proved by them for the adoption of some other. This was equivalent to the former state of things, so that all the agitation of the question had ended in a mere expression of opinion, and the broad-gauge party were not only left with all their former liberty, but were encouraged, and almost compelled, to push their system still farther wherever they could.

About the time of the passing of the Gauge Act, a Board of Commissioners of Railways was established, to whom the powers formerly possessed by the Board of Trade were transferred.

One of the first duties of the Commissioners was to provide for the due compliance with the order of the Board of Trade respecting the introduction of the narrow gauge, in conjunction with the broad, on the Oxford and Rugby Railway.

It was proposed to effect this either by laying a narrow gauge line concentrically between the two rails of the broad gauge, or by laying down only one additional rail between the two broad-gauge rails, making one of the latter serve for both broad and narrow gauges. Mr. Brunel recommended the second of these plans to be adopted on the Oxford and Rugby line.

After a careful consideration of the question, the Commissioners sanctioned the mixed gauge formed by the introduction of a third rail; this was accordingly laid down, and none of the dangers which were at the time prognosticated in reference to it were found to exist. It has been the plan almost exclusively used in the many cases where the combination of the two gauges has been required.

In 1846 the Great Western Railway Company had promoted a bill for a branch to Birmingham from the Oxford and Rugby line at Fenny Compton. The Act was obtained, but they were defeated on the question of the gauge. However, after the passing of the Gauge Act, the Company again attempted to carry the broad gauge to Birmingham. Their application was backed by a strong memorial from the districts interested, and in June 1847 an order was passed by the House of Lords directing the Board of Commissioners of Railways:—

> To inquire into the accommodation afforded by the several lines of railway
> now open, or in the course of construction, or projected, between London

and Birmingham; and to report to this House, early in the ensuing Session
of Parliament in what manner they are of opinion that the interests of the
public may be most effectually secured in regard to such lines; and whether
it is expedient that the broad gauge should be extended to Birmingham;
and if so, in what manner such an arrangement can be carried into effect
with the least interference with existing interests . . .

This, of course, opened up again the whole question of the comparative merits
of the two gauges.

The Railway Commissioners issued a series of queries addressed to the officials
of the Great Western and the London and North Western Railway Companies,
and others. On the part of the Great Western, answers were given by Mr. Brunel
and Mr. Daniel Gooch. Mr. Gooch also furnished the results, with tables and
diagrams, of a very comprehensive series of dynamometrical experiments, made
by him on a mile of straight and level line on the Bristol and Exeter Railway.
These experiments fully demonstrated the advantages of the broad gauge, and are
still the chief authority on train resistances.[12]

In their report the Commissioners adopted the opinion of the Gauge
Commissioners 'that a break of gauge was a most serious impediment in the
transport of merchandise, and that the broad gauge did not offer any compensating
advantage so far as that description of traffic was concerned.' In regard, however,
to passenger traffic, they found a case for further enquiry. They said:—

> It is notorious that higher speeds, with larger and heavier passenger
> trains, are regularly maintained on a part of the line of the Great Western
> Railway than on any other railway in the country. This fact is known
> and greatly appreciated by a very large portion of the public; and no
> opinion respecting the extension of the district within which the broad
> gauge should be adopted is likely to be received with confidence which
> is not founded on a full consideration of the circumstances to which the
> above fact is to be attributed, and of the extent to which, under differing
> circumstances, if attributable to the breadth of gauge, the gauge of the
> Great Western Railway offers this advantage (p.11).

They assumed that the greater speed was due to greater engine power, and
they admitted that the increase of gauge allowed of an increase in the size and
power of the locomotive. They arrived at the result that the broad-gauge engine
'can draw on a level an ordinary passenger train of 60 tons with as much facility
at sixty miles an hour as the narrow-gauge engines can at fifty,' the advantage,
however, diminishing with steep gradients. In their report the following passages
are to be found:—

> Such appear to the Commissioners to be the advantages which the broad
> gauge at present offers; and although they cannot consider them sufficient

to compensate the evils attendant on two gauges, if it were now possible to obtain uniformity of gauge, yet, as two gauges are established, it appears to them that it might be expedient, and for the public interest, on account of those advantages, to extend the broad gauge to Birmingham . . . (p.14.)

By introducing the mixed gauge on the Birmingham and Oxford Junction Railway, the line from Birmingham by Fenny Compton to London would probably offer, as a broad-gauge railway, as rapid a communication as the existing direct line;[13] and great as the advantages which the public have received by the rivalry between the gauges, in the rapid improvement in railway travelling, have been, it might even be expected that these would be further increased when the two systems are brought into direct competition, which as yet they have not been (p.16).

The report of the Railway Commissioners was presented in May 1848. Their decision was ratified by the passing of an Act in the same session for extending the broad gauge from Oxford to Birmingham; and the line was opened in October 1852.

Beyond Birmingham the Great Western Company purchased existing railways leading through Wolverhampton and Shrewsbury to Chester, and obtained access to Birkenhead and Manchester. It thus secured a communication with the great Lancashire towns and the manufacturing districts.

But all the lines north of Wolverhampton had been constructed on the narrow gauge, and therefore, unless the broad gauge had been laid down on these lines, there was a break of gauge between the northern districts and London.

The break of gauge was found to be a much more serious evil than had been anticipated by the Great Western Company when they were fighting their great battle in 1845. For passengers the inconvenience was unimportant; but for the goods traffic between the manufacturing towns and London it was serious, partly on account of the expense, but more especially in consequence of the loss of time. The delay of some hours by change of waggons, where great competition existed, was fatal.

For these reasons the abolition of the break of gauge became desirable. The number of narrow-gauge lines had by the year 1861 been so increased that there was no longer any hope of advantageously extending the broad gauge in the north. Therefore the mixed gauge was completed to London.

After the establishment of the narrow-gauge communication on the northern lines of the Great Western, and its prolongation to London, there was but little inducement to use the broad gauge north of Didcot.

So far as it extended, the broad gauge had exhibited in a marked degree the advantages Mr. Brunel claimed for it, and which were neither few nor unimportant.

It may be desirable before concluding this chapter to sum up those advantages:—

1. It gave the power of constructing more powerful engines, by which greater speed for passenger trains and greater tractive power for heavy goods trains were obtained.[14]

2. It gave more space for the convenient arrangement and beneficial proportions of the machinery, as well as for convenient access to it. In all these points difficulties had been found on the narrow gauge; and the compulsory restriction of so important a dimension as the width between the rails has been a bar to any improvements of great magnitude or comprehensive nature.

3. It gave, even with the overhanging carriage, the facility for obtaining large wheels, and consequently diminishing the axle friction without sacrifice of stability.

4. The greater width of base for the carriages to rest on gave increased steadiness and smoothness of motion, particularly at high speeds. It was the impulse given by the increase of speed and comfort obtained without difficulty on the broad gauge, which had led to the chief improvements introduced in railway travelling.

5. Greater safety was secured, particularly at high speed, from the greater stability of position due to the wider base, producing increased steadiness and diminishing the chance under exceptional circumstances of the derangement of any part of the train.

6. While the broad gauge was but little more costly than the narrow, the width of the works being determined not by the width of the rails, but by the width of the carriages, and the extra cost of rolling stock being very small,[15] the broad gauge could be worked more economically under parallel circumstances than the narrow.

7. It gave the facility of using broader vehicles with equal steadiness, in cases where the extra breadth would be useful, though the extra breadth was by no means an essential part of the scheme.

The truth of these assertions, as establishing the superiority of the broad gauge, was of course vehemently denied by the advocates of the narrow gauge.

One objection urged by them, the inconveniences of the break of gauge, has undoubtedly been proved by experience to be a very powerful one, so powerful indeed as to compel the abandonment of the broad gauge on the lines where any considerable quantity of goods traffic has to be carried in competition with other companies.

Had Mr. Brunel's original plan been carried out, and had the broad-gauge companies taken possession of all the western portions of England, and avoided extensions into the north, the points of contact would no doubt have been so unimportant that no great inconvenience would have arisen, or a few miles of double gauge would have removed any difficulty; but, under the actual circumstances of the case, the Great Western Company were forced to yield.

The advantages of the broad gauge were so much appreciated by the districts it served, that its abandonment was viewed with considerable displeasure, particularly in the neighbourhood of Birmingham; but the inconvenience of double traffic arrangements far outweighed the advantages derivable from the use of the broad gauge, to the limited extent it could be applied on those outlying portions of the Great Western system. For these reasons the Company came to the determination to work their northern lines on the narrow gauge only.

The broad gauge is therefore now confined to the district for which it was originally intended. Even in this district there are many points of contact with the narrow gauge; but the inconveniences of break of gauge are by no means so important as they were in the north, and do not, at present at least, menace the continued existence of Mr. Brunel's design.

1.  No copy of this report can be found; but documents of subsequent date sufficiently indicate the nature of the arguments Mr. Brunel used in it.

    Mr. Brunel had about this time given much attention to the principles of wheel carriages, as is manifested by an interesting article 'On Draught' written by him for the work on 'The Horse', published by the Society for the Diffusion of Useful Knowledge.

2.  With regard to this point, Mr. Brunel afterwards admitted that he had held a mistaken opinion. In speaking of his reasons for adopting the narrow gauge on the Taff Vale Railway in 1838, he said before the Gauge Commission:—'One of the reasons, I remember, was one which would not influence me now; but at that time I certainly assumed that the effect of curves was such, that the radius of the curve might be measured in units of the gauge, in which I have since found myself to have been mistaken.'

3.  See Mr. Brunel's report of August, 1838, printed in Appendix I.

    This plan was never adopted, as it was found desirable upon the broad gauge to use still wider carriages overhanging the wheels; but advantage was taken of the broader base to use wheels of greater diameter. However, in the saloon carriages, where ease of travelling was the chief object aimed at, the bodies were placed within the wheels.

4.  In the course of constructing the earth-works of a railway, the contractors were accustomed to lay down *temporary ways* or lines of rail, for the earth waggons to travel upon. When these were done with, the proper road for the trains was laid down; and this, to distinguish it from the former one, was called the *permanent way.*

5.  See Wood, *On Railways,* 3rd. edit. 1838, p.151.

6.  A full description of the original road of the Great Western Railway, communicated by Mr. Brunel, will be found in Wood's *Treatise on Railroads,* 3rd edit. 1838, p.708.

7.  At this time Mr. Brunel was confined to the house by the effects of his accident on board the 'Great Western' steam-ship (see p.181-82). Had he been on the spot, he would have been able to give the work careful consideration during its progress, and to judge of the expediency of proceeding with the plan.

8.  The continuity of the timbers diminishes the risk of trains leaving the line from small imperfections in the permanent way. And, should a train leave the rails, the

injury to the carriages and to the road is generally less serious than it is when the wheels of a carriage off the rails come into repeated and violent contact with the cross sleepers. Instances have frequently occurred where carriages which have left the rails have run considerable distances on the longitudinal timbers without injury.

9.  This experiment excited the greatest interest, and it was long afterwards related how Mr. Brunel, by the stroke of a hammer, had knocked to pieces the scientific deductions of Dr. Lardner, who, as was well known, had prompted Mr. Wood's decision in this matter.

    Mr. Brunel was so much impressed with the great influence which the operation of the blast-pipe had on the working of the locomotive that he afterwards investigated the whole subject, and made further experiments to determine whether or not it might be expedient to abandon the steam blast, and to maintain the draught in the chimney with a fan worked by a rotary steam jet.

10. The inconveniences of a break of gauge had already been brought into notice. One of the narrow-gauge companies, the Midland, worked two existing lines of railway, one between Birmingham and Gloucester, laid on the narrow gauge, and another between Bristol and Gloucester, on the broad gauge; and thus there was a break of gauge at Gloucester.

11. It should, however, be added, that the Commissioners had stated in the body of their report: 'We feel it a duty to observe here, that the public are mainly indebted for the present rate of speed, and the increased accommodation of the railway carriages, to the genius of Mr. Brunel and the liberality of the Great Western Company.'

12. These experiments will be found in the Appendix to the Report of the commissioners of Railways, respecting railway communication between London and Birmingham (ordered to be printed May 22, 1848).

13. This was fully borne out afterwards, the express trains running in the same time, 3 hours, over both routes, though the length of the broad-gauge line was 129 miles, as against 113 of the narrow. Similar favourable results have been since exhibited in the completion between the broad and narrow gauge lines to Exeter.

14. Among many important advances in railway travelling made on the Great Western Railway, it may be mentioned that it was on this line that express trains running long distances without stopping were first introduced; and that, in 1845, within about a year of the completion of the line to Exeter, express trains ran from London to Exeter, 194 miles, in 4½ hours. This rate of travelling, which was accomplished without difficulty by the broad gauge in its early days, has scarcely been exceeded since on any railway.

15. Even in the locomotives when of equal power, Mr. Brunel calculated that the extra weight was not more than about 500 lbs.

    The extra cost of the Great Western Railway was only, including land, from £300 to £500 per mile, or less than 10 per cent of the whole; although Mr. Brunel had taken advantage of the broad gauge to get carriage bodies 2 feet wider than was then usual.

    The wide carriages and wagons were found less costly than the narrow ones in proportion to the load they carried.

# VI

# THE ATMOSPHERIC SYSTEM

IN the year 1844 Mr. Brunel recommended the adoption of the Atmospheric System of propulsion on the South Devon Railway, a line of 52 miles in length, which he was then constructing between Exeter and Plymouth. This system had, under the management of Messrs. Clegg and Samuda, been in operation with success on the Dalkey line for some time before Mr. Brunel adopted their apparatus on the South Devon Railway. After it had been in use on the South Devon for about twelve months, it was abandoned, and the railway worked throughout by locomotives.

It is therefore as important as it is interesting to examine the causes of the failure of the Atmospheric System, and to consider the reasons which induced Mr. Brunel in the first instance to adopt it, and afterwards to recommend its abandonment.

Up to about the year 1843, the cost of railways, which was in a great measure due to the conditions imposed by the limited capabilities of the locomotive, had prevented their construction, except in cases where they would secure a large traffic, and at the same time traverse what was then considered a practicable country.

A curvature of one mile radius was regarded as the maximum generally admissible on a line where high speeds were aimed at, and auxiliary locomotives were required to work heavy gradients.

Nevertheless, by the growing wants of the public and the growing boldness of engineers, the railway system was gradually being forced into districts hitherto regarded as unsuitable for it; and no country was held to be impracticable where the gradients could be surmounted by the inconvenient and costly expedient of auxiliary power.

The south of Devon had for several years demanded railway accommodation, and at the period now under review, Mr. Brunel projected what was called the coast line. This line, while it best accommodated the population of the district, passed through a very difficult country. If it was to be constructed at a moderate cost, curves of a quarter of a mile radius had to be admitted; and above 30 miles of its entire length traversed a district involving the adoption of gradients steeper than had been elsewhere used for such considerable distances. The Act for this railway was obtained in the Session of 1844.

The South Devon Railway, on leaving Exeter, crosses the flat country on the right bank of the river Exe, as far as Starcross, a village nearly opposite to Exmouth. From this point it runs down to the coast and along the sea shore, by Dawlish, to Teignmouth; being protected by a sea-wall for the greater part of the distance, and passing through several headlands by short tunnels. Beyond Teignmouth it follows the left bank of the river Teign, which it crosses a short distance before reaching the station at Newton Abbott. The portion of the line from Exeter to Newton—21½ miles in length—is very nearly level, the steep inclines for which the railway is noted being west of Newton. Between Newton and Totnes for the first mile and a half the line is almost level, and in the next two miles it rises 200 feet, with gradients of 1 in 100, 1 in 60, and nearly a mile of 1 in 43. At the summit is a short tunnel; and thence the line descends 170 feet in a mile and three-quarters, with gradients of 1 in 40 and 1 in 48 for about three-quarters of a mile, and gradients of 1 in 57 and 1 in 88 for the rest of the incline. It then runs with more moderate gradients and about a mile and a half of level line to Totnes.

From the valley of the Dart at Totnes, the line rises at once by a rapid ascent of 350 feet in four miles and a half, with gradients varying from 1 in 48 to 1 in 90, more than a mile and a half averaging 1 in 50. Thence it runs, with easy up and down gradients, for a distance of 12 miles along the skirts of Dartmoor, crossing by lofty viaducts the deep valleys which penetrate the moor. It then descends to Plympton, in the valley of the Plym, falling 273 feet in a little more than two miles, with a gradient of 1 in 42½. From Plympton the line for two miles is level, and then rises on an incline of 1 in 80 for a mile and a half, and descends by a similar gradient into the Plymouth station.

The main characteristics of the railway are that, while it traverses a very heavy country, its principal changes of level are concentrated into four long and steep inclines. These four inclines were intended to be worked by auxiliary power.

Hitherto on gradients of unusual steepness a stationary engine with rope traction had been generally regarded as the only available expedient; but the special difficulties by which this system was encumbered rendered it unsuitable for high-speed passenger traffic, and practically inapplicable to an extended line. It had, however, been very successfully employed by Mr. Robert Stephenson on the Blackwall Railway, a line of about 3¾ miles in length.

Messrs. Clegg and Samuda, the projectors of the Atmospheric System, which was another mode of using stationary power, had, previously to this period, laboured to attract the attention and win the favourable opinion of engineers and the general public.

It is desirable, before proceeding further, to give a brief description of this system of traction, upon the merits of which distinguished engineers entertained widely different opinions.

Between the two rails of the line of way was laid a cast-iron tube, which on the Croydon and Dalkey railways and the completed or level portion of the South Devon Railway was fifteen inches in diameter. On the inclines it was proposed to use a twenty-two inch tube.

At intervals of about three miles along the line were erected stationary engines, working large air-pumps, by means of which air could be exhausted from the tube, and a partial vacuum created within it. A close-fitting piston was placed in one end of the tube, and the air being exhausted from it, the pressure of the external air on the surface of the piston which was towards the open end of the tube forced the piston through the tube towards the end where the air-pumps were working; so that if the piston were connected with a carriage running on the rails, it would draw the carriage with it. The connection between the piston and the carriage was arranged by Messrs. Clegg and Samuda in the following way:[1] Along the top of the tube was a slit about 2½ inches wide; this slit was closed by a long flap of leather, which was strengthened with iron plates, and secured to the tube at one side of the slit. One edge of the leather thus formed a continuous hinge; the other edge, where it closed on the tube, was sealed with a composition of grease, to render it air-tight. This flap was known by the name of the longitudinal valve.

When the valve was closed, the air could be exhausted from the tube in front of the piston, and a partial vacuum formed. Behind the piston, the air being at atmospheric pressure both within and without the tube, there was no objection to opening the longitudinal valve; and a bar, extending downwards from the under side of the carriage, entered the slit obliquely under the opened valve, and was connected to the rear end of a frame about ten feet long, the front end of which carried the piston. To allow the bar to pass along the slit, the valve was opened on its hinge, being pressed upwards by a series of wheels carried by the moving piston-frame inside the tube. The valve closed again after the passage of the train; and the tube was ready to be exhausted in preparation for the passage of the next train.[2]

The Atmospheric System was first tried in 1840. An experimental tube was laid down at Wormwood Scrubs on part of the short line now incorporated into the West London Railway, and then known by the title of the Bristol, Birmingham, and Thames Junction Railway. Its working was the subject of eager discussion among engineers.

In 1842 Sir Frederic Smith, R.E., and Professor Barlow, under an order from the Board of Trade, reported so favourably on the system with reference to the proposal for its application on the Dalkey branch of the Dublin and Kingstown Railway, that it was adopted there. In 1843 Mr. (afterwards Sir William) Cubitt determined to employ it on the Croydon Railway; and about the same time Mr. Robert Stephenson was desired by the Directors of the Chester and Holyhead Railway to report on the propriety of introducing it on that line.

Mr. Stephenson's report was based on a series of experiments on the working of the system at Dalkey. The view he took was adverse to its adoption, not only on the Chester and Holyhead, but on almost every railway whatsoever; and this on the ground that, though it was quite capable of being developed into a practical working system, yet on lines with ordinary gradients the atmospheric traction must be considerably more costly than locomotive traction, and on steep gradients than rope traction; in other words, that, as a mechanical appliance it was, though practicable, not economical.

Mr. Stephenson's report had no sooner appeared than the correctness of his conclusions was disputed on his giving evidence before a Committee of the House of Commons, in the spring of 1844, on the Croydon and Epsom Railway Bill.

Mr. Brunel also was summoned as a witness. Previously to this time he had taken a great interest in the various attempts which had been made to introduce the Atmospheric System, and he had himself conducted experiments at Wormwood Scrubs and at Dalkey. As early as July 1840 he had considered its applicability to the Box Tunnel incline on the Great Western Railway. He had also considered it in reference to various projected lines; and in 1843 he recommended it for adoption in a long tunnel on one of the steep inclines of the Genoa and Turin Railway, the success of the system being then (he wrote) sufficient to justify its use on a part of the line protected from weather. It was not, however, applied on this railway.[3]

Mr. Brunel's views at this time are indicated in the following letter:—

April 8, 1844.

Any part I have taken in examining into the system has been purely from the desire which I always feel to forward good inventions; and when I have formed a decided opinion, no fear of the consequences ever prevents my expressing it. My great anxiety, however, is to see a line of railway and all its appurtenances made expressly for the Atmospheric System, and worked accordingly; until this is done the results will be comparatively unsatisfactory.

Although unwilling to express general opinions, Mr. Brunel spoke strongly before the Croydon and Epsom Railway Committee in favour of the advantages of the Atmospheric System under certain circumstances, and approved of its use on that line.

A few months later Mr. Brunel recommended the Directors of the South Devon Railway Company to adopt the Atmospheric System, and they resolved to act on his advice.

His report was as follows:—

August 19, 1844.

I have given much consideration to the question referred to me by you at your last meeting—namely, that of the advantage of the application of the Atmospheric System to the South Devon Railway.

The question is not new to me, as I have foreseen the possibility of its arising, and have frequently considered it.

I shall assume, and I am not aware that it is disputed by anybody, that stationary power, if freed from the weight and friction of any medium of communication, such as a rope, must be cheaper, is more under command, and is susceptible of producing much higher speeds than locomotive power; and when it is considered that for high speeds, such as sixty miles per hour, the locomotive engine with its tender cannot weigh much less than half of the gross weight of the train, the advantage and economy of dispensing with the necessity of putting this great weight also in motion will be evident.

I must assume also that as a means of applying stationary power the Atmospheric System has been successful, and that, unless where under some very peculiar circumstances it is inapplicable, it is a good economical mode of applying stationary power.

I am aware that this opinion is directly opposed to that of Mr. Robert Stephenson, who has written and published an elaborate statement of experiments and calculations founded upon them, the results of which support his opinion.

It does not seem to me that we can obtain the minute data required for the mathematical investigation of such a question, and that such calculations, dependent as they are upon an unattained precision in experiments, are as likely to lead you very far from the truth as not.

By the same mode M. Mallet and other French engineers have proved the success of the system; and by the same mode of investigation Dr. Lardner arrived at all those results regarding steam navigation and the speed to be attained on railways, which have since proved so erroneous.

Experience has led me to prefer what some may consider a more superficial, but what I should call a more general and broader, view, and more capable of embracing all the conditions of the question—a practical view.

Having considered the subject for several years past, I have cautiously, and without any cause for a favourable bias, formed an opinion which subsequent experiments at Dalkey have fully proved to be correct; viz. that the mere mechanical difficulties can be overcome, and that the full effect of the partial vacuum produced by an air-pump can be communicated, without any loss or friction worth taking into consideration, to a piston attached to the train.

In this point of view the experiment at Dalkey has entirely succeeded. A system of machinery which even at the first attempt works without interruption and constantly for many months, may be considered practically to be free from any mechanical objection.

No locomotive line that I have been connected with has been equally free from accidents.

That which is true for one railway of two miles in length is equally true for a second or third, although they may be placed the one at the end of the other; the chances of an accident are only in the proportion of the number, or in other words, the length, a proportion which holds equally good with locomotives, except that a locomotive may be affected by the distance it has previously run, while a stationary engine and its pipes cannot in like manner be affected by the previous working of the neighbouring engine and pipes.

In my opinion the Atmospheric System is, so far as any stationary power can be, as applicable to a great length of line as it is to a short one.

Upon all these points I could advance many arguments and many proofs, but I shall content myself with saying that, as a professional man, I express a decided opinion that, as a mechanical contrivance, the Atmospheric apparatus has succeeded perfectly as an effective means of working trains by stationary power, whether on long or short lines, at higher velocity and with less chance of interruption than is now effected by locomotives.

I will now proceed to consider the question of the advantage of its application to the South Devon Railway.

It will simplify the discussion of the question very much if it is considered as a comparison between a double line worked by locomotives in the usual manner, and a single line of railway worked by stationary power, the only peculiarity of the present case being that upon four separate portions of the whole 52 miles stationary assistant power would under any circumstances have been used, these four inclines forming together one-fifth of the whole distance.

It is necessary to consider it as a question of a single line on account of the expense, the cost of the pipe for each line being about £3,500 per mile.

An addition of £7,000 per mile, or of about £330,000 in the first construction could not be counterbalanced by any adequate advantage in the saving in the works on the South Devon Railway, and probably not by any subsequent economy or advantage in the working; but the system admits of the working with a single line, without danger of collision, certainly with less than upon a double locomotive line. And I believe also that, considering the absence of most of the causes of accidents, there will even be less liability of interruption and less delay in the average resulting from accidents than in the ordinary double locomotive railway.

By the modification of the gradients and by reducing the curves to 1,000 feet radius where any great advantage can be gained by so doing, and by constructing the cuttings, embankments, tunnels, and viaducts for a single line, a considerable saving may be effected in the first cost.

In the permanent way and ballasting, the reduction will be about one-half. I should propose to make the rails about 52 lbs. weight and the timber 12 x 6; the quantity of ballast would probably be rather more than half but at the present prices of iron and timber the saving could not be less than £2,500 per mile.

From a careful revision of the works generally, I consider that a reduction may be effected in the following items, and to the amount specified in each, viz., ballasting gradients and curves:—

|  | £ | s. | d. |
|---|---|---|---|
| Reduction in earth work | 16,500 | 0 | 0 |
| "   in length of principal tunnel | 14,000 | 0 | 0 |
|  | 30,500 | 0 | 0 |

*Saving by single line*

| | | | |
|---|---|---|---|
| Earth work | 25,000 | 0 | 0 |
| Tunnels | 11,500 | 0 | 0 |
| Viaducts | 15,000 | 0 | 0 |

*Permanent way and ballast*

| | | | |
|---|---|---|---|
| To allow for sidings, say 50 miles, £2,500 | 125,000 | 0 | 0 |
| | 207,000 | 0 | 0 |

*Per contra*

| | £ | | | |
|---|---|---|---|---|
| Pipe on 41½ Miles[4] | 138,500 | | | |
| Increase on inclined planes, 10½ Miles | 6,500 | £ | s. | d. |
| | 145,000 | 0 | 0 | |
| Engines for the 41½ Miles | 35,000 | 0 | 0 | |
| Patent right, say | 10,000 | 0 | 0 | |
| | 190,000 | 0 | 0 | |
| The difference in first cost therefore is | 17,000 | 0 | 0 | |
| To this must however be added the cost of the locomotive power, with its attendant expenses of engine-houses, &c., which cannot, I think, be put at less than | 50,000 | 0 | 0 | |
| Making a saving of | 67,000 | 0 | 0 | |

I have not included in the expense of the Atmospheric apparatus that of the telegraph, because at its present reduced cost of £160 per mile I am convinced its use would repay the outlay in either case.

It would appear, then, that the line can be constructed and furnished with the moving power, in working order, on the Atmospheric System, for something less than the construction only of the railway fitted for the locomotive power, but without the engines; and that taking into consideration the cost of locomotive power, a saving in first outlay may be effected of upwards of £60,000.

But it is in the subsequent working that I believe the advantages will be most sensible.

In the first place, with the gradients and curves of the South Devon Railway between Newton and Plymouth, a speed of thirty miles per hour would have been, for locomotives, a high speed, and under unfavourable circumstances of weather and of load, it would probably have been found difficult and expensive to have maintained even this; with the Atmospheric, and with the dimensions of pipes I have assumed, a speed of forty to fifty miles may certainly be depended upon, and I have no doubt that from twenty-five to thirty-five minutes may be saved in the journey.

Secondly, the cost of running a few additional trains so far as the power is concerned is so small, the plant of engines, the attendance of engine-men, &c.,

remaining the same, that it may almost be neglected in the calculations; so that short trains, or extra trains with more frequent departures, adapted in every respect to the varying demands of the public, can be worked at a very moderate cost. I have no doubt that a considerable augmentation of the general traffic will be thus effected, by means which with locomotive engines would be very expensive, and frequently unattainable, particularly as regards one class of short trains, whether for passenger or goods, which from the inconvenience of working them by locomotives are hardly known—I refer to trains between the intermediate stations.

By many means, which the easy command of a motive power at any time, at every part of a line, must afford of accommodating the public, I believe the traffic may be increased.

It appears to me also that the quality of the travelling will be much improved; that we shall attain greater speed, less noise and motion, and an absence of the coke dust, which is certainly still a great nuisance; and an inducement will thus be held out to those (the majority of travellers) who travel either solely for pleasure, or at least not from necessity, and who are mainly influenced by the degree of comfort with which they can go from place to place.

Lastly, the average cost of working the trains will be much less than by locomotives.

With the gradients of the South Devon Railway, and assuming that not less than eight trains, including mail and goods trains, running the whole distance, and certainly one short train running half the distance, be the least number that would suffice, I think an annual saving of £8,000 a year in locomotive expenses, including allowance for depreciation of plant, may very safely be relied upon.

For all the reasons above quoted, I have no hesitation in taking upon myself the full and entire responsibility of recommending the adoption of the Atmospheric System on the South Devon Railway, and of recommending as a consequence that the line and works should be constructed for a single line only.[5]

In this report Mr. Brunel rested his recommendation principally on two assumptions, which he held to be indisputable—(1) That stationary power, if freed from incumbrances such as the friction and dead weight of a rope, was superior to locomotive power; and (2) That the Atmospheric System of traction was theoretically a good and economical method of applying stationary power, and that it was also a practical and working system, as had been shown in its first and somewhat crude application at Dalkey.

The superiority of stationary as compared with locomotive power depends on two principles—(a) That a given amount of power may be supplied by a stationary engine at a less cost than if supplied by a locomotive. (b) That the dead weight of a locomotive forms a large proportion of the whole travelling load, and thus inherently involves a proportionate waste of power—a waste which is enhanced by the steepness of the gradients and the speed of the trains.

A detailed examination of these principles is given in the note to this chapter.

It is there shown that at the time referred to stationary power could be obtained for one farthing per horse power per hour, while locomotive power cost more than one penny per indicated horse-power per hour, the cost of the locomotive power being more than four times that of the stationary power. On a level line at a speed of 60 miles per hour, for each horse-power usefully employed, a locomotive, in consequence of its own dead weight and the friction of its machinery, is obliged to expend more than one and a half horse-power; in this case, therefore, the useful work done costs nearly seven times as much as if it had been performed by a stationary engine. Again, on so moderate an ascent as one in 75, for each horse-power usefully employed, the locomotive has to expend at 40 miles per hour more than two horse-power, and at 60 miles per hour three horse-power; so that the useful work done costs in the one case nearly ten times, in the other thirteen times, as much as if it had been performed by a stationary engine.

This great advantage of stationary over locomotive power was a sufficient justification for introducing a system which promised to realise it to any considerable extent; although many difficulties might have to be encountered.

As has been already stated, Mr. Brunel had satisfied himself that the Atmospheric System was theoretically economical, and that its trial at Dalkey had shown that it was practically free from mechanical objections. Mr. Stephenson, indeed, had admitted that the mechanical details had been brought to a remarkable degree of perfection; but this admission was not any qualification of the radical difference of opinion which existed between him and Mr. Brunel, to which Mr. Brunel drew attention in his report of August 1844.

The subsequent abandonment of the Atmospheric System led many to believe that Mr. Brunel had been rash in rejecting the detailed investigations and conclusions of Mr. Stephenson, and that the adverse conclusions which Mr. Brunel had refused to entertain were subsequently established.

But, in fact, the failure of the Atmospheric System was due to failure in some of its mechanical details, and was not due to those inherent defects in its principle which Mr. Stephenson in his report considered that his experimental data had established.

Mr. Brunel's opinions were again brought prominently into notice in the evidence given by him before a Select Committee of the House of Commons, which was appointed in the Session of 1845, in consequence of the number of projected lines which it was proposed to work by the Atmospheric System.

The following letter, written by Mr. Brunel to one of the members of the Committee, explains what he considered to be his position at this time:—

April 3, 1845.

I am summoned to attend your Committee on Friday, and as it is known that I have expressed opinions favourable to the Atmospheric System, and that I am actually applying it upon a line of some length, it would be considered an absurd affectation, and would, moreover, be useless to attempt, to avoid giving evidence when called upon; but I am a most unwilling witness. I think it rather hard upon a professional man, who wishes to be cautious and prudent, that he should be called upon to express general opinions, which, if written even in the most studied and careful language, cannot be so worded as to be applicable to every case that may hereafter arise, or to be proof against the unfair and unscrupulous attack of the paid writers on these controversies. I mention these difficulties, which I feel that you may make some allowances for, if my feelings should appear in my evidence, or if that evidence should appear to fall very short of the opinions I am known to entertain, and which I must entertain to induce me to apply the system extensively, as I am doing. I find it difficult to define the points upon which it would be desirable to examine engineering witnesses, and I really believe that, entering freshly upon the subject and feeling as one of the public, you are more likely to elicit the useful points than one who, like myself, has been turning his whole attention (lately, at least) solely to the mechanical construction. However, I enclose a copy of a letter I addressed to a party interested in the patent, which refers to my opinions on the several points—opinions, however, expressed without that caution to which I referred as so necessary.

(Enclosure.)

March 31, 1845.

I object very much to giving evidence upon the abstract point of the applicability of a particular system, and thus furnishing general opinions which others are to apply as they may choose to particular cases; and if I could, I would refuse to give evidence at all before the present Committee, whatever might be the consequence to the promoters of the Atmospheric System. Circumstances, however, render such a refusal impossible; but I am equally anxious not to be drawn into becoming an advocate of a system.

When I gave evidence last year,[6] although it was then very much against my inclination, it was in support of a particular case; and it was only incidentally that my opinion was advanced as to a system. The evidence is now avowedly sought in support of a system, and I do not, as I before stated, intend to become an advocate of this or any other system. I mention these my views to prevent disappointment. If the following facts and opinions are likely to be of use to the Committee, I can give evidence on them.

I made experiments upon the portion of railway laid on Wormwood Scrubs. These experiments were made for my own private satisfaction, and not made public in any way.

They satisfied me of the mechanical practicability of the system.

In 1843 and 1844 I made several experiments upon the Dalkey railway.

The result of my observations and of those experiments is an opinion that the mechanical difficulties attending the application of this system may be overcome, and the whole as a machine made to work in a very perfect manner; that is, as a mechanical power for locomotion it will generally, but not in every case, be more economical than what is strictly called locomotive power, that high speeds may be more easily attained, that, from the absence of the locomotive engine, the rails may be constructed and maintained in more perfect order, and as a consequence the carriages may be constructed and worked in a more perfect manner, and so as to run more smoothly, and that in all respects the travelling may be rendered more rapid, more luxurious, and more safe. As regards the last, viz. safety, collisions may be rendered altogether impossible, or most remotely possible; while all other sources of danger, now very small, may be almost entirely removed by the increased perfection of the rails.

As regards first cost, a single line may be made to answer all the purposes of a double locomotive line for most railways, except main lines in immediate connection with the metropolis, or forming trunk lines for others with important branches not under the same control; and a single Atmospheric line will generally cost as little, often less, including the working power, than a double locomotive line without the engines. I am now constructing a line of 52 miles in length entirely for the Atmospheric System. I already see many advantages to be attained in the setting out and constructing of a railway, if originally designed for this system; the principal advantages can only be attained, and, above all, the principal difficulties in the system can only be properly provided against, where the line is originally designed for the system; the choice of gradients and curves and levels, the position, and, above all, the arrangements of stations, will generally be totally different in the two systems, and the difficulties to be avoided will equally differ.

It is unnecessary to give any extracts from Mr. Brunel's evidence,[7] as it only repeats the opinions and calculations embodied in his report of August 1844.

The Committee, while they allowed that experience could alone determine under what circumstances of traffic or of country the preference to the Atmospheric or the Locomotive System should be given, reported very strongly in favour of the general merits of the Atmospheric System.

As soon as it had been decided that the South Devon was to be constructed as an Atmospheric line, the dimensions of the cuttings, embankments, and tunnels were arranged for a single line, except on the long incline west of Totnes, and on that east of Plympton. In the approaches to the principal summits the gradients were made somewhat steeper, so as to reduce the excavations.

In December 1844, Mr. Brunel prepared a specification and drawing of a steam-engine and vacuum pump, and a copy was sent to the most eminent engine-builders, together with a letter inviting tenders for six pairs of engines. The letter concluded as follows:—

> Any party whose tenders may be accepted, shall, if required, furnish forthwith a more detailed drawing and specification of the engines as they propose to furnish them; the specification and drawing now sent being

expressly made very general, in order that each manufacturer may, so far as is consistent with the general requisites and conditions, adopt his own methods of construction, or use any existing patterns.

The economical character of the engines which Mr. Brunel desired to obtain is sufficiently indicated by the requirement that they were to be high-pressure condensing engines, fitted with double-seated expansion valves, and having boilers proved to 100 lbs. per square inch, and guaranteed to work with safety valves loaded to 40 lbs. per square inch.

The tenders accepted were those of Messrs. Boulton and Watt, Messrs. Rennie, and Messrs. Maudslay and Field.

Mr. Brunel left it to the contractors to prepare their designs without any interference on his part, deeming it best to rely on their unfettered judgment.

The task of manufacturing and fitting the cast-iron tube was one of some difficulty, the longitudinal slit allowing of considerable distortion in the casting. This work was undertaken by Mr. George Hennett, by the aid of a set of very effective tools devised by Mr. T.R. Guppy.

The tubes were supplied at the rate of one mile per week, and by the middle of 1846 nearly the whole line was laid to Newton, and the valve was ready to be fixed.

In the autumn of 1846, Mr. Joseph Samuda went to Dawlish, taking with him a staff of assistants trained in the working of the system at Croydon; and every effort was made to advance the completion of the engines and the other parts of the apparatus.

Owing, however, to vexatious delays in the erection of the engine-houses and engines, it was not until the commencement of 1847 that a piston-carriage was able to traverse the first six miles out of Exeter. And, though repeated experimental trains continued to be run, no passengers had been conveyed by Atmospheric trains prior to the general meeting of the shareholders, at the end of August. Mr. Brunel's report on this occasion was as follows:—

August 27, 1847.

It is a subject of great regret, and to no one more than to myself, that we have as yet been unable to open any portion of the line to the public with the Atmospheric apparatus, although a considerable distance has for some months been in a state to admit of frequent experiments being made upon it. This delay has arisen principally, if not entirely, in that part of the whole system which it might have been expected would have been the least exposed to it—namely, the construction and completion of the steam-engines.

It is due to Mr. Samuda that I should say that, so far as regards the mere pipe and valve, and other details which may be said to constitute the Atmospheric apparatus, we might long since have commenced. But the engines, although designed without any interference with their plans, and furnished by the best makers of the country, and although differing so slightly from the ordinary construction of steam-engines, have proved sources of continued and most vexatious delays, both in the unexpected length of time occupied originally in their erection, and in subsequent correction of defects in minor parts. While the engines were imperfect, it would not only have been unwise to have commenced working the line, even had it been practicable, but the frequent interruptions to the continuous working of all the engines rendered it impossible to complete and test the different portions of the Atmospheric apparatus. There are still some defects to be remedied in one or two of the engines, and I am using every endeavour, by persuasion and by every other means in my power, to urge on the manufacturers in their work of completion. Within the last week or two only have we been able to work at all continuously between Exeter and Teignmouth, so as to have the opportunity of trying the different parts, and getting the various details requisite for actually working trains tested and brought to sufficient perfection to ensure efficiency and regularity.

Since the beginning of last week, however, four trains per day have been run regularly, stopping at the stations, and keeping their time as if working for traffic. The tube and valve appear in good order, and the whole has worked well, but the running in this manner can alone show the deficiencies which may still exist in the details necessary for stopping, and starting quickly from the stations, and all the other minor operations incidental to working the traffic in the ordinary course; and, until all these arrangements are completed, and the engines in more perfect order, I think it would be much better to defer at least the substitution of the Atmospheric for the locomotive working. Trains, in addition to those now running may perhaps be advantageously worked for the public, after a further short continuance of the present practising.

The two engines completing the number to Newton are nearly ready for trial, and it is to be presumed that, after the experience of the past, the makers will be enabled to put them at once into an efficient state.

The delays and difficulties attending the bringing into operation the Atmospheric System upon this portion of the railway have been beyond all anticipation, and beyond what any previous experience would have justified anybody in anticipating. The difficulties have all been seriously aggravated by the necessity (consequent, certainly, upon the original delays) of working the line with locomotives during the construction and completion of the Atmospheric apparatus. Not only has the constant occupation of the line interfered with the progress of the work, but it has been necessary to devise all the arrangements so as to admit stations, sidings, and line being worked either by locomotive or by Atmospheric in succession, or even at the same time.

These difficulties, added to those always consequent upon the introduction of any new system, have been most wearying and incessant, and I am not surprised that the public and the proprietors should have been impatient. I trust the ultimate result will remove any grounds for disappointment.

The stress of personal anxiety and personal fatigue, experienced by Mr. Brunel and by all who were engaged in the work, was very severe, and continued so to the end. Not only was the progress in the completion of the work slow, but in spite of every exertion the results were incessantly marred by unfortunate contingencies which involved further delay, discouragement, and expenditure. Moreover, the reaction which followed the railway mania had set in; calls were ill responded to, and great difficulty was experienced in raising the money requisite for the completion of the line.

Under these circumstances it was resolved, on September 1, not to incur any new expenses in relation to the Atmospheric System beyond Totnes, and to limit any expenditure already contracted for, until its working between Exeter and Totnes had been fairly tried, except to provide assistant power up the two inclines.

On September 8 the Atmospheric trains began to take their share in the passenger duty of the line, four trains running each way daily; and, except when occasional mishaps caused delay, the new mode of traction was almost universally approved of. The motion of the train, relieved of the impulsive action of the locomotive, was singularly smooth and agreeable; and the passengers were freed from the annoyance of coke dust and the sulphureous smell from the engine chimney.

In other respects the record of progress is but a chequered one, and exhibits, in spite of great and able efforts and brightening intervals of occasional improvement, indications of growing difficulties deepening into ultimate defeat.

In examining the chronicle of events which correspondence and memoranda supply, it is inevitable that references to failure and disaster should be found relatively in far greater abundance than records of success; and this for the simple reason that there was at that time great use in taking note of the unfavourable incidents that occurred, almost none in mentioning successful work.

There is therefore some danger of falling into a mood of unjust depreciation, such as Mr. Brunel had in energetic terms urged the Directors to guard themselves against. He protested against their requiring (as they once intended to do)—

> continuous and detailed reports—if true and honest, of course containing nothing but accounts of mishaps—of a system which (he says) we are struggling to render perfect. Why, a daily account of our locomotive mishaps would ruin the locomotive system, if it were new! I will undertake to say that the mishaps of yesterday or to-day on the Great Western Railway were as great as that of Tuesday on the South Devon . . .

The Atmospheric System was vaguely credited with every delay which a train had experienced in any part of its journey; though, in point of fact, a large proportion of these delays was really chargeable to that part of the journey which was performed with locomotives. It often happened that time thus lost was made up on the Atmospheric part of the line, as is shown by a record of the working, which is still extant. In the week, September 20–25, 1847, it appears that the Atmospheric trains are chargeable with a delay of 28 minutes in all; while delays due to the late arrival of the locomotive trains, amounting in all to 62 minutes, were made up by the extra speed attainable on the Atmospheric part of the line.

Not unfrequently, however, casualties occurred; due indeed to remediable causes, but yet of discouraging aspect in themselves, and deriving additional weight from the manner in which they reacted on the cost of working. Such, for instance, were the frequent and occasionally very serious breakages in essential parts of the pumping-engines. Again, the cupped leathers of the travelling piston, which made it air-tight, were often destroyed while it was passing the various inlet and outlet valves. Improvements in the valves were introduced to meet this difficulty; but the remedy could not be applied at once throughout the line, and much inconvenience was thus experienced, and a considerable expense incurred. Another source of inconvenience was the water which at times accumulated within the tube.

In many respects the results which had been calculated on were realised, and the new arrangements necessary to the working of the system were successfully brought into operation.

The speed of the trains corresponded fully with the degree of vacuum obtained; that is to say, the train resistances proved to be what had been anticipated.[8]

After the trial of a great variety of air-pump valves, a form was adopted which was found to answer exceedingly well.[9]

In the Atmospheric tube, the system of self-acting inlet and outlet valves, by which the piston was enabled to leave the tube on approaching a station and enter it again on recommencing its journey, were, on the whole, successfully adapted to their duty.

Again, an arrangement for starting the train rapidly from the station, without the help of horses or of locomotives, had been brought practically into operation. This arrangement consisted of a short auxiliary vacuum tube containing a piston which could be connected with the train by means of a tow-rope, and thus draw it along till the piston of the piston carriage entered the main Atmospheric tube. Some accidents at first occurred in using this apparatus, but its defects were after a time removed; and it is hardly to be doubted that the various minor difficulties of the Atmospheric System could soon have been effectually mastered.

It now remains to show how it was that, in spite of much that was hopeful, a vigorously sustained contest ended in defeat, instead of being prolonged into victory.

In working the Atmospheric System on the South Devon Railway grave difficulties were throughout encountered, for which to the last only imperfect remedies could be found.

As regards the power consumed, the engines of each pumping station worked up to about 236 indicated horse-power, and their regulated duty for each train, including the anticipatory pumping, was equivalent to 5 minutes of work for every mile of the length of tube they had to exhaust. As the running speed averaged 40 miles per hour, or a mile in 1.5 minutes, the 236 horse-power during the 5.5 minutes of pumping must be regarded as equivalent to 865 horse-power during the actual passage of the train. Now, making full allowance for piston friction and extra friction on curves, for the power expended in getting up speed, for the excess of air-pump resistance due to the changes of temperature experienced by the air under exhaustion, and even for the very large actual amount of friction in the engines employed, the work done should have been represented by an expenditure of 240 horse-power during the passage of the train. If to this is added an allowance for leakage, such as the experiments at Dalkey indicated would be amply sufficient with the longitudinal valve in good condition, it may be said that Mr. Brunel had a right to expect that the duty would be performed with an expenditure of 300 horse-power; whereas it actually required 865 horse-power, or nearly three times the amount.

The explanation of this waste is simple.

Serious and unexpected causes of failure developed themselves in the longitudinal valve, and led to an excessive amount of leakage. A great part of the normal duty of the engines was, as has been stated, to exhaust the tube previous to the entry of the train; and when, owing to leakage, the amount of air to be so pumped out was greatly increased, it became necessary that the operation should be commenced much earlier. There was thus a longer time during which the leakage could take place, and a still greater amount of air to be pumped out. It therefore followed that a large increase of leakage involved waste of power in an enormously increased proportion.

The length of time occupied by the anticipatory pumping was often increased by the difficulty of arranging proper telegraphic communication on the South Devon Railway, and by the absence of it on the Bristol and Exeter Railway. The Electric Telegraph was in its infancy, and though Mr. Brunel had been the first to apply it in connection with railways, namely, between London and Slough on the Great Western Railway in the year 1839, it was not brought into perfect working order on the South Devon Railway till the Atmospheric System was on the point of being abandoned. The result of the defects in the telegraph was that, when a train was late, warning was not received at the several engine-houses;[10] and thus, when this was the case, the pumping-engines which had been started at the right interval of time before the train was due, had to be kept at work for a needlessly long period pumping out the air, which was all the while leaking in through the deteriorated valve. This inefficiency of the telegraphic apparatus would have been of trifling importance but for the defects of the valve. Had the valve been as perfect as it was expected to be, the vacuum, after it had been formed, could have been maintained by an expenditure of power very moderate in comparison with that which was actually required.

As regards the relative cost of the power consumed, it appears that, owing to imperfections in the engines, their expenditure of fuel per indicated horse-power was more than double that of the best of the Cornish pumping-engines, to which they were analogous; while the cost of working was more than three times as great.

The defects in the engines were for the most part such as might have been remedied; and this would have been done, had not the excessive duty imposed on them by the leakage of the longitudinal valve prevented their being stopped for repairs and alterations. In this way the defects of the different parts of the apparatus mutually aggravated each other.

It appears then that, chiefly owing to the defective longitudinal valve, the engines were expending nearly three times the power which they should have done for a given tractive duty, according to previous experience, and the results obtained on the Dalkey line, and that they cost per horse-power at least three times as much as was expected.

The cost of traction, nearly nine times as much as had been calculated on, was between two and a half and three times what it would have been with locomotive power; and this was on a level part of the line, where the comparative advantages of the Atmospheric System were not exhibited as they would have been in the part which had steep gradients.

The imperfections of the longitudinal valve have now to be described. By its condition the Atmospheric System had to stand or fall. With an efficient valve, the defects of the other parts of the apparatus would have been of minor importance, and time would have been given for remedying them. When the leakage became considerable, the defects of the telegraph and the defects of the engines alike assumed a formidable aspect.

The failure of the valve was due, partly to the composition which was used to seal the joint where it opened, and partly to the material of which the valve consisted. The difficulty of obtaining a suitable composition was the first which had to be encountered. On the South Devon a lime soap was eventually found to answer the purpose well. Its surface, however, from exposure to light and air, formed into a hard skin; and to remedy this a thinner and more fluid material, a compound of cod-oil and soap, was laid on to keep it soft. This answered satisfactorily, but it required frequent renewal, as it was apt to be drawn into the tube by the rush of air when the valve was opened. The renewal of the various compositions, and the careful examination and repair which the valve constantly required, was a cause of great anxiety and expense.

But in the materials of the valve lay the source of the more serious difficulty.

The ready affinity of leather for oil and grease, and its suppleness and closeness of grain when saturated with substances of that nature, had long been known and utilised. It had not been anticipated how readily, with air-pressure on one side and a partial vacuum on the other, the oily matters with which the leather was charged would escape from it, especially in the presence of water. Although, while the leather was saturated with water, the valve was remarkably air-tight; when frost supervened the water became frozen, and gave a fatal stubbornness of texture, which rendered the valve incapable of closing properly. Again, in long-continued drought the leather became intractable from its dryness; and the stiffening, whether from frost or drought, rendered it liable to be torn. An immediate application of seal-oil penetrated the leather, and relaxed its stiffness; but the remedy often could not be applied in time, and, moreover, was expensive.

A still more grave defect was all the while becoming matured, and was undermining every hope that a suitable dressing could be discovered, and that the longitudinal valve might be made perfect.

Under the joint action of water in the leather, and of the affinity of iron for tannin—and on the enduring presence of tannin within its texture the consistency of leather depends—a destructive decomposition had long been at work; the oxide, established in the iron plates of the valve by continued contact with damp leather, had been steadily abstracting the tannin; thus the leather had become converted into an ink-stained and comparatively decomposed tissue. Large portions of it became torn, and incurably pervious to air.

It was not until early in June 1848 that Mr. Brunel discovered the condition which the valve had assumed. He then instituted a careful examination throughout the line, and the extent of the disorder was realised.

The state of things which existed when this discovery was made in effect involved the renewal of the valve the whole distance from Exeter to Newton; so that, as the cost of the valve was £1,160 per mile, an immediate outlay of some £25,000 became essential to the maintenance of the system, and this at the time when the real difficulties of the valve question had become most apparent. By galvanising the iron plates of the valve the mutually destructive action of the iron and the leather might have been prevented; but a remedy was also required for the other serious defect which leather, as the material of the valve, was found to exhibit, namely, its tendency to become permeable to air after long-continued use under air-pressure, owing to the inward escape of the material with which it had been dressed.

These difficulties were not only such as had not been anticipated, but such as no one was justified in anticipating.

It now became necessary for Mr. Brunel to consider what course, under the circumstances, it was most advisable for the Company to adopt.

A Committee of the Board was appointed to examine the whole question; and, at their desire, Mr. Brunel made a report upon it, which was as follows:—

August 19, 1848.

You have called upon me to report to you upon the present state of the Atmospheric apparatus, and particularly upon the circumstances connected with the partial destruction of the longitudinal valve which has lately occurred, and the probability of remedying this serious defect, and of keeping the valve in repair and in good working order.

Such a report involves necessarily the consideration of the whole question of our experience of the working of the Atmospheric System; because, to arrive at any clear appreciation of the present state of the apparatus, I must refer to the circumstances which have affected our working up to the present time, and particularly to the several difficulties which we have had to encounter and their effects.

The first difficulty, and one which was as unexpected as it was serious, was in the working of our stationary engines. Upon the efficiency of these machines must of course ultimately depend the economy and efficiency of the working of the whole system, however perfect in itself might be the Atmospheric apparatus. Accordingly, great precautions were taken—precautions which I still think such as to justify the expectation that we should secure the best engines that could be made.

The three first manufacturers of the day were employed—Messrs. Maudslay (who had had some experience in this particular branch, having made the engines for the Croydon railway), Messrs. Boulton and Watts, and Messrs. Rennie. They prepared their own designs and I knew that they each bestowed much thought in the preparation of these designs, and took considerable interest in the results.

Mr. Samuda, a man of considerable mechanical abilities, having all the experience that could be had upon the subject, and deeply interested in the success of the engines, was also employed to superintend their manufacture.

Notwithstanding all these precautions, notwithstanding excellent work-manship, these engines have not, on the whole, proved successful; none of them have as yet worked very economically, and some are very extravagant in the consumption of fuel, burning nearly double the quantity of others, while the average is very considerably more than it ought to be.

The apparent causes of this excess are various in the different engines, but all resulting more or less apparently from the want of experience in this particular application of power, and from the circumstance of the form of the engines being somewhat novel, and involving slight differences in the proportion and arrangement of the parts; and the consumption of steam being greater than was calculated upon, it has been obtained by a more wasteful expenditure of fuel, and the evil has been aggravated.

The difficulty of remedying this state of things has been increased by the consequence of defects in the Atmospheric apparatus, which, causing a much greater demand upon the working of the engines, has delayed, or has entirely prevented, our throwing an engine out of work, to introduce the requisite improvements.

Still, so far as this defect in the engines is concerned, there is no doubt that it is susceptible of considerable, if not complete remedy, and that a reduction of one-third may be effected in the consumption of fuel.

In the Atmospheric apparatus itself our difficulties have been more numerous.

We have suffered from extreme cold, particularly when it followed quickly upon wet.

We have suffered from extreme heat, and also from heavy falls of rain. These difficulties have in turn been encountered and gradually overcome, and

I think the effects of all these causes upon a valve in good condition may now be obviated, if not entirely, yet so much so as to render their operation unimportant.

The same remedy applies to all three—keeping the leather of the valve oiled and varnished, and rendering it impervious to the water, which otherwise soaks through it in wet weather, or which freezes in it in cold, rendering it too stiff to shut down; and the same precaution prevents the leather being dried up and shrivelled by the heat; for this, and not the melting of the composition, is the principal inconvenience resulting from heat. A little water spread on the valve from a tank in the piston-carriage has also been found to be useful in very dry weather, showing that the dryness, and not the heat, was the cause of leakage; but a new difficulty has arisen, and a new defect has been discovered, one much more serious in its extent and its possible consequences, and one which renders the operation of each of the previously mentioned causes of difficulty much more powerful and mischievous.

Within the last few months, but more particularly during the dry weather of last May and June, a considerable extent of longitudinal valve failed by the tearing of the leather, at the joints between the plates; the leather first partially cracked at these points, which causes a considerable leakage, particularly in dry weather; after a time it tears completely through, and that part of the valve is destroyed, and requires to be replaced.

A considerable extent has thus been replaced, but the whole of the valve is more or less defective from this cause; the amount of leakage is considerable, and the working altogether inefficient. I have examined carefully portions of the valve that have been removed, and I find that at the part which has given way the texture of the leather seems to be destroyed—it is black, and has evidently been acted upon by the iron of the plates.

Upon some parts of the line the injury seems to be more general than upon others; but it is very difficult to examine the valve in place, so as to form any correct opinion of the extent of the evil.

As regards the cause of this defect, Mr. Samuda, who under his contract is at present liable for the repair of the valve, urges that the valve was kept for a length of time in cases after it was delivered to the Company, and that, exposed to damp, and the oil in the leather not being renewed on the surface; the iron may have rusted, and the leather have been injured; and he refers to instances lately observed, in which valves taken out of the top of a case which had been exposed to wet do show similar signs of injury.

Supposing, however, this assumption to be correct, it would not seem to affect the question of his liability. He suggests also, as a cause, that the valve remained for a length of time in place without being used and even worked over by locomotive engines, which prevented its being properly oiled and attended to; that the evil has been aggravated by an attempt to reduce too much the use of oil to the leather; and, lastly, that the piston-gear has been allowed to get out of adjustment, so that the leather of the valve has been strained.

I shall not, however, here enter into the discussion of this question of liability, but confine myself to the consideration of the evil, and the possibility of remedying it.

Of the extent of the evil, for the reason I have given, it is impossible to form any accurate opinion; it is impossible, therefore, to say that it does not extend more or less over the whole distance, excepting, of course, that which has been already replaced. That which is injured cannot be repaired in place, but must be removed, and the remedy can only be applied in the new valve.

It is quite possible that a valve made in the same manner as the present, if properly attended to from the first, and with our present experience, might not be subject to this destruction, and Mr. Samuda states that such is the case at Dalkey; but I do not think that I could rely upon this result. By painting, but, better still, by zincing or galvanising the iron plates, and making them overlap a short distance, both the chemical and the mechanical action of the plate upon the leather appears to be prevented, and I believe, therefore, that this evil may be remedied at a small increased cost in any new or repaired valve that might be laid down: but of the existing valve I can say no more than I have done. It is not now in good working condition, and I see no immediate prospect of its being rendered so.

From the foregoing observations, it will be evident that I cannot consider the result of our experience of the working between Exeter and Newton such as to induce one to recommend the extension of the system.

I believe that if the longitudinal valve were restored, the working expenses might be immensely reduced; that the quantity of fuel consumed which is the great item of expense, may be diminished by one-third; that the price of the fuel, which now costs 18s. per ton at the engine-houses, ought to be reduced at least 12 per cent; and that the total cost may thus be brought down to a moderate amount, such as I had originally calculated upon. But the cost of construction has far exceeded our expectations, and the difficulties of working a system so totally different from that to which everybody, traveller as well as workmen, is accustomed, have proved too great; and therefore, although, no doubt, after some further trial, great reductions may be effected in the cost of working the portion now laid, I cannot anticipate the possibility of any inducement to continue the system beyond Newton.

With respect to the future working of the apparatus between Exeter and Newton, I feel in great difficulty as to expressing any opinion, seeing that a very large expense has been incurred, and believing, as I do, that the cost of working may be so very much reduced; but that reduction can only be effected by the almost entire renewal of the valve, and by some expenditure in the engines. And unless Mr. Samuda or the patentees undertake the first, and extend considerably the period during which they would maintain it in repair, and unless they can offer some guarantee for the efficiency of that valve, I fear that the Company would not be justified in taking that upon themselves, or incurring the expense attending the alteration of the engines.

I believe that for the inclined planes, as an assistant power, the apparatus will be found applicable and efficient; and as the engines and the pipes are nearly ready at Dainton, it may be found desirable to try it there, provided a satisfactory arrangement can be entered into for the maintenance and efficiency of the valve.

I have not referred to our great disappointment in not obtaining the assistance of the telegraph in the working of the engines, and the greatly increased consumption of coal consequent upon the working the engines unnecessarily, because this evil is now nearly removed; but some further reductions may still be made by using the telegraph by night as well as day, which has not yet been in our power to do, but which I trust will be commenced this week.

The Committee to whom this report was made, and who had been also in constant communication with Mr. Brunel, placed the result of their investigation before the Board. The Directors, after carefully considering the information given them, reported as follows:—

> Your Directors, without pronouncing any judgment as to the ultimate success of the Atmospheric System, and while they are prepared to afford to the patentees and other parties interested in it the use of their machinery for continuing their own experiments, have arrived at the conclusion, with the entire concurrence and on the recommendation of Mr. Brunel, that it is expedient for them to suspend the use of the Atmospheric System until the same shall be made efficient at the expense of the patentees and Mr. Samuda.

At the meeting in August, the proprietors adopted the Directors' report, and the line was worked throughout by locomotives on and after September 9.

In the following November Mr. Thomas Gill, the chairman of the Board of Directors, published an 'Address to the Proprietors,' in which he strongly deprecated the abandonment of the Atmospheric System, and proposed that the Company should embark on a further experiment. Mr. Gill's pamphlet was referred to three of the Directors, Mr. Thomas Woollcombe, Mr. Charles Russell, and Mr. James Wentworth Buller. With Mr. Brunel's assistance, and to a great extent from memoranda written by him, they prepared a statement which went very fully into all the points raised by Mr. Gill.

After combating Mr. Gill's propositions, they observe:—

> Of the two men who are most deeply concerned in the further trial of any reasonable experiment to perfect the Atmospheric System we find that one, Mr. Brunel, disapproves of the proposal for the purpose as insufficient and unsatisfactory; the other, Mr. Samuda, had not sufficient confidence in the result, or in Mr. Gill's estimates for its accomplishment, to offer the only security which would justify the Company in endeavouring to effect it.

In conclusion they express an opinion that the suspension of the Atmospheric System in the previous September was a prudent and necessary step, and that nothing had since occurred to justify its resumption.

The proprietors adopted the view taken by the Committee, and no further attempt was made to work the railway on the Atmospheric System.

Under these circumstances, it cannot be a matter of surprise that Mr. Brunel was much censured for having advised the South Devon Railway Company to work their line on the Atmospheric System.

The reasons which led him to recommend the use of the Atmospheric System on the South Devon, and the causes of its failure, have been very fully described, and it has been also shown that the most important of these were the defects of the pumping-engines, and the deterioration of the longitudinal valve.

When the formidable character of these difficulties had fully declared itself, the South Devon Railway Company were not in a position to spend any more money upon a system which, as the event had proved, was, in one of its most important details, still in the experimental stage.

There can be no doubt that the abandonment of the Atmospheric System was the wisest step which, under the circumstances, could be adopted; and it was recommended to the Directors by Mr. Brunel with a simple and self-sacrificing disregard of every consideration except that which was always paramount with him, the interests of those by whom he was employed.[11]

## NOTE
### Comparison of Stationary and Locomotive Power

In order clearly to set forth the reasons which justify the statement made by Mr. Brunel,[12] that stationary power if freed from the weight and friction of any medium of communication, such as a rope, must be cheaper than locomotive power, it is desirable to consider, (1) the waste of power which arises from the locomotive having to move itself as well as the train; and (2) the excess of cost at which a given power was supplied by a locomotive, as compared with that at which it could have been supplied by a stationary engine.

On the first point, the best information can be obtained from experiments made by Mr. Daniel Gooch during the gauge controversy. The results are very suitable for use in the present investigation, as the South Devon was to be a broad-gauge railway. Moreover, as the broad-gauge engine with which these experiments were tried was one of a class more powerful for their weight not only than the contemporary narrow-gauge engine, but also than the engines Mr. Brunel had experience of when he wrote his report three years previously, the results may be considered to represent very favourably the then existing case for the locomotives.

The engine employed in the experiments weighed, with its tender, about fifty tons. The maximum power it was capable of delivering by the pressure of steam in its cylinders was represented as a tractive force of 4,900 lbs. at a speed of 60 miles an hour, equivalent to 784 indicated horse-power; and at 40 miles an hour 5,200 lbs., equivalent to 555 indicated horse-power.

It is next to be considered how this power would, when running at the speeds mentioned, be employed in overcoming the elements of resistance. These are:—

(1) The working friction of the machinery.

(2) The rolling resistance of the engine and tender.

(3) The air resistance due to the engine frontage.

(4) The rolling resistance of the train.

(5) The air resistance on the portion of the train unprotected by the tender.

(6) The resistance due to gradient.

The following symbols and quantities may be conveniently made use of to denote the various terms of the equation between force and resistance.

| | |
|---|---|
| Total available tractive force in lbs | F |
| Weight of engine and tender (superfluous load) in tons | 50 |
| Weight of train (useful load) in tons | W |
| The sum of the resistances of machinery, rolling resistance, and air resistance of engine and tender | R |
| Rolling resistance of train in lbs. per ton | K |
| Gradient | G |
| Speed in miles per hour | V |

Resistance of air (according to the received empirical formula)

$$= \frac{1}{400} \ \text{(frontage area)} \times V^2$$

| | |
|---|---|
| Frontage area of train in square feet | 63 |
| Frontage area of portion of train unprotected by the tender, in square feet | 24 |

For a locomotive train therefore

$$F = R + WK + \frac{24}{400} V^2 + (50+W)\ 2240\ G$$

For a system that dispenses with the locomotive

$$\text{Tractive force} = WK + \frac{63}{400} V^2 + W\ 2240\ G$$

Therefore

$$W (K + 2240\ G) + \cdot 1575\ V^2$$

= the useful tractive force, and

$$R + 112000\ G - \cdot 0975\ V^2$$

= the tractive force wasted by the use of the locomotive.

Therefore

$$F = \{R + 112000\ G - .0975\ V^2\} + \{W(K + 2240G) + .1575\ V^2\}$$

and the useful load

$$W = \frac{F - R - 112000\ G - .06\ V^2}{K + 2240\ G}.$$

The values which Mr. Gooch's experiments give for the two selected speeds are as follows[13]:—

| Miles per Hour | R (lbs.) | K (lbs. per ton) | F (lbs.) |
|---|---|---|---|
| 40 | 1500 | 12.5 | 5200 |
| 60 | 2100 | 18.6 | 4900 |

Using these values, the results in the following table are obtained, being the conditions appropriate to the two speeds at successive ascending gradients:—

| Miles per hour | Ascending Gradient | Useful Load in tons | Superfluous Load in tons | Gross Load in tons | Useful Horse-power | Waste Horse-power | Gross Horse-power | Ratio of Waste to Useful Horse-power |
|---|---|---|---|---|---|---|---|---|
| 40 | 0 | 288 | 50 | 338 | 411 | 144 | 555 | ·35 |
| | 1/200 | 128 | 50 | 178 | 352 | 203 | 555 | ·58 |
| | 1/100 | 71 | 50 | 121 | 292 | 263 | 555 | ·90 |
| | 1/75 | 50 | 50 | 100 | 252 | 303 | 555 | 1·20 |
| | 1/50 | 23.8 | 50 | 73.8 | 173 | 382 | 555 | 2·21 |
| | 1/40 | 11.7 | 50 | 61.7 | 113 | 442 | 555 | 3·91 |
| | 1/36.3 | 7 | 50 | 57 | 82 | 473 | 555 | 5·77 |
| 60 | 0 | 139 | 50 | 189 | 504 | 280 | 784 | ·56 |
| | 1/200 | 68 | 50 | 118 | 415 | 369 | 784 | ·89 |
| | 1/100 | 35.7 | 50 | 85.7 | 325 | 459 | 784 | 1·41 |
| | 1/75 | 22.5 | 50 | 72.5 | 265 | 519 | 784 | 1·96 |
| | 1/52.3 | 7 | 50 | 57 | 160 | 624 | 784 | 3·90 |

Thus, on a level line, the engine, working up to 555 horse-power, could just draw 288 tons of train at the rate of 40 miles per hour, wasting on its own resistance only one-third of the power usefully employed on the train; but when the speed was increased to 60 miles per hour, it could not, though working up to 784 horse-power, draw more than 139 tons of train, wasting on its own resistance more than half the power usefully employed on the train. And again, at 40 miles per hour, though, as just stated, it could draw on the level 288 tons, it could only draw 24 tons of useful load at that speed up 1 in 50; while at 60 miles per hour, though it could draw, as stated, 139 tons of train on the level, it could only draw 23 tons of useful load up 1 in 75; and at the respective speeds of 40 and 60 miles per hour, it could only take one carriage (7 tons) up the respective gradients of 1 in 36, and 1 in 52.

Hence to maintain a minimum speed of 40 miles per hour with locomotive power on a line with long gradients of 1 in 40 involved on those parts of the line a wasted power of nearly 4 times that usefully employed; and if a minimum limit of 60 miles per hour were contemplated, a locomotive of the most powerful class in existence three years subsequent to Mr. Brunel's report advising the adoption of the Atmospheric System would only have been able to take a single carriage up an incline of 1 in 52. So heavily at high speeds on steep gradients is the performance of a locomotive taxed by the resistance due to its own dead weight.[14]

A comparison has now to be made between the cost of power as developed by a locomotive and as developed by a stationary engine.

From the well-known experiments made for the information of the Gauge Commissioners in December 1845, taking the high speed trials as the basis of calculation, it appears that 4.5 lbs. of coke per horse-power per hour may be taken as the average consumption of the engine.[15]

It will be well, however, to allow for the improvement which was at the time anticipated in locomotive working, and to assume an expenditure of 4 lbs. of coke per indicated horse-power per hour, as representing the case then for the locomotive engine.

Coke may be taken to have at that time cost 21s. a ton, or .0094s. per lb. Moreover, a careful analysis of the Great Western Railway half-yearly reports, for 1844 and 1845, shows that for every shilling expended in coke, 1.44 shillings were expended on the average in wages, oil and waste, repairs, etc.

Putting the results together, it appears that for each single indicated horse-power delivered by a high-speed locomotive, the cost per hour was 0.00915s. or 1.098d.; that is to say, about 1 1/10d. per hour.

Let this now be compared with the cost per horse-power per hour at which the best Cornish pumping engines had long been known to perform the work. This comparison is manifestly a rational one—with reference to the kindred employment of engine power in atmospheric pumping-engines.

The performances of nearly all the pumping-engines in Cornwall were for many years so systematically and exactly reported, and the reports of each were so critically scrutinised by the rival makers, that the data they supply may be relied on without hesitation. It was well known that the best of the engines continuously performed useful work with a consumption of coal at the rate of 2.33lbs. per delivered horse-power per hour, or, counting coal at 16s. per ton (a fair price on the South Devon), at the cost of 0.2d., or one-fifth of a penny per horse-power per hour.

But it was not in its consumption of fuel alone that stationary power was the more economical; the expenditure in wages, oil, and tallow on one of the pumping-engines above referred to, when doing 200 horse-power of useful work, did not exceed 20s. for the twenty-four hours, or one-twentieth of a penny per horse-power per hour, while the cost of repairs was merely nominal.

Thus if fuel, wages, oil, and tallow be brought into one item, it is seen that the cost of one horse-power in stationary engines such as the then existing Cornish engines was only 0.25d. per hour, or less than one-fourth of its cost when developed by a locomotive, which as been shown to have been 1.098d. per hour.

1.  The apparatus patented in 1839 by Mr. Samuel Clegg and Messrs. Jacob and Joseph Samuda, and improved from time to time by them, was that adopted in almost all the attempts made in this country to introduce the Atmospheric System. In reckoning up the force which was available for mastering the practical difficulties of the undertaking, the death, in 1844, of Mr. Jacob Samuda must be considered to have been to his brother, and to all others concerned, a great and irreparable loss.

2.  A considerable amount of engine power was necessarily consumed in exhausting the tube before the passage of the train commenced; and it might at first sight appear that this work was wasted, and that it was only the work which the engine performed during the passage of the train which was useful in traction. This, however, was not the case; for, as was admitted by the more scientific of the opponents of the Atmospheric System, the power employed in anticipatory pumping was work legitimately stored up and re-delivered in relief of the engines during the passage of the train. A waste of power incidental to the Atmospheric System was indicated by the heat of the air which was delivered by the exhausting pumps. This waste, however, amounted on the average to only 10 per cent of the total work done. A further source of waste of power was the friction of the air passing along the tube to the exhausting pumps; this waste was found to amount, on the average, to from 10 to 15 per cent of the total work done.

3.  This was almost the only case in which Mr. Stephenson approved of the application of the System.

    Before the Croydon and Epsom Committee, in answer to the question, 'Does the Atmospheric railway give you any power of using practically and usefully steeper inclines than the locomotive railways?' Mr. Stephenson said, 'Yes, I think it does, but still at a very inordinate loss of power; still it is within the scope of the Atmospheric System under particular circumstances. I remember a case where it might be advantageous. Mr. Brunel went to Italy for the purpose of laying out a line there, and from Genoa over the Apennines he had to form a line; it would probably rise 15 or 20 miles at 1 in 100 or 1 in 60 or 70. Where there is that continuous line of ascent, where no stoppages are required, where the locomotive is totally inapplicable, there I can conceive nothing more eligible than the Atmospheric plan'.

4.  The length of the line was 52 miles, but as it was considered that auxiliary stationary power would in any case be necessary on the 10½ miles of very steep inclines, the cost of the Atmospheric apparatus is taken on 41½ miles.

5.  It must be remembered that beyond the South Devon Railway was the projected railway through Cornwall, which, with its long and heavy gradients, was, in all its features, even more suitable than the South Devon for the application of the Atmospheric System. Had that system succeeded, and been introduced on the Cornwall Railway, a very great saving might have been made in the cost of the works of this line.

6.  i.e. before the Croydon and Epsom Committee. See above, p.108

7.  It will be found on pp.35–52 of the Minutes of Evidence taken before the Atmospheric Committee (ordered to be printed April 24, 1845).

8.  This appeared with sufficient clearness from the general comparison between vacuum, weight of train, and speed. The exact appropriation of the force employed was shown by some dynamometric experiments made on the line.

The highest speed recorded was 65 miles per hour, with a train of 28 tons, the speed averaging 64 miles per hour for four level miles of the line, the vacuum being 16 inches. This speed should have exhibited a resistance of about 21 lbs. per ton, or 588 lbs., as the running resistance or friction, and 645 lbs. for the resistance of the air; in all 1,233 lbs. Now, the pressure due to 16 inches of vacuum on the piston is 1,390 lbs., which gives 157 lbs. as the friction of the piston; a result which corresponds sufficiently well with a direct dynamometric experiment.

Going to the other extreme, there are numerous records of trains of 100 tons which attained, on a level of four miles in length, average speeds of from 30 to 35 miles per hour, with 16 inches of vacuum, one train of 103 tons going 324 miles per hour with 16.9 inches of vacuum.

9. This valve consisted of a number of long delicate blades of spring steel, arranged parallel to each other, as in a musical box, but with wider intervals. These plates rested on a series of truly faced bars, which crossed the end of the air-passage. The slightest pressure outwards lifted the springs; and as the area of opening was large, a very free passage was given to the air. On the current ceasing, the blades instantly, yet without shock, replaced themselves in contact with the bars, clipping them tightly under a very small reverse pressure, and effectually closing the passage. Their merit consisted in their being almost without weight, and thus promptly re-closing the aperture by a delicate elastic reaction.

10. Trains frequently arrived late on the Atmospheric portion of the South Devon Railway, owing to its being at the end of the long trunk line from London to Exeter, and having at its other end a locomotive line contending with very heavy gradients.

11. It may be mentioned, that, from the date of the abandonment of the Atmospheric System, he refused to receive any remuneration for his professional services as engineer beyond a nominal retaining fee.

12. See above, p.108.

13. These quantities are the result of the experiments made in September 1847. They agree with what is now the received opinion of authorities on train resistances, and represent favourably the case for the locomotives at the time of Mr. Brunel's report in August 1844. At the time when Mr. Brunel wrote his report of August 1844, the weight of a locomotive, as has been said, bore a higher ratio to its power.

14. It must be borne in mind that all the inconveniences attending the use of auxiliary locomotives must be encountered, or else the excessive dead weight of an engine powerful enough to take a train up the steepest gradient in a hilly district must accompany it for the whole length of that part of the line.

15. No dynamometer was used in these experiments, but all other requisite data were recorded with the greatest exactness, and the horse-power employed may be deduced by means of the scale of resistance which the subsequent dynamometric trials supply. Moreover, the result above arrived at for the consumption of coke is verified by an examination of published indicator diagrams taken off the same engine on another occasion.

# VII

# Railway Bridges and Viaducts

In Chapter IV a general history has been given of the railways of which Mr. Brunel was the engineer; but the bridges and viaducts designed by him are so numerous and important that it has been thought advisable to devote a separate chapter to their consideration.

The bridges selected for mention have been grouped according to the nature of the material used in their superstructure. This arrangement is the most convenient one for giving a concise description of the most remarkable of Mr. Brunel's bridges, and for stating the circumstances which guided him in the determination of the particular form of construction used in each case.

The works are therefore divided into four groups, namely, brickwork and masonry, timber, cast iron, and wrought iron.[1]

## *Brickwork and Masonry Bridges*[2]

The viaduct which carries the Great Western Railway over the valley of the river Brent near Hanwell is the first of Mr. Brunel's important railway works.[3] It is a handsome brickwork structure, 65 feet high, with eight semi-elliptical arches, each 70 feet span and 17 feet 6 inches rise. The spandrils of the arches are lightened by longitudinal spandril-walls; the piers are also hollow, and the structure is throughout made as light as possible. It is on this account interesting, as showing the care taken by Mr. Brunel from the commencement of his practice to distribute the material in the simplest and most effective manner.[4]

The great bridge over the Thames at Maidenhead contains two of the flattest, and probably the largest arches that have yet been constructed in brickwork. The river, which is about 290 feet wide, flows between low banks; in the middle of the stream there is a small shoal, of which Mr. Brunel took advantage in building the centre pier.

It was originally intended that the foundation of the bridge should be on the chalk, which was at a short distance below the surface; but it was found to be very soft, and Mr. Brunel therefore decided to place the foundations of the bridge on a hard gravel conglomerate overlying the chalk. The main arches are semi-elliptical, each of 128 feet span and 24 feet 3 inches rise. They are flanked at each end by four semicircular arches, one of 21 feet span, and three of 28 feet span, intended to give additional water-way during floods. The radius of curvature at the crown of the large arches is 165 feet, and the horizontal thrust on the brickwork at that point is about 10 tons per square foot.

In the interior of the structure immediately landward of the large arches, Mr. Brunel constructed flat arches loaded with concrete. The centerings of these were struck, and an active thrust opposed to the main arches before their centerings were eased.[5] The line of pressure of each main arch was diverted downwards by the thrust of the flat arch adjoining it without the necessity of employing a great mass of brickwork in the abutment.

The woodcut (fig. 1) shows the form of the main arches and the flat arch referred to.[6]

The Maidenhead bridge is remarkable not only for the boldness and ingenuity of its design, but also for the gracefulness of its appearance. If Mr. Brunel had erected this bridge at a later period, he would probably have employed timber or iron; but it cannot be a matter of regret that this part of the Thames, although subjected to the dreaded invasion of a railway, has been crossed by a structure which enhances the beauty of the scenery.

There are two other large brick bridges over the Thames, one at Gathampton and another at Moulsford, that at Moulsford crossing the river obliquely at an angle of 45°. In each of these bridges there are four arches, of 62 feet span on the square.

Other good examples of brick bridges are the turnpike road bridge, 60 feet high, with three arches, across the deep cutting at Sonning Hill, and the bridge, with one opening of 60 feet and four side arches of 18 feet span, over the river Kennet at Reading.

The bridge over the Avon at Bathford, of 87 feet span, and the bridge crossing the same river at Bath, with an arch of 88 feet span, are handsome Bath-stone structures with semi-elliptical arches. Near Bristol there is an ornamental bridge of masonry with three Gothic arches, the centre arch having a span of 100 feet.[7] Another bridge of Gothic design, with two arches of 56 feet span, carries the railway over the Floating Harbour.[8]

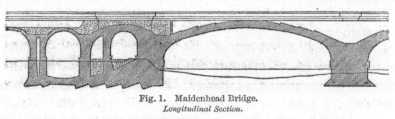

Fig. 1.   Maidenhead Bridge.
*Longitudinal Section.*

*Scale of feet.*

The Royal Albert Bridge

The bridges which have hitherto been noticed are all on the Great Western Railway. On the Bristol and Exeter Railway there is a large stone bridge over the New Cut at Bristol, built in 1840, which has a single segmental arch of 120 feet span, and 20 feet rise. Owing to some imperfect workmanship in the interior masonry of the arch, and possibly to some unequal yielding of the abutments, the crown sunk much more than had been expected.

On his later railways Mr. Brunel did not build large arches of brickwork or masonry, though he constructed several lofty and extensive viaducts of these materials with spans varying from 40 to 60 feet.

Mr. Brunel seldom employed artificially piled foundations to support masonry. When the ground was soft, he preferred to rely on a large extent of bearing surface, and ensured uniformity of settlement by an accurate distribution of the load. Several of his large viaducts and bridges, standing on ground of a soft and spongy nature, were constructed on this principle.

A class of bridge of striking outline was used in the cuttings on the Bristol and Exeter Railway, and on the other railways subsequently made. Bridges of this class were called flying bridges. Instead of arches resting on piers and abutments, the bridge has a single arch, reaching from one side of the cutting to the other, and springing from the slopes, which it helps in some measure to support. A flying bridge of large dimensions near Weston-super-Mare carries a road across the cutting at a height of 60 feet above the line of rails, with a clear span of 110 feet.

The quantity of masonry in these bridges is much less than in those of the ordinary construction; and lofty and expensive centering is not required, as the bridge can be built before the cutting is excavated to its full dimensions.

This class of bridge, by the avoidance of abutments and counterforts, simplifies the construction of skew arches, while on sharp curves it presents but little obstruction to the view along the line.

A curious use of arches of this construction, as applied by Mr. Brunel, may be seen on the South Wales Railway near Llansamlet, between Neath and Swansea. A deep cutting through the coal measures showed a tendency to slip, and a large amount of excavation would have been required to flatten the slope, as a hill rose immediately above the side of the cutting. Four of these flying arches were thrown across the cutting at short intervals, and weighted with heavy copper slag, so that the sides of the cutting are kept apart by the thrust of the loaded arches.

Among the skew bridges on Mr. Brunel's railways, there are a few of extreme obliquity. Of these may be mentioned two large road bridges near Berkeley, over the Bristol and Gloucester Railway, one being 48° and the other 53° off the square. Both the bridges are of brickwork, and in the arch of the first one, which was set in Roman cement, hoop iron was introduced in the manner successfully employed by Sir Isambard Brunel. On the South Devon Railway, near Plympton, there is a skew bridge 63° off the square.

On the Great Western Railway, in the neighbourhood of Bath and Bristol, there are skew bridges of ashlar masonry built on the mechanically correct principle of spiral tapering courses, the bed-joints in every part of the arch being made at right angles to the lines of pressure. By this method the arch does not depend for its stability on the friction and cohesion of the materials, as it does to a great extent in very skew bridges, built in the usual way with spiral parallel courses, especially when the arches are semi-circular or semi-elliptical.

Mr. Brunel's bridges of masonry and brickwork were well known for the comparatively small quantity of material used in them; and, though it was requisite that the materials and workmanship should be of superior quality, their cost was comparatively small.

The specifications he prepared for all his works, and on which the contracts were based, were noted for the completeness with which they were drawn up, and for their not requiring a standard of perfection higher than that which was actually to be carried out. The confidence with which Mr. Brunel was regarded enabled him to insist with effect on the work being executed according to his interpretation of the contract.

In connection with the design of engineering works, and especially of brickwork and masonry bridges, the following letter from Mr. Brunel to one of his assistants, who was abroad, will be found interesting:—

December 30, 1854.

Let me give you one general piece of advice—that while in all works you endeavour to employ the materials used in the most economical manner, and to avoid waste, yet always put rather an excess of material in quantity. You cannot take too much pains in making everything in equilibrio; that is to say, that all forces should pass exactly through the points of greater resistance, or through the centres of any surfaces of resistance. Thus, in anything resembling a column or strut, whether of iron, wood, or masonry, take care that the surface of the base should be proportioned that the strain should pass through the centre of it. Consider all structures, and all bodies, and all materials of foundations to be made of very elastic india-rubber, and proportion them so that they will stand and keep their shape: you will by those means diminish greatly the required thickness: *then add 50 per cent.* So in trussed framework of wood or iron, experience shows that you cannot refine too much upon the perfection of the designing of every little detail by which all strains are carried exactly through the centres of the rods or struts and the centres of the bearing surfaces. And remember, always in retaining walls to give plenty of batter; never build an upright wing-wall, or retaining wall. To a man who has an instinctively mechanical mind—and no other can be an engineer—the advice I have given you above is all I need say; but this advice is the result of a good deal of experience, purchased by failures of my own, and by looking at those of others, and is, I assure you, valuable advice, to be followed literally and strictly, and not to be considered as a mere theoretical refinement, to be neglected in practice. Practically too much attention cannot be paid to these precautions. I have found that there is not a single substance we have to deal with, from cast-iron to clay, which should not practically be treated strictly as a yielding elastic substance, and that the amount of the compression or tension, as the case may be, is by no means to be neglected in practice any more than in theory. Bear in mind also that which is too often neglected and involves serious consequences, that masonry or brickwork has not half the strength which is generally calculated upon until the mortar is hard, and that you cannot keep centres or shores up too long.

### Timber Bridges and Viaducts

Mr. Brunel's timber bridges and viaducts are remarkable on account of the extensive scale on which he employed that material, and the simple and efficient type of construction which he adopted in the largest structure as well as in the smallest.

In 1841 Mr. Brunel constructed a timber bridge of five spans to carry a public road over the Sonning cutting of the Great Western Railway, a short distance east of Reading. The total width of the space across which the road had to be carried was 240 feet. The superstructure rests on four tall frameworks or trestles of timber forming the piers. Two of these piers are on either side of the railway, and the others are about halfway up each slope.

The road rests on a platform of timber planking, carried on three longitudinal beams, which are supported at nearly equal distances by timber struts radiating from points on the piers about 12 feet below the level of the carriage road. The system of arrangement of these struts will be best understood by a reference to the woodcut given below (fig. 5, p.142) of one of the Cornwall viaducts, of which the Sonning bridge may be regarded as in some measure the prototype.[9]

The skew timber bridge on the Great Western Railway near the Bath Station, carrying the line over the river Avon, was constructed about the same time as the Sonning bridge. It has two spans of 36 feet each on the square, but the obliquity is so great that the span on the skew is 89 feet. Each opening has six laminated arched ribs parallel to the line of the railway. These support the platform of the bridge, and are built up in five layers of curved Memel timber, six inches thick, bolted together. The thrust is counteracted by iron ties connecting the ends of the ribs. The inner spandrils are filled in by cross-ties and braces, and those of the outer ribs by ornamental cast ironwork.

The two bridges already described are almost the only timber bridges of importance on the main line of the Great Western Railway from London to Bristol. Shortly after the completion of this railway Mr. Brunel began to make an extensive use of timber in his designs, and in so doing took full advantage of the largeness of the material, in order to avoid intricacy of construction.

A well-known arrangement for forming beams of greater strength than could be obtained by single pieces of timber was adopted by Mr. Brunel after a careful investigation of its merits. This arrangement consists in joining together two beams of timber placed one above the other, by means of bolts and joggles, so as to form a beam nearly equivalent in strength to a single piece of timber of the same depth as the two pieces united.[10] By this plan, the length which could be spanned by simple beams, without the introduction of trussed framework, was nearly doubled.

The distance between the piers of railway bridges is generally too great to allow of the superstructure being constructed of simple beams, and in such cases Mr. Brunel adopted forms of framing similar in the arrangement of their parts to the common designs of king and queen trusses employed in roofs.

One of Mr. Brunel's early timber viaducts was that erected in 1842 at Stonehouse, on the Bristol and Gloucester Railway. It consisted of a series of five openings of queen trusses 50 feet span, resting on piers formed of timber trestles.

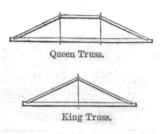

Queen Truss.

King Truss.

In the Bourne viaduct, across the Stroudwater Canal, on the Cheltenham and Great Western Union Railway, there was a span of 66 feet, with three timber trusses, for the two lines of way. Each of these trusses may be described as a king truss with an internal queen truss. The inclined timbers or principals rested in iron shoes upon the piers, and were connected together by bolts and joggles.

The upper horizontal or collar beam of the queen truss carried the roadway planking, which was continued upon beams supported by the principals. The timbers carrying the roadway received support from struts radiating from the feet of the queen posts, which were connected with the apex of the king truss by iron ties. The horizontal tie bars were of wrought iron. The arrangement of the truss is shown in the woodcut (fig. 2).

The side openings consisted of four spans of 30 feet, with trusses of the Stonehouse viaduct type, of one span of 25 feet and ten spans of 20 feet, with double beams.

The St. Mary's viaduct, across the canal in the Stroud Valley, was constructed with one span of 74 feet, with trusses similar to those at the Bourne viaduct.[II]

In the year 1846 Mr. Brunel made an elaborate series of experiments on the strength of large timber. Some account of these is given in the note to this chapter.

Fortified by the information thus obtained, he was able to proceed with confidence to an extensive use of timber in the viaducts of the South Devon, the Cornwall, and other railways.

Between Totnes and Plympton, the South Devon Railway, running along the skirts of Dartmoor, crosses four deep valleys, by lofty viaducts, all of the same design.

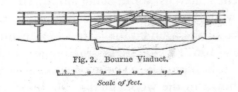

Fig. 2. Bourne Viaduct.

Scale of feet.

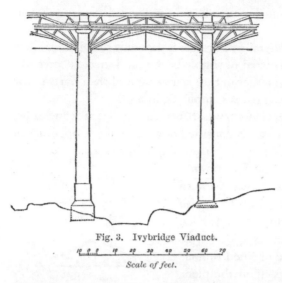

Fig. 3. Ivybridge Viaduct.

Scale of feet.

Three of them can be seen at one time, and they form striking and elegant features in the landscape.

The viaduct at Ivybridge is the highest of these. It is on a curve, and has eleven openings of 61 feet each; the extreme height is 104 feet.

The piers are of masonry, each consisting of two slender and slightly tapered shafts about 7 feet square, rising to the level of the rails. The superstructure was originally designed for a railway on the Atmospheric System, and was therefore only intended to bear the load of a train of carriages. The framework was placed below the level of the rails, and, as will be seen in the woodcut (fig. 3), it consists of a polygonal frame, with a few subsidiary struts, the feet of the main timbers being tied together by wrought-iron rods. There are two of these frames, one at each side of the bridge, to support the planking of the roadway. Before the construction of the viaducts was proceeded with, a complete span of the superstructure, consisting of a pair of the frames with the planking, was erected at Bristol, and tested to ascertain the efficiency of every part.

When it became necessary to strengthen the superstructure to enable it to carry the weight of locomotives, a strongly trussed parapet was added above the trusses, as shown in the woodcut. After the lapse of twenty years, the timber having begun to decay, wrought-iron girders have been inserted, which rest on the stone piers; the framing, however, has not been removed.

Shortly after the completion of the viaducts on the South Devon Railway, those on the South Wales Railway were constructed. The most important on this line are those at Landore and Newport.

The viaduct at Landore, near Swansea, is 1,760 feet long, as the railway here crosses a wide valley. It has 37 openings, and there are a variety of spans, one of 100 feet, two of 73 feet, two of 64 feet, two of 50 feet, and the rest of about 40 feet each. Most of these consist of a superstructure of queen trusses. The piers are of different materials, some being almost entirely of masonry, some partly of masonry and partly of timber, and others entirely of timber, according to the nature of the foundation.[12] The chief feature is the centre span, with an opening of 100 feet, the superstructure of which is a very fine piece of timber-work.[13] It has four trusses, one on either side of the two lines of rails, of the form shown in the woodcut (fig. 4). The truss consists of a four-sided frame placed within a five-sided frame, the angles of each polygon

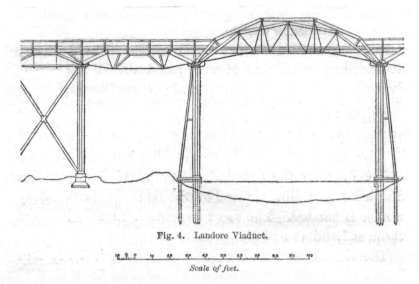

Fig. 4. Landore Viaduct.

Scale of feet.

being connected by bolts and struts with the middle of the sides of the other polygon.

The planking of the roadway rests on double beams, supported at several points in the manner shown in the woodcut, each point having suspension-rods to connect it with the nearest angles of the frames. The arrangement of the double polygonal frame and of the tie-rods enables the transverse strength of the timbers to exercise considerable resistance to any distortion of the shape of the truss by a rolling load. To prevent any tendency of the top of the frame to yield sideways under the compressive strain, the tops of the trusses are connected by transverse struts or braces, the two outside trusses being steadied by raking ties attached to outriggers projecting from below the flooring of the bridge. The thrust of the polygonal frames is resisted by wrought-iron tie-bars at the level of the roadway beams. All the tie-rods in this bridge are double, with one bar on each side of the timbers, to avoid the necessity of making large bolt-holes.[14]

The viaduct at Newport consists of eleven spans with queen trusses, resting on piles. The main span, over the river Usk, is 100 feet, and was constructed with timber trusses very similar to those at Landore. Shortly before it was finished, the viaduct was burnt down. In rebuilding it, wrought-iron trusses were employed for the main span.

The works of the Cornwall Railway were commenced in the year 1852. The district through which the line passes is very deficient in the materials requisite for the construction of a railway. The granite of the country is for the most part only applicable for ashlar; and the slate, which is flat-bedded and so far fit for rubble masonry, is frequently inferior in quality.

In consequence of the number of valleys that the railway had to cross, the aggregate length of the viaducts, thirty-four in number, exclusive of the Saltash bridge, is upwards of four miles on a line of sixty miles. By the use of timber, a great saving was effected in the first cost of the works; and though it is a material which in time requires renewal, its use on the Cornwall Railway enabled the line to be made with the capital at the command of the Company; while, allowing for the cost of subsequent repairs, the total expenditure did not differ much from what it would have been had the superstructure of the viaducts been of more durable materials. The comparatively small cost of these structures enables them to be, in certain places, economically substituted for embankments, as was done on the Cornwall Railway.

The viaducts are to be found over the whole length of the line, but they are most frequent between the Liskeard and Bodmin Road stations, where the railway crosses numerous branches of the Glynn valley.

Most of these viaducts are of one type of construction.

The piers are formed of plain walls, built up to thirty-five feet below the level of the rails, those of the more lofty viaducts being strengthened by buttresses. In the woodcut (fig. 5) is shown a portion of the St. Pinnock viaduct, from which the form of these piers will be understood.

This viaduct is the loftiest on the Cornwall Railway, the rails being at a height of 153 feet above the ground. A description of the superstructure will serve to explain the design of the principal viaducts on the line.

The roadway planking rests on three beams, which run longitudinally throughout the whole length of the viaduct. Each of these beams consists of two pieces of timber, one above the other, fastened together by bolts and joggles.

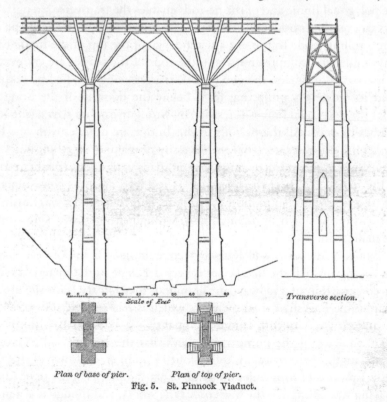

Scale of Feet

Transverse section.

Plan of base of pier.          Plan of top of pier.

Fig. 5.   St. Pinnock Viaduct.

The piers are 66 feet apart, centre to centre, and the longitudinal beams are supported, at four nearly equi-distant points in this space, by straight single timbers radiating from the tops of the piers. The feet of the timbers, which rest on the masonry in cast-iron shoes, are connected together by wrought-iron tie-bars; and the framework is made rigid by iron diagonals.

It will be observed in the transverse section (fig. 5) that the whole weight of the superstructure is concentrated immediately over those points in the piers where the three buttresses meet. The diagonal braces which are attached to each set of the main timbers give transverse stability to the superstructure.[15]

It was desirable, both in first construction and in subsequent repairs, to have a uniform dimension for the spans, and the subdivision of 66 feet was determined on as being suitable for the economic construction of the greater part of the work. The subdivision of this length was such as to allow of single whole timbers being sufficient for the direct supports of the longitudinal beams; and as these beams were supported at intervals of 15 to 20 feet, no intermediate trussing was required. As the inclined timbers met the tops of the piers at a moderate inclination, the outward thrust caused by unequal loading of the spans of the viaduct was inconsiderable, and was easily counteracted by light iron ties.

The stone for the piers was for the most part procured in the neighbourhood, the design of the masonry being such as to enable stone of the country to be used; and, as the timber superstructure was built in pieces of moderate size,[16] and easily obtained, the expenditure was probably not far from the minimum under the existing conditions.[17]

On the South Devon and Tavistock Railway, the viaducts are six in number, and from 62 to 132 feet in height. In these the piers were made of a somewhat simpler form than those just described. At the lofty viaducts, the buttresses were made with a uniform batter throughout their height. The Walkham viaduct, near Tavistock, 132 feet high, with fifteen openings of 66 feet span each, may be considered to exhibit the most matured design of Mr. Brunel's timber viaducts.

On the West Cornwall Railway a type of viaduct similar to that described above was adopted; but as the general height was not so great, the spans were 50 feet each, and the longitudinal beams were supported at three points in each span, instead of at four as on the Cornwall Railway. In consequence of the nature of the foundations, the piers of the nine viaducts on this line were for the most part formed of upright timbers well braced together, standing upon masonry footings. The viaduct at Angarrack, 98 feet high, with 16 spans, which was constructed in 1851, was remarkable for its light appearance, owing to the small number of timbers in the superstructure and piers.

Mr. Brunel paid great attention to the preservation of the material of the timber bridges and viaducts. As early as 1835 he had been in communication with Mr. Faraday as to the best method of testing the extent to which the Kyanising solution penetrated into wood. Mr. Brunel made a careful trial of all the different methods of preserving timber, and employed the more successful of them on a very considerable scale. He was so impressed with the importance of preserving processes being properly applied, that he on several occasions preferred to keep the operation of preserving the timber in the hands of the Company, in order that it might be done thoroughly, and under his own supervision. He also minutely attended to the details by which timber structures may be protected from decaying influences.

### Cast-Iron Bridges

Mr. Brunel did not make an extensive use of cast iron for the superstructure of bridges. His views as to the employment of this material in girders are clearly expressed in the following extract from a letter to one of the Directors of the Great Western Railway:—

April 18, 1849.

Cast-iron girder bridges are always giving trouble—from such cases as the Chester Bridge, and our Great Western road bridge at Hanwell, which, since 1838, has always been under repair, and has cost its first cost three times over, down to petty little ones, which, either in frosty weather or from other causes, are frequently failing. I never use cast iron if I can help it; but, in some cases it is necessary, and to meet these I have had girders cast of a particular mixture of iron carefully attended to, and I have taught them at the Bridgewater foundry to cast them with the flange downwards instead of sideways. By these means, and having somebody always there, I ensure better castings, and have much lighter girders than I should otherwise be obliged to have. The number I have is but few, because, as I before said, I dislike them, and I pay a price somewhat above ordinary castings, believing it to be economy to do so.

I won't trust a bridge of castings run in the ordinary way, and at foundries where I have not a person always watching; and, even if I did, the weight requisite in a beam of ordinary metal and mode of running would more than make up for the reduced price.

The bridge at Hanwell referred to in this letter was one on the main line of the Great Western Railway, over the Uxbridge road. In 1847 the planking caught fire, and the cast-iron girders were destroyed by the heat.

The researches of Mr. Eaton Hodgkinson had drawn attention to the importance of a proper proportionment of the top and bottom flanges of cast-iron girders, and Mr. Brunel now made some experiments on this point. As part of this investigation, eight girders, 30 feet long and 16 inches deep, were tested by weights until they gave way. The comparative areas of the top and bottom flanges were varied until a correct proportion between the two was arrived at. The general result of these large-scale experiments showed a lower breaking than that deduced from Mr. Hodgkinson's formula.

When Mr. Brunel afterwards had occasion to use cast-iron girders, which was chiefly for road bridges over railways, they were made of the form which his experiments had shown to be the best;[18] but he repaired the Hanwell Bridge with wrought iron.

At about the same time the necessity for spanning wide openings had led to larger girders being required than could be manufactured in single castings, and Mr. Brunel had a large cast-iron girder made, 46 feet long and 4 feet deep, of five pieces bolted and keyed together. It was tested until it gave way with a load of 92 tons on the middle. The result showed that the several parts had been well connected, and that the strength of the beam was not much less than the calculated strength of a beam of the same size in a single piece. Mr. Brunel did not, however, use girders of this construction, as the rapid introduction of wrought iron rendered it unnecessary.

Cast iron was introduced, though not for girders, in many of the brick and stone bridges on the Great Western Railway. It was used in the form of troughs sunk into the crown of the arch in bridges where the headway was very limited. The rails were laid along the bottom of the trough within a few inches of the soffit or underside of the arch.

Although, after the careful experiments and investigations he had made, and the experience he had obtained, Mr. Brunel did not make use of cast iron for large girders, he looked forward to the possibility of such improvements being introduced into the manufacture as would enable sound castings of considerable size to be made of homogeneous material.

He expressed this opinion in a letter to the Secretary of the Commission on the Application of Iron to Railway Structures. This Commission (which Mr. Brunel called 'The Commission for stopping further improvements in bridge building') was appointed 'for the purpose of inquiring into the conditions to be observed by engineers in the application of iron in structures exposed to violent concussions and vibration.' Mr. Brunel, in common with most engineers, thought it would be very inexpedient that any *règles de l'art* should be laid down, and took up the cudgels boldly on behalf of the liberty of the profession:—

March 13, 1848.

At present cast iron is looked upon, to a certain extent, as a friable, treacherous, and uncertain material; castings of a limited size only can be safely depended upon; wrought iron is considered comparatively trustworthy, and by riveting, or welding, there is no limit to the size of the parts to be used. Yet, who will venture to say, if the direction of improvement is left free, that means may not be found of ensuring sound castings of almost any form, and of twenty or thirty tons weight, and of a perfectly homogeneous mixture of the best metal? Who will say that beams of great size of such a material, either in single pieces or built, may not prove stronger, safer, less exposed to change of texture or to injury from vibration, than wrought-iron, which in large masses cannot be so homogeneous as a fused mass may be made and which when welded is liable to sudden fracture at the welds?[19]

### Wrought-Iron Bridges

Notwithstanding the cost of wrought iron, but a short time elapsed between its introduction into bridge building and its use in structures of great magnitude. Mr. Brunel had been long familiar with the application of riveted wrought-iron work, and he was the first to encourage its use on a large scale in shipbuilding by recommending its adoption in the 'Great Britain' steam-ship in 1838.

### Girder Bridges

The strains on girders made of homogeneous material have been carefully and ably calculated by mathematicians; and the investigations thus made have directed inquiry into the right channels for determining the nature of the stresses on the several parts of the built-up structures now so much in use. Principles have by degrees been laid down, and lines of thought have been suggested and followed out which were unknown at the time when wrought-iron girders were first introduced in the construction of railway bridges.[20]

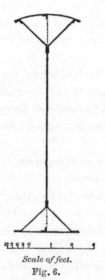

Scale of feet.
Fig. 6.

Experimental Girder
*Transverse Section*

Shortly after Mr. Brunel began to use wrought iron for bridge girders, he made an experiment in order to determine the weak points of a large wrought-iron plate-girder. Mr. Edwin Clark, in his work on the 'Britannia and Conway Tubular Bridges,' vol. i. p.437, gives a description of what he justly terms 'this magnificent experiment.' The girder was of the section shown in the woodcut (fig. 6), 70 feet in length, and of ¼-inch plate throughout. It was weighted gradually, and gave way with a load of 165 tons on the centre, by the tearing apart of the vertical web plate near the ends of the girder. When this portion had been strengthened, and the girder again loaded, it gave way with a load of 188 tons by the simultaneous failure of the top and bottom flanges, that is to say, of the plates forming the triangles shown in the woodcut.[21]

The superior tensile strength of wrought iron to that of cast iron, and the facility with which pieces could be joined together by riveting, enabled girders of great size to be made. The thin wrought-iron plates were arranged so as to form the top and bottom flanges of the girders as well as the upright web connecting them. The metal in the top of a girder being in compression, it was important so to dispose it that it should resist the tendency to yield sideways under the strain. This requirement was met in the experimental girder by the triangular section of the top flange; and the convenience of this form for joining together a number of plates, without difficulty or the use of long rivets, led Mr. Brunel to use the triangular section also for the bottom flange.

Subsequent improvements in the facilities for bending wrought-iron plates enabled him to use a form of cross section of wrought-iron girder, the top flange of which was a nearly circular tube, the best shape of strut to resist longitudinal compression. It is shown in the woodcut (fig. 7), and was used in many of his bridges.

This form was afterwards modified to that shown in fig. 8. The semicircular top plate is stiffened by occasional cross diaphragms, and while it was a good form to resist compression, it was more easily painted than the closed-in top flanges shown in figs. 6 and 7.

The forms of wrought-iron girder already referred to are those known as plate girders, with continuous webs made of plates riveted together, and therefore analogous to the beams of cast iron which they almost entirely superseded. On Mr. Brunel's railways there are a great number of bridges of these forms of girder, where the spans do not exceed 100 feet. For larger spans he used wrought iron, in large and deep trussed frames, by which means a great degree of economy was attained in the employment of the material.

Fig. 7.

Girder on
South Wales Railway,
70 feet span
*Transverse Section*

The care which he had taken to satisfy himself of the action of the strains in plate girders was of service in all the greater structures he designed, as in all of them he employed wrought-iron girders to carry the roadway, of a type somewhat similar to those already described, the girders being supported at frequent intervals by the main framework or truss.

### Opening Bridges

The first large opening bridge which Mr. Brunel constructed was a roadway swing bridge, 12 feet wide, across the new lock at the Bristol Docks. The length of the overhanging end is 88 feet, and the other, or tail end, which is 34 feet long, rests upon two wheels, which travel on a circular rail. The weight of the overhanging end is rather more than counterbalanced by large blocks of cast iron, forming part of the pavement of the tail end. Almost the whole weight is borne on a centre pivot, assisted by four wheels in fixed bearings, upon which runs an inverted circular rail attached to the underside of the bridge. On the pivot, which rests on a large cast-iron bed-plate, are two discs, one of steel and the other of brass, which can readily be lubricated, or taken out and renewed.

On the sides of the bridge are longitudinal wrought-iron plate-girders. The top flange is pear-shaped, and the bottom flange triangular, having three curved plates. The flanges are connected together by a vertical plate web of wrought iron. The section is shown on the woodcut (fig. 9). It admits of very simple riveting, without the use of angle irons. The form of the bottom flange is suited to the compressive

Fig. 8.

Girder on Eastern
Bengal Railway,
92 feet span
*Transverse Section*

Scale of feet.

Fig. 9.    Cumberland Basin Swing Bridge.

strain it has to bear when the bridge is being moved. The top flange has also wrought-iron tie-bars within the tube. When the bridge is across the lock and open for traffic, the overhanging end rests on cams, which are tightened up so as to lift and support the ends of the girders. As the bridge rests almost entirely on a pivot of small diameter, it turns with great ease.

Near Gloucester there are two skew swing bridges somewhat similar to each other in arrangement. Almost all the weight while turning is supported on the piston of a hydraulic press, and the bridge therefore turns round on the water in the cylinder. The first bridge is on the main line of railway leading to South Wales, across a branch of the River Severn, and is for two lines of way. It has three girders, 125 feet long, of the form shown in fig. 7 (p.147). The water pivot is in the middle of the length of the bridge, which spans two openings of 50 feet on the square. Before being turned the bridge was intended to be lifted slightly off its bearings by the hydraulic press, and steadied by four wheels, on which a portion of the weight was to be made to rest by long springs within the girders, the range of which was to be limited in one direction by a fixed stop. The central pier consists of five cylinders of cast iron, each 6 feet in diameter, filled with concrete, surmounted by a cast-iron ring or roller path. The railway company was obliged to make this an opening bridge in order to provide for the free navigation of the river should the old stone bridge lower down be altered. This has not been done, and the railway swing bridge, constructed in 1851, has not yet been opened.

The other swing bridge at Gloucester is on the Dock branch, for one line of way, with an opening of 50 feet on the square, the overhanging length of the girders being 70 feet. While raised from its bearing and turning on its water pivot it is steadied by two tail wheels, like the bridge at the Bristol Docks.

On the Bullo Pill branch of the South Wales Railway there is a small wrought-iron drawbridge, for one line of way, of 30 feet span. It is a lifting bridge on the *bascule* principle, like many bridges over canals in this country and in Holland. The opening part turns on a horizontal axle, and is lifted by rods attached to the ends of two large beams or levers, turning vertically, which are supported above the railway on a timber framework. At the other ends of these beams is a counterbalance weight. The bridge is opened or shut by pulling down either end of the beams with a small chain.

The other bridges are on the main line of the South Wales Railway, and are four in number, each for two lines of way.

One at Loughor is a wrought-iron swing bridge, of 30 feet opening, of the ordinary construction, with girders 90 feet in length, resting upon 36 rollers, which are secured in a ring concentric with the pivot. The opening and closing is effected by means of a crab, fixed clear of the bridge, near the centre. A chain passes from the overhanging end of the bridge to this crab, and taking one or two turns round the barrel, to ensure a sufficient amount of friction, is led to the tail end. The bridge can thus be opened or shut by turning the crab handle in opposite directions. The overhanging end, when across the river, is raised upwards to a small extent by weighted levers, and wedges are then drawn in under it to give it a solid bearing.

At Kidwelly and at Haverfordwest there are wrought-iron lifting bridges, the former of 20 feet, and the latter of 30 feet span. Each of these turns on a horizontal axle like the Bullo Pill bridge; but, instead of being lifted by levers overhead, it has a narrow, heavily-weighted tail end, beneath the planking of the viaduct, which is pulled down with a chain worked by a crab. The portion which carries each line of way is made to open independently. In this form of bridge no wedges or adjusting arrangements are required for the bearings of the overhanging end.

Over the river at Caermarthen is a skew bridge of three girders, each 116 feet long, for a double line of way. It occupies two spans and rolls back, so as to leave a 50-feet opening for the navigation. The swing bridge at Bristol, already described, was at first intended to be a rolling bridge, and to be furnished with wheels to run back on fixed rails, but the difficulty of forming a good foundation for the wheel path led to the design being altered. At Caermarthen the same difficulty was overcome by putting wheels turning in fixed bearings on the pier and abutment of the bridge. The undersides of the girders carry inverted rails, and run back on the wheels. The bridge, when shut, is on an incline of 1 in 50. When about to be opened it is made to assume a horizontal position by turning a supporting cam to lower the overhanging end, and the tail end then rises sufficiently to pass clear above the part of the railway over which it runs back.

By this arrangement the bridge, while in motion, moves along a level path. It is opened and closed by hydraulic machinery.

All these opening bridges have worked satisfactorily since they were constructed.

## Trussed Bridges

When the timber viaduct over the river Usk, at Newport, was burnt down,[22] Mr. Brunel decided to form the new superstructure of the centre opening with three iron trusses, for the two lines of way.

These are bow and string girders, of 100 feet span, and were made of considerable height, not only to reduce the strain on each of the members of the framework, but also in order that the rib or upper portion of each truss might be braced diagonally to the corresponding portion of the other trusses, and headway left for the locomotive chimneys to pass underneath. This bracing counteracts any tendency of the ribs to bend sideways under the compressive strain. The form of the trusses is shown in fig. 1, (p.153). Each truss is a wrought-iron polygonal arch of triangular section, from which is suspended a horizontal girder supporting the roadway. This girder also forms the tie which connects the feet of the arch and counteracts its thrust. The diagonal braces shown on the elevation of the bridge prevent the arch from being distorted by the unequal loading caused by a passing train. The middle truss is twice the strength of each of those at the outside, being made so by increasing the thickness of the plates. One of the outside trusses was tested with a distributed load of 1½ tons per foot-run of its length.

At about the same time that the Newport viaduct was reconstructed, Mr. Brunel designed the bridge over the Thames on the Windsor branch of the Great Western Railway. This is a very large example of the bow and string girder, the span being 202 feet, and the height of the truss 23 feet. The trusses are three in number, for two lines of way, the middle one being twice the strength of the outside trusses. The elevation of the Windsor bridge is shown in fig. 2 (p.153). The bridge is oblique to the river, being 20° off the square. To steady the arched ribs sideways a system of diagonal bracing extends over the whole of the top of the trusses, except at the ends, where headway has to be left for the trains.

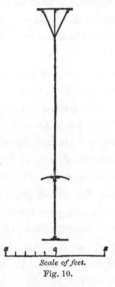

A section of the arched ribs and of the roadway girders in the centre of one of the trusses is given in the wood cut (fig. 10). The arched rib, to resist compression, is of triangular section.[23]

The borings to ascertain the nature of the ground at the foundations of the piers were made in 1846, but it was not until 1848 that the works were commenced. Each abutment consists of six cast-iron cylinders, 6 feet diameter, which were sunk by excavating the gravel from their interior by hand dredging and by placing weights on the top so as to force them down.

*Scale of feet.*
Fig. 10.

Truss of
Windsor Bridge
*Transverse Section*

When each cylinder had been by this means sunk low enough to ensure a good foundation, it was filled with concrete in the following manner. A mixture formed of Thames ballast and Portland cement, in the proportions of 8 to 1, was put into a canvas bag; this was lowered inside the cylinder to the bottom, and, by pulling a rope, the mouth of the bag was opened, and the concrete deposited under water in the bottom of the cylinder. Whenever the work was interrupted, great care was taken before recommencing it to clean off any deposit, in order that the new concrete might adhere well to the old. When the cylinder had been filled to such a height that there was no danger of its floating up when emptied, the water was pumped out. The inside was then filled with concrete in the ordinary manner. On the top were placed oak platforms, which support the trusses of the bridge.

One of the outside trusses was tested at Bristol in July 1849, by loading it gradually with iron rails, beginning from one end, until the whole truss was uniformly weighted with 270 tons, or 1½ tons per foot-run, observations on the deflection of different points of the bottom girder being made both during the loading and unloading. The results of this test were perfectly satisfactory.

The superstructure of the bridge was erected on scaffolding, and the line was opened on October 8, 1849.

It will be desirable here to notice one or two important features in this as in almost all Mr. Brunel's bridges.

The ordinary permanent way was laid over the bridges with ballast of sufficient thickness to enable the road to be kept in repair in the same manner as the other parts of the line. As there was no change in the nature of the support given to the rails, no concussion was caused on a train entering or leaving a bridge. The ballast took off from the structure the vibration of the train; and, in the event of carriages or even engines getting off the line, it helped in a great measure to prevent their ploughing through the flooring. Where the flooring was of timber the ballast protected it from fire. Also in long bridges there was no necessity for any contrivance of sliding rails to allow for the effects on the structure of changes of temperature. On the other hand, the ballast added to the weight on the bridge. With the timber viaducts this was an advantage, since it kept the various parts of the framework in close contact, and prevented sudden jars being brought on them by the rapidly applied load of a passing train. Even on the large bridges the cost of the extra material requisite to support the weight of the ballast was more than compensated for by the advantages above referred to.

Mr. Brunel employed timber flooring, as being the safest in the case of carriages getting off the line, and also as being the cheapest. This flooring in the iron bridges was generally laid diagonally on wrought-iron cross girders, which were placed not at right angles to the line, but obliquely, in order that the two wheels of the same axle of an engine or heavy waggon might be on different cross girders at the same time. By this arrangement the cross girders could be made of less strength, and a saving effected in their cost and weight.

The bridge over the Wye at Chepstow, and the Royal Albert Bridge over the Tamar at Saltash, are the largest and most important of Mr. Brunel's bridges.

They are remarkable not only for their dimensions, but also for the economical character of the designs, the form of their superstructures, and the methods by which the foundations of the piers were made.

At the part of the river Wye where it is crossed by the Chepstow Bridge, a cliff of limestone rock rises on the left bank to a height of 120 feet above the bed of the river, forming the precipitous edge of a broad table-land; while on the right bank the ground slopes gently for a considerable distance, rising only a little above high water, and is composed partly of clay and partly of loose shingle interspersed with large boulder stones. As it was necessary to leave a clear headway of 50 feet above high water for the navigation, the line on one side of the river is on an embankment of great height, and on the other side it penetrates the cliff about 20 feet below the top. The whole space to be bridged over, 600 feet wide, was divided into a river span of 300 feet, and three land spans of 100 feet each (see fig. 3, p.153) At one end of the great span a secure abutment was offered by the cliff of limestone rock; but at the other end, and under the piers of the smaller spans, the ground throughout was soft, and full of water. There was, however, rock at a depth of 30 feet below the bed of the river.

To reach this foundation with masonry, by means of a coffer dam, was almost impracticable, as it was 84 feet below high water.

The plan of building a stone pier on a foundation of piles was considered, and abandoned on account of the expense.

The method of sinking the cast-iron cylinders of the Windsor bridge has been already described. The pneumatic process of sinking cylinders had been introduced with great success at the Rochester bridge.

In this process the cylinder is closed at the top and air forced in by pumps until the water is expelled at the bottom. Workmen in the interior excavate the ground and remove any obstacles which prevent the cylinder from sinking, weights being added to force it down. As the air within is at high pressure, the workmen enter, and the materials are passed in and out, through an intermediate chamber, called an 'air lock,' fitted with air-tight doors. The pneumatic method was ultimately employed at Chepstow, to assist in sinking the cylinders.

Before he decided on the plan for the foundations, Mr. Brunel had an experimental cylinder made of cast iron, 3 feet in diameter, at the bottom of which was an exterior screw flange 12 inches broad, and 7 inches pitch, making one complete turn. This screw cylinder penetrated the ground like an ordinary screw pile. In one instance it was rapidly sunk to a depth of 58 feet, through stiff clay and sand, in 142 revolutions;[24] yet, on another trial, when boulders were encountered, there did not appear to be sufficient penetrating power. In one of these trials, the screw, having got into a bed of running sand, had no hold, and failed to descend. Mr. Brunel then had the cylinder partly raised, and another screw added at some distance above the lower one. It was then successfully screwed down.

Mr. Brunel, however, ultimately decided on forming the piers of cast-iron cylinders forced down by loading and afterwards filled with concrete, and the work was commenced in the spring of 1849.

With this form of construction all uncertainty of obtaining a secure foundation was removed, as the pneumatic method was in reserve, in case of excessive influx of water, to sink the cylinders to the rock, if it could not be reached by simpler means; and additional cylinders could be added, so as to obtain any amount of area of base that might be thought necessary.

The land piers for the 100 feet spans consist each of three cylinders, which are 6 feet in diameter, joined together in lengths of 7 feet. The main pier, which supports one end of the great truss, consists of a double row of cylinders, six in all, the lower parts of which are 8 feet in diameter, joined together in lengths of 6 feet. The bottom of each cylinder was made with a cutting edge, so as to penetrate the ground easily.

Most of the cylinders were sunk by the process of excavating the ground within them and weighting the top, the water being kept down by pumping. As the ground consisted chiefly of wet sand and shingle, danger was apprehended from

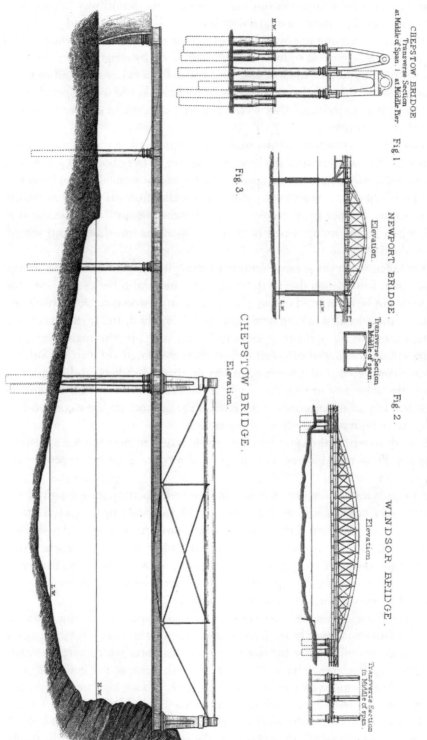

IRON BRIDGES.

CHEPSTOW BRIDGE
Transverse Section
at Middle of Span. | at Middle Pier

Fig. 1.

Fig. 3.

NEWPORT BRIDGE.
Elevation.

Fig. 2.

Transverse Section
in Middle of span.

CHEPSTOW BRIDGE.
Elevation.

WINDSOR BRIDGE.
Elevation.

Transverse Section
in Middle of span.

its tendency to run in from the outside, while the excavation was in progress. This would have diminished the lateral stability of the cylinders; and great care was taken not to excavate too near the bottom, but merely to loosen the ground round the cutting edge and to force the cylinder down by weights. Stiff clay was sometimes used to prevent the wet sand and gravel from being squeezed in from the outside. When the cylinders had been sunk to the rock, and it had been dressed off to form a level foundation, they were filled with concrete in the same manner as at the Windsor bridge.

In sinking the cylinders of the main pier, much greater difficulties were encountered than with those of the land piers, owing to large boulders and pieces of timber being met with near the bottom. When still at some distance from the rock, a length of one of these large cylinders cracked, from its having met with an obstruction. Timber struts were then fixed within it, until the obstacle was passed, when it was strengthened by a strong wrought-iron hoop, and forced down to the rock.

In April 1851, when the greater number of the cylinders had been sunk, it was apparent that, from delays due to the influx of water and other causes, some of them could not be completed by the time that the superstructure would be ready. Mr. Brunel then decided to employ the pneumatic method, and by means of this apparatus some of the remaining cylinders were sunk. In the main pier four auxiliary columns, formed of 7-feet cylinders, were placed close to the others. They were connected to the 8-feet cylinders by strong brackets, and supplied a great additional bearing surface. Any slight inaccuracy of position in the cylinders was corrected by adjusting cones at the level of the ground; on these cones 6-feet cylinders were built up to the level of the railway.

The depth to which the cylinders were sunk and their position are shown in fig. 3, p.153. From this drawing also the general form of the superstructure will be understood.

The bridge is for two lines of way; each line is carried between two longitudinal girders 7½ feet deep, of the section given in the woodcut, fig. 11 (p.155). Each girder has a triangular top flange with a plate iron vertical web, and a slightly curved plate for the bottom flange. The roadway girders over the three land spans of 100 feet are in one piece, and are therefore continuous girders, 300 feet long, supported at two intermediate points. Those across the main span are also 300 feet long, and are supported by the main truss.

The truss for each line of way consists of two suspension chains, one on each side of the roadway, hung from either side of the ends of a horizontal circular tube, arched slightly for the sake of appearance, which rests on piers rising about 50 feet above the level of the rails. The pier at the land end is of masonry, and the upper part of the middle pier is of cast iron, resting on the cylinders already mentioned. Each pier has two archways for the trains to pass through. The chains carry the roadway girders at four points, and the tube is supported at two intermediate points in its length by upright standards resting on the chains. Thus, while the weight of the structure is supported somewhat in

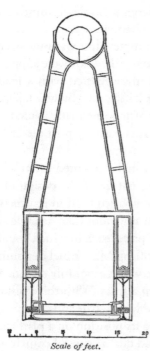

Scale of feet.

Fig. 11.   Truss of Chepstow Bridge.
Transverse Section.

the same manner as in a suspension bridge, the inward drag of the chains is resisted by the tube. To prevent the framework from being distorted by unequal loading, it is made rigid by diagonal chains connecting the upper and lower ends of the two upright standards.

The main truss may be described as an inverted queen truss. The tube which has to resist the compressive strain due to the inward pull of the chains is 9 feet in diameter, and is made of boiler plate ¾ and ⅝ of an inch thick, stiffened at intervals by diaphragms. The chains are like those of suspension bridges, each formed of 12 and 14 links alternately, these being 10 inches deep, and varying from ¾ to ¹¹⁄₁₆ of an inch thick.[25]

At the ends of the tube, where the chains are connected to it, there are several thicknesses of plate; between which the links of the chains are introduced, and a round pin, 7 inches in diameter, passes through both plates and links. The strain is thus conveyed from the chains to the ends of the tube.

Though the trusses for the two lines of way are completely distinct, the tubes are braced together horizontally, to increase their stiffness sideways.

The woodcut (fig. 11) represents a transverse section of the truss for one line of way, and shows the circular tube with the internal diaphragms, the upright standards which support it, the roadway girders, and the chains.

In consequence of the great depth of the truss, which is about 50 feet, or one-sixth of the length, the strains on the several parts are comparatively small for such a large span.

The weight of wrought-iron work in each of the trusses of the main opening is 460 tons, inclusive of the longitudinal and cross girders, which weigh 130 tons.

At the points where the roadway girders are intersected by the inclined chains, they are not fixed to the chains, but rest upon them, rollers and saddles being placed between; and at the ends of the short horizontal links, in the middle of the span, there are screws for adjusting the level of the girders.

These arrangements were made in order that the roadway girders might not be strained by the slight alteration in the form of the truss which takes place when a load comes on the bridge.

The continuous roadway girders were, in the case of the large span, supported at six points, and in those over the three land spans at four points. As the strains on continuous beams, supported at so many points, had not at that time been fully investigated, Mr. Brunel had the subject carefully enquired into both by calculation and experiment, and was thus enabled to proportion the section of the girders to the strains at each point in their length. Some account of this investigation is given in the note at the end of this chapter.

As soon as the ironwork for the first truss was completed, it was put together parallel to the river bank close to the site of the bridge. The ends were supported on temporary piers, and the structure was uniformly weighted with a load of 770 tons, or 2¾ tons per foot run. In unloading it, the weight was taken off from one end of the truss, so as to test its strength when unequally loaded. The testing having been satisfactorily completed, the truss was taken to pieces, and preparations were made for erecting it.

It was necessary that the river traffic should not be interrupted for any long period; this circumstance materially influenced the nature of the design of the superstructure, which was such that no scaffolding was required in its erection, nor was there any interference with the navigation for more than a single tide. The truss was made so that it could be divided into parts, each of which could be lifted separately and quickly. For the operation of lifting Mr. Brunel determined to use chain purchases worked by crabs.[26] The tube was temporarily stiffened by portions of the main chains, arranged so as to form a truss. With this assistance it was able to carry its own weight when suspended by the two ends.

The preliminary operation of slewing the tube to its position on a platform at right angles to the river, was a work requiring a good deal of careful contrivance. When this had been accomplished, a pontoon, consisting of six wrought-iron barges, was placed opposite the end of the tube, and all was ready for floating it across the river.

The floating took place on Thursday morning, April 8, 1852. The tube had been rolled forward on two trucks till its end overhung the pontoon; and, as the tide rose, the pontoon floated with the end of the tube resting on it. In order to guide it in a straight line across the river, hawsers were attached to points on the bank up and down the stream, and were led to crabs on the pontoon, so that by hauling on either hawser the tube was kept in its right course. As spring tides at Chepstow rise 40 feet, there is a rapid current except for a very short time.

The operation of drawing the tube across was commenced at a little after nine o'clock, and by a quarter to ten the pontoon had reached the other side safely, and the tube spanned the river. All proceeded with perfect quiet and regularity under the management of Mr. Brunel, who was assisted by Mr. Brereton and Captain Claxton. As soon as the pontoon reached the further shore, the chains of the lifting tackles were attached to the tube. The tube was lifted in the course of the day to the level of the railway, and afterwards to its place on the top of the piers, when the suspension chains and the rest of the truss were attached to it. The bridge was opened for a single line of way on July 14, 1852. The second tube was floated in a similar manner to the first, and the bridge was completed shortly afterwards.

The total cost of the Chepstow bridge was £77,000.[27]

The Royal Albert Bridge, which carries the Cornwall Railway across the River Tamar at Saltash, is the last and greatest of Mr. Brunel's railway works.

A railway into Cornwall, crossing the river Tamar, was proposed as early as 1844. Mr. Brunel at one time thought of carrying the trains across on a steam ferry similar to those which had been successfully introduced by Mr. Rendel.

In 1845 a company was formed and an application made for an Act to construct the railway either with a steam ferry at Torpoint or by a bridge at Saltash. The latter plan was sanctioned by Parliament.

The height of the line shown on the section at the crossing of the Tamar was 80 feet above high water. The Admiralty, however, required that this height should be increased. No further steps were taken till the beginning of 1847, when some preliminary borings and sections were made, in order to prepare definite plans for the bridge. The facts then ascertained were so encouraging as to strengthen Mr. Brunel in his opinion that the difficulties to be encountered would not be found greater than had been anticipated.

The river at Saltash is 1,100 feet wide, with a depth in the middle of about 70 feet at high water. It had at first been intended to construct the bridge with one span of 255 feet, and six of 105 feet, with superstructures of timber-trussed arches.[28] In compliance with the requirements of the Admiralty, the design was altered to two spans of 300 feet, and two of 200 feet, with a clear headway of 100 feet. This arrangement would have required three piers in deep water. Mr. Brunel subsequently decided to have only one pier in deep water, and to have two spans of 465 feet each. It was afterwards found that these could be reduced to 455 feet.[29]

Twenty years before, while engaged with his father on the Thames Tunnel, he had conceived the idea of working under a diving-bell of great dimensions. Sir Isambard approved of the suggestion, and thought of applying it in sinking the shafts of the Tunnel. Drawings were prepared, but the circumstance of a patent for a similar idea having been taken out by Lord Cochrane partly deterred him from carrying out the project, though some sketches were afterwards made for constructing a lighthouse by means of this arrangement. When the construction of the Cornwall Railway had to be considered, Mr. Brunel thought that his old idea would be applicable to the difficulties to be encountered at Saltash.

Although the plan of using a large diving-bell was one which was nearly certain to be successful, Mr. Brunel thought it probable that a large cylinder of wrought iron could be constructed to serve as a coffer-dam, and that after sinking it through the mud, the bottom edge might be sufficiently water-tight to admit of the water being pumped out, and the masonry of the pier built in the ordinary manner.

A trial cylinder, 6 feet diameter and 85 feet long, was made, partly to ascertain whether or not this plan was practicable, but mainly for the purpose of thoroughly examining the site of the centre pier, where the surface of the rock was 80 feet below high water.

A strong framework was fitted on two gun-brig hulks, with powerful tackle for lowering and raising the cylinder. After it had been lowered to the bottom, five borings were taken within it, reaching through the mud to the rock. The cylinder was then shifted and similar borings made.

The positions of the borings, one hundred and seventy five in all, were carefully recorded; and thus a minute and accurate survey was obtained of the surface of the rock.

The site of the pier was afterwards determined by means of a model constructed from these observations. In January 1849, when sufficient information had been obtained, the water was pumped out of the trial cylinder, and the mud excavated down to the rock. A short piece of masonry was then built, to demonstrate the practicability of building a pier in such a situation.

The expenditure of the Company for works of all kinds was shortly afterwards curtailed as much as possible, and no further progress was made for upwards of three years. However, information had been gained which proved that a masonry pier could be built in the middle of the river, on a good rock foundation which was there covered by a thickness of about 16 feet of mud.

During the suspension of the works, all the plans were revised, with the view of reducing the first cost wherever practicable; and Mr. Brunel decided not to make the bridge for a double line, even if there were money forthcoming to do so. His reason for this is given in the following report to the Board of Trade, made in 1852:—

> This bridge had been always assumed to be constructed for a double line of railway as well as the rest of the line. In constructing the whole of the line at present with a single line of rails, except at certain places, the prospect of doubling it hereafter is not wholly abandoned, but with respect to the bridge it is otherwise.
>
> It is now universally admitted that when a sufficient object is to be attained, arrangements may easily be made by which a short piece of single line can be worked without any appreciable inconvenience . . . This will make a reduction of at least £100,000.

In the summer of 1852 the designs of the bridge were matured, and by the beginning of 1853 the Admiralty had approved of them; the work of constructing the great cylinder for the centre pier was then commenced.

It was determined to provide for the possibility of having to employ the pneumatic process. The cylinder had a diameter of 35 feet at the bottom, and about 20 feet above the lower end of it a dome was made to form the roof of the diving-bell; from the centre of the dome rose a tube 10 feet in diameter to the level of the top of the great cylinder. As a diving-bell of this size, under 80 feet of water, might have proved unmanageable, an annular space, forming a gallery or jacket of 4 feet in width and 20 feet high, was formed round the inner circumference of the bottom of the cylinder below the dome. This annular space was divided by radial vertical partitions into eleven compartments, and was connected at the top by an air-passage with a 6-foot cylinder, which was placed eccentrically inside the 10-foot cylinder already mentioned, and served as a communication between the outside and the annulus. On the top of the 6-foot cylinder were placed the air-locks of the pneumatic apparatus which had been used at Chepstow. Thus air might be pumped into the annular space, the water expelled, and the work carried on without having to use air pressure under the whole of the dome. In that part of the 10-foot cylinder which was not occupied by the 6-foot cylinder a powerful set of pumps were fixed to keep down the water in the central space, and diminish the pressure under which the men worked, thus utilising whatever advantage could be gained from the great cylinder acting as a coffer-dam. As it had been ascertained that the surface of the rock dipped to the south-west to the extent of about 6 feet in the width of the pier, the bottom of the cylinder was made oblique, so as to fit the surface of the rock. These arrangements are represented in the transverse section of the great cylinder (p.161).[30]

The great cylinder, having been constructed on the river bank, was moved down to low water on launching-ways, and floated off by the rising tide. Guided between four pontoons, it was finally sunk in correct position in June 1854.

Some delay in penetrating the mud was caused by a bed of oyster shells, which had to be cut through by one edge of the cylinder. In consequence of some irregularities of the surface of the rock, the cylinder at first deviated considerably from an upright position; and it was necessary to use the pneumatic apparatus to gain access to the rock, and excavate it. The height of the annulus below the dome was such that it was not quite filled by the mud when the cylinder rested on the bottom. The work of getting the mud out of the annular space was much facilitated by the division of it into compartments.

By February 1855, the cylinder had been sunk to its full depth in an upright position, and it then rested everywhere on the rock, its lowest point being 87 feet 6 inches below high water.

Much trouble was given by a spring of water issuing out of a fissure in the rock, in one of the compartments, but the flow was stopped by driving close sheet piles into the fissure. The rock in the annulus was dressed, and the space filled by a ring of granite ashlar masonry which was built to a height of about 7 feet all round. The state of the work at this time is that represented in the section of the cylinder, (p.161).

The rock consisted of greenstone trap, so hard that tools could with difficulty be got to work it. When the ring of masonry was completed, it was expected that the bottom might be sufficiently water-tight to act as a coffer-dam, and allow of the mud being taken out from the central part of the cylinder, below the dome. But the pumping power was not at first sufficient for this purpose, and it was thought that it would be necessary to employ the pneumatic process in this space also.

However, by rapid and incessant pumping the water was lowered so as to allow of the mud and rock being excavated, and the masonry in the central space built without having again recourse to the use of air pressure. The leakage water was conveyed to two wells, formed of cast-iron pipes built into the masonry, from which the water was pumped. The inner plates of the annulus were cut out, and the work in the centre which consisted of granite ashlar set in cement was thoroughly bonded into the ring of masonry already built. When the work was carried up to the level of the dome, both the dome and the internal 10-foot cylinder were cut out and removed. When the building had been carried up some height, the pump wells were filled with cement concrete, and the influx of water stopped. Finally, about the end of 1856, when the masonry was completed to the cap of the pier, the upper part of the great cylinder was unbolted and taken ashore, it having been made in two halves with that object. Thus the most difficult part of the undertaking was successfully completed.[31]

The centre pier of the Saltash bridge is, like many great engineering works, out of sight, and little regarded by any but professional men. The rest of the bridge forms a striking feature in a beautiful landscape, and its appearance is well known.

The whole length of the bridge is about 2,200 feet, and is divided into two great spans over the river of 455 feet each and seventeen side spans, varying from 70 to 90 feet, which are on sharp curves. The piers of the side spans, as well as the two large piers carrying the land ends of the main trusses, are of masonry. The masonry of the centre pier is 35 feet in diameter, and is carried up about 12 feet above high water level. On it stand four cast-iron octagonal columns, rising up to the level of the railway. The piers which support the ends of the great trusses are constructed with arched openings, through which the trains pass.

The transverse elevation of the centre pier (p.161) shows the octagonal columns connected by cast-iron open-work, and the arched opening. The upper part of the centre pier is a cast-iron standard, and that of the land piers is of masonry cased with cast iron.[32]

The elevation shows the great height of the structure, the rails being 190 feet and the highest part of the truss 260 feet above the lowest point of the foundations.

The railway is carried over each of the smaller openings between two longitudinal girders, and over the main spans it is carried between similar longitudinal girders, which are suspended at intervals from the main truss.

Each truss consists of a wrought-iron oval tube, which forms an arch, and of two suspension chains,[33] one on either side of the tube, connecting its two ends. The rise of the arched tube above its abutments on the top of the piers is the same as the fall of the suspension chains below the same level. At eleven points in the length of the truss the chains are connected to the tube by upright standards, which are braced together by diagonal bars, in order to resist the strains due to unequal loading. The roadway girders are suspended from the truss at the upright standards already mentioned, and at an intermediate point between each of them.

The truss has the great depth of 56 feet in the centre; this conduces materially to the economy of the construction, as it diminishes the strain upon the principal parts, the tube and the chains, and so enables them to be made of smaller dimensions.

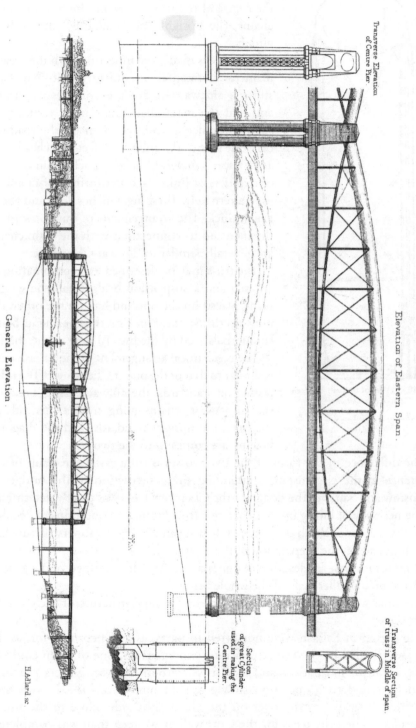

THE ROYAL ALBERT BRIDGE.

Transverse Elevation of Centre Pier.

Elevation of Eastern Span.

General Elevation.

Transverse Section of truss in Middle of span.

Section of great Cylinder used in making the Centre Pier.

H. Adlard sc.

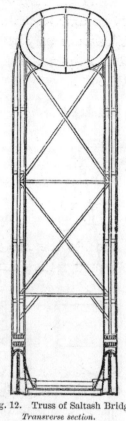

Fig. 12. Truss of Saltash Bridge.
*Transverse section.*

*Scale of feet.*

The woodcut (fig. 12) is a transverse section in the centre of the truss, showing the oval tube, the chains, the upright and standards, the roadway girders.

The tube is made oval in section with the greater diameter horizontal, in order that it may have stiffness sideways under the compressive strain, and that the main chains may hang vertically at such a distance as to leave room for the roadway between them. The tube is 16 feet 9 inches broad and 12 feet 3 inches in height. Each chain consists of two tiers of links, each tier formed of 14 and 15 links alternately. These are 7 inches deep and about 1 inch thick. The arrangements of the ironwork of the tube and its connections with the main chains are generally similar to those at Chepstow.

The truss may be described as a combination of an arch and a suspension bridge, half the weight being placed on the one and half on the other, the outward thrust of the arch on the abutments being counterbalanced by the inward drag of the chains.

The mechanical arrangement of the Saltash truss is similar to that of the one at Chepstow.[34] The tube, resting on standards, the railway passing beneath, the suspension chains hung from either side of the tube, the upright standards, and the diagonal bracing are common to the two structures.

The difference in the form of the two trusses is principally the result of the difference in the circumstances attending the construction of the bridges at Chepstow and Saltash. The design of the Chepstow truss was chiefly determined by the necessity of lifting up the separate parts of it under conditions of peculiar difficulty; while at Saltash the mode of floating and lifting the superstructure had great influence in the preparation of the design.

On p.161 is given an elevation of one span of the Saltash bridge, and a general elevation on a smaller scale of the whole bridge.

The total weight of wrought ironwork in the superstructure of each span is 1,060 tons.[35]

The trusses at Saltash were not lifted in parts, as at Chepstow; for, as the river was divided by the centre pier into two openings, one of them could be left clear for the navigation, and each truss, with its roadway girders attached, could be raised to its position slowly and in one piece. The trusses were constructed parallel to the river on the Devonshire side, close to the site of the bridge. When the truss for the Cornwall or western span was completed, temporary piers were erected to support the ends, and the scaffolding having been removed, the roadway was loaded with 1,190 tons, uniformly distributed.[36]

This test having proved satisfactory, preparations were made for floating the truss. Docks were made underneath it near the two ends, and in each of these docks two iron pontoons were placed. Valves were then opened to admit water, and the pontoons were allowed to sink on timbers prepared to receive them.

Upon each pair of pontoons was erected an elaborate framework of timber to carry the weight of half the truss, or between 500 and 600 tons. The framework consisted of stout timber props, some of them 40 feet long, extending from the pontoon to the arched tube, and was attached to the tube by iron suspension rods, so that when the operation of floating was completed, the pontoons would be free to pass from underneath the truss.

Mr. Brunel had previously taken part in operations of this nature. When Mr. Robert Stephenson was about to undertake the floating of the tubes of the Conway and Britannia bridges, he asked his friends Mr. Brunel and Mr. Locke to give him their assistance. They were present at all, or nearly all, these difficult operations, and Mr. Brunel had an active share in the work, especially in the floating of the first Conway tube. By Mr. Brunel's advice Mr. Stephenson had obtained the services of Captain Claxton to superintend the nautical part of the work; and Captain Claxton was, as a matter of course, with Mr. Brunel in a similar capacity at Chepstow and Saltash.

At Saltash fortunately there was not so swift a tide as there had been at the Britannia and Conway floatings.

In order to haul the truss out, warps were laid from the pontoons to a gun-brig hulk near the centre pier, and to a barge higher up the river. On this barge were also placed ready for use the ends of four warps, leading to capstans and crabs on board vessels moored at various points. To keep the truss from being drifted up or down the river while being moved out, radius lines were laid from the pontoons to moorings, with arrangements for hauling in on them if required.

In order to ensure his directions being clearly understood and promptly attended to, Mr. Brunel assembled a number of his assistants, one of whom was placed as 'Captain' in each of the vessels containing the hauling capstans, to superintend the men, and to execute orders. These orders were given by signals.

It was most important that the attention of the captain should not be diverted by looking out for the signals, and that there should be no chance of a signal not being seen by him because he was attending to some other of his duties. There was, therefore, in each vessel an assistant whose sole duty it was to watch for the signals, to give the appropriate interpretation to the captain, and to acknowledge the signal by a flag corresponding to that by which it was given.

Mr. Brunel directed the operations from a platform in the centre of the truss. The signals were given from a smaller platform immediately above, and were made by red and white flags, held in front of black boards, which were turned towards the vessel signalled to. Printed papers containing instructions were distributed to all engaged; the signalling was carefully rehearsed, as also was every other part of the operations which could be tried beforehand.

September 1, 1857, was the day fixed for the floating. During the morning, the men, about 500 in number, assembled at their stations on the vessels and pontoons. Captain Claxton had command of the arrangements afloat, and as a reserve force to act in any emergency several boats were lent from H.M.S. 'Ajax' and the Dockyard. With Mr. Brunel were Mr. Brereton and Captain Harrison, the commander of the 'Great Eastern.' Mr. Robert Stephenson was expected, but a serious attack of illness prevented him from being present.

At about one o'clock in the afternoon signals from the tops of the temporary piers on which the truss rested, showed that the ends had been lifted three inches clear. Mr. Brunel then gave the signal for the men in the pontoons to haul on the warps, and the great structure glided slowly out to the centre of the river. A pause was then made, while the warps which were to swing the truss round into its place were being attached to the pontoon which was farthest from the centre pier. When this was done, the different ropes were hauled upon in obedience to signals, so as to keep the other pontoon close to the centre pier, upon which, as a pivot, the truss swung round in a quarter circle till it occupied the whole of the western half of the river, and was brought close to its appointed resting-place. It was finally adjusted to its exact position by strong tackles attached to the piers. Water was then admitted into the pontoons; and, as the tide fell, they were allowed to drift away, leaving the truss resting on the piers, the roadway girders being but a few feet above the water.

The whole operation was conducted with the most perfect order and regularity. The beauty of the scenery and the changing effect produced by the truss in the various positions it assumed as it was being moved forward and swung round into its place, rendered the operation as interesting to the spectators as its results were satisfactory to Mr. Brunel and to those who assisted him.[37]

In the task of lifting the truss, as well as in that of floating it, Mr. Brunel had the great advantage of the experience gained at the Britannia Bridge. There the piers were built first, and the tubes hauled up with link chains by hydraulic presses placed on the tops of the towers. The design of the piers at Saltash did not allow of this plan being adopted, and they were built up under the truss as it was lifted. Under each end of the truss were three hydraulic presses; the two outside presses combined, or the middle one by itself, were sufficient to lift the weight. Mr. Brunel had also at first intended to have strong screw-jacks, which were to be kept screwed up underneath the truss, and so to support the weight, if by any accident the presses failed. A modification of this plan was adopted; the rams of the presses had a screw thread cut on them, and a large nut on each was kept screwed up hard against the top of the press as the ram emerged from it. As an additional precaution, timber packing, in thin layers, was placed in the space between the completed portion of the pier and the end of the truss as it was lifted. Great care was thus taken to guard against any mishap.

The tube was lifted 3 feet at a time at each end. The operation went on slowly, in order to allow the masonry of the land pier to set after it had been built up underneath the truss. The work was carried on with great system and care under the immediate superintendence of Mr. Brereton. Mr. Brunel was only able to be present during one of the lifting operations, as he was then engaged in the launch of the 'Great Eastern.'

By July 1858, the first truss had been lifted to its full height, and the second truss was ready for floating. The arrangements were generally similar to those on the previous occasion; the course, however, to be traversed by the pontoons was more intricate than on the previous occasion, as the land pier on the Devonshire side of the river, over which the first truss had passed, had been built up to receive the end of the second.

This truss had, therefore, to be moved first outwards till the pontoons were clear of the docks, then it had to move endways up the river, and to swing round into position. Mr. Brunel was obliged to remain abroad from ill-health, and Mr. Brereton conducted the operations. Although the weather was not favourable, and the wind high, the truss was safely landed on the piers; and was afterwards raised in the same manner as the first one.[38]

The general elevation, p.161, shows the proportions of the bridge. On the Devonshire side, the side spans pass over fields, and on the Cornwall side over the town of Saltash.

The general effect of the bridge is in no way heightened by an expenditure of money on architectural ornament; for, with the exception of a few unimportant mouldings, the bridge is absolutely unadorned. The total cost was £225,000—a very moderate expenditure, especially when the difficult work at the centre pier is taken into account. This result is due not only to the careful manner in which all the details of the design were prepared, but also to the great attention given throughout to the construction.

His Royal Highness the Prince Consort, as Lord Warden of the Stannaries, permitted the bridge to be called the Royal Albert Bridge, and consented to open it in person. The ceremony was performed on May 3, 1859. Mr. Brunel was compelled to be absent on the Continent, for the sake of his health, and was represented on the occasion by Mr. Brereton.

After Mr. Brunel's return to England, he paid a hurried visit to the Cornwall Railway, and, for the first and last time, saw in its completed state the great work on which he had expended so much thought and care.

NOTE
*Experiments on Matters connected with Bridge Construction*

No account of the structures designed by Mr. Brunel would be complete without a reference to the elaborate care he always took, wherever it was practicable, to satisfy himself by experiment of the qualities of the materials employed, and of the correctness of the principles followed. It would not here be possible to give a detailed record of all his experiments, but an account of some of the methods employed by him will be interesting.

Some of the larger of Mr. Brunel's experiments on cast and wrought-iron girders have already been mentioned.[39] He scarcely ever made any large girder or framework without having it fully tested, and he made extensive and elaborate experiments, most of them on a very large scale, on the strength of some of the materials and component parts of his different structures.

Among the large scale experiments tried by Mr. Brunel, were those on the compressive strength of yellow pine timber, which were made at Bristol in 1846, and were on specimens from 10 to 40 feet in length, and from 6 to 15 inches square. A framework of four upright pieces of whole timber, nearly 50 feet high, contained four strong bars of wrought iron, placed vertically, and attached at their lower ends to the cylinder of a hydraulic press. Along these bars, a casting could be moved, and fastened at different heights by keys, in such a manner as to have its undersurface, which was planed, perfectly horizontal. The ends of a specimen having been made exactly square to its length, it was put in this apparatus, with the upper end bearing against the lower surface of the movable casting, and the lower end resting on the top-surface of the ram of the hydraulic press, which was also planed and adjusted so as to be horizontal. The keys, which attached the movable casting to the bars, were now driven tight, and the pump of the press worked, weights being placed on the end of the lever, to correspond with increments of pressure on the ram. These weights were added gradually, until the specimen gave way. The accuracy of this mode of measuring the pressure was tested by direct loading of the ram with rails, which was repeated several times during the course of the experiments, so as to guard against any change in the amount of friction of the press. For each increment of weight, the compression of the specimen was measured on its four faces, and its deflection, or amount of bending, on two adjacent faces. The transverse stiffness of long specimens was also tried, by supporting them at each end, and loading them in the middle. The deflection in the middle thus observed corresponded very closely with what might have been expected from the observations on the direct compression; and from the constants so obtained, the strength of those specimens, whose length was very great as compared with their transverse dimensions, could be obtained by Euler's theory, but for the stouter specimens the strength per square inch was found to be nearly constant. From these experiments, a complete practical knowledge of the properties of yellow pine timber, when subjected to end pressures, was obtained, knowledge new at the time, and almost essential to Mr. Brunel in designing the many viaducts which he afterwards constructed.

Mr. Brunel also made experiments on the strength of pine timber when exposed to pressure on the side or at right angles to the fibre. By this means he determined the area which it was desirable to provide for the washers of bolts, and the weight which might safely be placed on transverse timbers or sills of viaducts.

Mr. Brunel's experiments on riveting were also important. Most of these were made with specimens 20 inches wide, and half an inch thick. They were compared with specimens of solid iron, of the same quality and thickness as the riveted specimens, and also the same width minus the rivet holes, so as to have equal efficient sectional areas. Double covering plates and double riveting were used in all cases, the variation being in the widths of the covering plates, and the number and arrangement of the rivets. The experiments were continued until thirty in all had been made, and the strongest form of joint was considered to have been arrived at.

In connection with the lifting of the parts of the Chepstow Bridge, an elaborate series of experiments was made on ropes, chains, and wire-rope, so as to ascertain which of these it was desirable to employ, as possessing the greatest advantages. The experiments made were of two kinds, one to determine the absolute strength of the specimen when subjected to a straight pull, and the other to observe what took place when it was worked over a sheave. In the first set the specimen was held at each end in the jaws of a pair of wrought-iron clamps, which were tightened up by means of screws. One of the clamps was attached to a fixed beam, and from the other was suspended a large cylindrical tank, which was gradually filled with water until the specimen gave way, the breaking strain being the weight of the tank and water. This weight was ascertained by actual weighing with a steelyard when the water in the tank was at different heights. Observations on the extension, shrinkage of the circumference, and change in the pitch of the spiral of the rope were made with different loads, and the strength of a sufficient number of the yarns of which the rope was composed was tried to ascertain the loss of strength by combining the yarns into a rope. In the second set the specimen, clamped as before, was passed over a sheave, the axle of which rolled horizontally on planed cast-iron plates, in order to diminish friction. To each clamp was attached a cylindrical iron tank. Water being admitted to the highest tank until downward motion commenced, its influx was stopped, and the tank descended, the other one rising. Water was then admitted to the now highest tank until motion again commenced, and this process was repeated until the specimen gave way, the tanks getting fuller of water at each movement, at which times the difference of weight of the two tanks was observed. This, minus the slight friction of the apparatus, represented the rigidity of the rope. The extensions were observed as in the first set of experiments.

The specimens consisted of hemp, manilla, shroud laid and hawser laid ropes, from 8 to 10 inches in circumference, round and flat wire ropes, and chains of different sizes of about the same strength as the ropes. The sheaves also were of different diameters. These experiments resulted in Mr. Brunel deciding to use chains for lifting the bridge, and this mainly from the circumstance that chains work more satisfactorily over a sheave than either hemp or wire ropes.

Some small scale experiments made by Mr. Brunel are deserving of notice. These were made to verify calculations on the longitudinal girders of the Chepstow truss, which are virtually continuous beams of five unequal spans. It was desirable to test the results of analysis by experiment, in order to be assured that no errors had been committed in its application. Mr. Brunel accordingly devised the following simple form of experiment for this purpose. A deal rod, exactly half an inch square and 38 feet long, quite free from knots, was supported on props of equal height, above the perfectly horizontal and planed surface of a large beam of timber. The props were placed so as to correspond relatively to the actual spans, and the rod was loaded uniformly by means of a chain. It was thus bent into an elastic curve, the ordinates of which were very carefully measured, at every foot along the length, by a finely divided scale and magnifier. The pressure on each prop was also determined, by removing any particular one, and suspending the point of the rod immediately over it to a steelyard, the weight being observed when the point of the rod was exactly at the same level as before the prop was removed. The obvious condition, that the sum of the pressures on the props should be equal to the weight of the rod and its load, furnished a satisfactory means of testing the results of these weighings. The rod being turned over on each of its four sides, the experiments were repeated, and the average taken, in order to eliminate the effects of initial curvature, or of unequal elasticity. Diagrams of the elastic curves were then made, showing the correspondence of theory with experiment, and this was so close as to leave no doubt that a true knowledge of the nature of the strains had been arrived at. One of these diagrams is given by Mr. Edwin Clark in his work on the 'Britannia and Conway Tubular Bridges,' vol. i. p.462.

By modifications of the plan Mr. Brunel adopted in this experiment, the strains on continuous beams of varying section may be ascertained with considerable accuracy.

1.    It would of course be impossible here to give a description of all Mr. Brunel's bridges, or even to refer to the most important of them with that minuteness which would be required if this were a book written for professional use. The following publications may be consulted:—Bourne's *History and Description of the Great Western Railway*, 1846; Brees' *Railway Practice*, 1837; Simms' *Public Works of Great Britain*, 1838; *Proceedings of Institution of Civil Engineers*, vols. 14, 25, and generally; Molinos et Pronnier, *Construction des Ponts Metalliques*, 1857; Humber's *Cast and Wrought-Iron Bridge Construction*, 1861; and Humber's *Record of Engineering* for 1866. At the end of this chapter a note has been given of publications in which many of the bridges have been referred to.

2.  In the early days of the Great Western Railway special designs were made for every one of the ordinary bridges over and under the railway but when, in consequence of the rapid extension of the Great Western system, the number of bridges to be designed became very large, Mr. Brunel had a set of 'standard drawings' prepared and engraved, which embodied the experience gained, and contained designs suitable for various situations. The contract drawings were made by adapting to the particular circumstances of each case the standard drawing which was most applicable to it. This system, besides securing uniformity of construction, introduced a considerable amount of economy; since, the standard drawings being based upon the results arrived at in an extensive practice, the proper structural arrangements and dimensions were indicated with far greater accuracy than could be attained in a reasonable time by an independent calculation in each individual case.

3.  It was called the 'Wharncliffe Viaduct,' in acknowledgment of the services rendered to the company by the late Lord Wharncliffe as Chairman of the Committee in the House of Lords. Drawings of this bridge are given in Simms' *Public Works of Great Britain*, 1838, pl. 54, 55, and 56; and in Bourne's *History and Description of the Great Western Railway*, 1846.

4.  At the back of retaining walls, such as the abutments and wing walls of bridges which were subject to the pressure of earth behind them, Mr. Brunel introduced what were termed 'sailing courses,' projecting shelves corbelled out at the back of the wall. The weight of earth resting on these shelves virtually increased the weight of the back of the wall, and assisted it in resisting the forward pressure of the earth.

5.  During the construction of the bridge a part of the crown of the eastern arch proved defective, in consequence of the cement in the middle of the brickwork not having set sufficiently at the time when the centering was eased. Apprehensions which had been entertained by some as to the safety of the structure were groundless, for when the defective part was taken out and replaced, no further trouble was experienced. The bridge has stood well, and has shown none of those symptoms which an overstrained structure exhibits.

6.  Simms' *Public Works of Great Britain*, pl. 57, 58; Bourne's *History and Description of the Great Western Railway*, p.36.

7.  When Mr. Brunel for architectural effect employed Gothic or pointed arches, he occasionally made the main part of the arch of a form different from the curve visible on the face, but he more frequently made it of the same pointed form throughout. In this case he did not obtain equilibrium by loading the crown, but he kept the line of pressure sufficiently within the thickness of the arch by strengthening the haunches.

8.  Brees' *Railway Practice*, pl. 42.

9.  There is an illustration of this bridge in Bourne's *History and Description of the Great Western Railway.*

10. Joggles are small pieces of hard wood or cast-iron of rectangular cross section, placed between two beams, and fitted carefully into notches cut across them. The beams are then bolted firmly together. The object is to stop the slipping which

would occur between the two surfaces if the beams were merely laid one on the other and loaded, and so to make two pieces of equal size act as one of double the depth, and therefore of four times the strength of a single piece. In 1841 Mr. Brunel made experiments to satisfy himself of the strength gained by this method; and he afterwards effected the arrangement by ensuring an exact fit in the notches by tightening up the joggles with wrought-iron wedges.

11.  A drawing of this bridge is appended to the *Report of the Commission on the Application of Iron to Railway Structures,* 1849.

12.  Mr. Brunel, by taking care in his timber structures to distribute the load uniformly, was frequently able to dispense with costly foundations. On one occasion a timber viaduct 30 feet high was placed upon a broad platform resting on an old embankment upward of 50 feet in height. A similar arrangement was adopted by him in some cases on slips in embankments which it would have been uncertain and expensive to make up with earthwork.

13.  For a description of this viaduct, see *Proceedings Inst. C.E.,* vol. xiv. For 1854–5, p.492.

14.  Mr. Brunel about this time introduced a great improvement in the manufacture of wrought-iron bolts for bridges, when these, as is usually the case, are screwed at the ends. A screw cannot be made on a bolt with out the metal being cut into to the extent of the depth of the thread, and the strength thereby considerably reduced. The section, taken at the bottom of the thread, is much smaller than that of the bolt; and, moreover, owing to the abrupt change at the commencement of the thread, the strength is not so great as that due to the reduced sectional area. The improvement consisted in swelling out the iron of the end of the bolt where the thread was to be made, so that the diameter, at the deepest part of the thread, should be fully equal to that of the bolt. The saving of metal by this improvement, especially in the case of long bolts, is very considerable.

15.  An arrangement was introduced by Mr. Brunel to prevent any disturbance in the permanent way by a settlement of the embankments at the ends of the viaducts. There were no large abutments and wing walls, but the end of the viaduct was formed with a queen truss in the parapet, which rested on a platform on the top of the embankment slope. In any slight settlement of the earth, the end of the viaduct sunk with it, and the permanent way was not disturbed. Wedges were provided to raise the ends of the trusses and readjust their level. This was an important provision, as it applied to about 60 cases on the Cornwall Railway.

16.  The simple arrangement of the timber-work was specially arranged with a view to giving facility for replacing portions of it, should they decay.

17.  The viaducts on the Cornwall Railway between Plymouth and Truro are thirty-four in number. Of these there are nineteen which have from six to twenty openings, and are from 80 to 153 feet in height. The aggregate length of these nineteen viaducts is nearly 2¾ miles.

18.  A drawing showing this form of section is appended to the *Report of the Commission on the Application of Iron to Railway Structures,* 1849.

19.  The remainder of this letter is printed in Chapter XVI, p.356

20. Mr. Brunel was thoroughly conversant with the principles of mathematical analysis, but at the same time he preferred, when it was possible, to use geometrical methods of solution for engineering problems.

21. In the note at the end of this chapter mention is made of some experiments made by Mr. Brunel on riveted joints.

22. See above, p.141.

23. A description of this bridge is given in Humber's *Bridge Construction,* vol. i. p. 228, and vol. ii. pl. 70, 71, 72; Molinos et Pronnier, *Construction des Ponts Metalliques,* p.328, pl.20, 21, 22.

24. See *Proceedings Inst. C.E.,* for 1847–8, vol. vii. p.138

25. After some perseverance, Mr. Brunel succeeded in getting these unusually large links, which were 20 feet long, rolled in a single piece without welding on the eyes. He had to go down himself to the manufactory in order to get the men into the way of doing the work.

26. See note at the end of this chapter for experiments on ropes and chains. The crabs were designed and made specially for this duty. Each had two barrels, grooved to receive the chain, which was passed several times round both barrels, so as to get sufficient grip; and it was in this way possible to wind in with the crabs any length of chain without having to stop to fleet, as would have been the case had a single-barrelled crab been used.

   These crabs were subsequently used at the floating of the Saltash trusses, and at the launch of the 'Great Eastern.'

   A similar arrangement was applied in the paying-out machinery of the Atlantic cable, and is still used for the picking-up gear in the 'Great Eastern.'

27. See *Encyclopædia Britannica,* 'Iron Bridges,' vol. xii. p.320, pl.23, 24, 25.

28. These would have been magnificent specimens of timber-work, and a design for somewhat similar trusses had at one time been prepared for the Chepstow Bridge. It is worthy of mention that Sir Isambard Brunel at one time designed a timber arched bridge of 800 feet span to cross the Neva at St. Petersburg.

29. Before the work was begun Mr. Brunel made calculations to determine whether or not it might be desirable to cross the river with one span of 850 feet, in order to avoid the great depth at the centre pier.

   A few extracts from letters relating to bridges of large span will be interesting:

January 31, 1852.

I have revised my calculation as to a span of 1,000 feet, and find that even with the loads and limitations of strains which I adopt—namely a proper thickness of ballast, and a possible load of a train of engines without tenders, and a limitation under such a load of 5 tons' strain per square inch, that a span of 1,000 feet may be made in England of the very best workmanship, and sent out and erected for I should say safely £250,000, of course a single way—another £250,000 ought, I should think, to cover the rest of the bridge.

I should like to explain to you the mode I should propose for raising such a bridge, weighing 7,000 tons.

December 1, 1852

As you ask me my opinion of the advisability of patenting your bridge, I give it you; though you will probably be the first person who will have followed such advice if you do so, and might safely patent such a novel mode of using advice.

In my opinion you cannot patent the bridge. Without detracting in the least from your merit of invention, the form has been so frequently and exactly applied that no patent could hold. The Saltash bridge now just advertised for letting is exactly on the same principle as regards form; and this is so old to my knowledge that I can claim no invention, and the use of cast iron for such purpose is also incapable of being patented. There is much that is good in your bridge, and you deserve credit, but you would find innumerable claimants to dispute, and successfully, your attempt to claim a monopoly by a patent—I myself for one. Besides, I see it published in a book.

May 30, 1854.

As to your present enquiry, I do not think that what I am doing at Saltash would be applicable in this case; but without being guilty of great presumption, I think I may say that if the same plan will not do, it is fair to assume that the same brains which concocted the plan to suit the difficulties of the Tamar might very likely find the means of overcoming those of the Severn.

If I should be able to suggest a feasible plan, and there should be found people ready to make it, I shall have the satisfaction of bridging the Severn as well as the Tamar.

30. It has of course been impossible to refer to the points on which Mr. Brunel was aided, in his different works, by the suggestions of his assistants; but it may be mentioned here that, as appears from one of Mr. Brunel's letters, the plan of working under a diving-bell had been proposed by Mr. William Glennie, his assistant, before he knew that Mr. Brunel had decided to adopt it.

Mr. Brunel also mentions in another letter that the 'very simple and effectual manner' of applying the pneumatic apparatus, by forming the annular space round the circumference of the bottom of the cylinder, was suggested to him by Mr. Brereton, when the method of constructing the cylinder was being finally settled.

31. In the *Proceedings of the Institution of Civil Engineers,* vol. xxi. 1861–2, will be found a paper by Mr. Brereton, giving a detailed description of the means employed for the construction of the central pier.

32. Shortly after Mr. Brunel's death some of his friends on the Board of the Cornwall Railway placed the following inscription, in raised letters, over the land archways— I.K. BRUNEL, ENGINEER, 1859.

33. A portion of the chains used were those which had been made for the Clifton Suspension Bridge, Mr. Brunel's earliest design.

34. Though it is convenient to explain the nature of the strains in the Saltash bridge as an arch and suspension bridge combined, it is not intended to imply that there is any virtual difference between this truss and the one at Chepstow, for in both

the strain on the tube counteracts the strain on the chains, though the one tube is curved and the other straight.

35. Humber's *Bridge Construction,* vol. i. p.231; vol. ii. pl. 78, 79, 80.

36. This load amounted to two and three quarter tons per foot run, in addition to the weight of the truss. Under this load the central deflection was about 5 inches.

   The strain on the iron of the tube and of the chains with a load of one ton per foot run, in addition to the weight of the truss, flooring, and ballast, is under four tons per square inch.

37. The difficult operations of floating and lifting the superstructures of the Chepstow and Saltash bridges were carried out entirely by Mr. Brunel and his assistants, there being no contractor engaged in, or responsible for, the work in either case.

38. A photograph taken shortly after the floating of the second tube forms the frontispiece of the first volume of Humber's *Bridge Construction.*

39. See above pp. 144 and 146.

# VIII

# THE 'GREAT WESTERN' STEAM-SHIP

IT will readily be conceded that Mr. Brunel's railway works, which have formed the subject of the five preceding chapters, would have given him ample employment for the thirty years of his professional life. Nevertheless, during almost the whole of that period—namely, from 1835, the year of the passing of the Great Western Railway Bill, to his death in 1859—he was also engaged in the accomplishment of undertakings which had for their object the systematic development of Ocean Steam Navigation.[1]

The 'Great Western,' the first steam-ship which made regular voyages across the Atlantic, the 'Great Britain,' the first large iron steam-ship, and the first large ship in which the screw propeller was used, and, lastly, the 'Great Eastern,' were Mr. Brunel's works, built under his direction, in the midst of his other engrossing occupations, and at the sacrifice of his health and life.

The history of these projects will contain records of many disappointments as well as of success; for no great and novel undertaking can be perfected at once and without changes of plan and arrangement. As engineer to the Companies which built these steam-ships, Mr. Brunel advised the adoption of measures strongly in opposition to current popular opinion, and far bolder and more daring than even his recommendation of the broad gauge and the atmospheric system. The results obtained have verified his calculations, and the conclusions he sought to establish are now so generally accepted that it is difficult to believe that they were ever questioned. No one now has any doubt that large vessels can with safety be built of iron, or that the screw propeller can be advantageously employed in ships of war and the mercantile navy; no one can now deny that it is practicable for steam-ships to make long voyages across the ocean with regularity and speed.

A detailed account will now be given of the ships whose performances first demonstrated the truth of these propositions.

Although the 'Great Western' was the first steamer which was built for regular voyages between Europe and America, the first attempt to use steam in the direct voyage across the Atlantic was made by an American ship of 300 tons burden, called the 'Savannah,' and built at New York. Her engines were of small power, with paddles made to ship and unship. She made only two voyages to and from Europe: in the first of these she left the port of Savannah on May 25, and anchored at Liverpool on June 20, 1819.

No further advance in Ocean Steam Navigation seems to have been attempted until 1835. In the October of that year, at a meeting of the Directors of the Great Western Railway Company, at Radley's Hotel, in Bridge Street, Blackfriars, one of the party spoke of the enormous length, as it then appeared, of the proposed railway from London to Bristol. Mr. Brunel exclaimed, 'Why not make it longer, and have a steamboat to go from Bristol to New York, and call it the "Great Western?"' This suggestion was treated as a joke by most of those who heard it; but at night Mr. Brunel and Mr. T.R. Guppy, one of the Directors, talked it over, and afterwards consulted three of the leading members of the Board—Mr. Scott, Mr. Pycroft, and Mr. Robert Bright. They took up the idea warmly, and a committee was formed to carry out the project.

As a preliminary measure, Mr. Guppy and Captain Christopher Claxton, R.N., made a tour of the great ship-building ports of the kingdom, in order to collect information. The results of their inquiries were embodied in a report, dated January 1, 1836, which describes at great length the advantages to be gained in large vessels. The manuscript was submitted to Mr. Brunel previously to its publication, and he inserted the following passage:—

> The resistance of vessels in the water does not increase in direct proportion to their tonnage. This is easily explained; the tonnage increases as the cubes of their dimensions, while the resistance increases about as their squares; so that a vessel of double the tonnage of another, capable of containing an engine of twice the power, does not really meet with double the resistance. Speed therefore will be greater with the large vessel, or the proportionate power of the engine and consumption of fuel may be reduced.

This was an important addition to the report, for it enunciates the principle which governed Mr. Brunel in determining the dimensions and power, not only of the 'Great Western,' but also of the 'Great Britain' and 'Great Eastern' steamships.

Immediately after the publication of this report a Company was formed in Bristol called 'The Great Western Steam-Ship Company,' Mr. Peter Maze being the Chairman, and Captain Claxton the Managing Director. Captain Claxton's exertions in the service of the Company from its formation to its dissolution were unremitting and invaluable. He was also, from the date of Mr. Brunel's first connection with Bristol, one of his most intimate friends, and his confidential adviser on all points on which nautical experience was of value.[2]

Mr. Patterson (an eminent ship-builder of Bristol) was selected to superintend the building of the first ship, under the direction of a 'Building Committee' consisting of Captain Claxton, Mr. Guppy, and Mr. Brunel. Whenever railway business called Mr. Brunel to Bristol, which at this time was at least once in every week, the Committee and Mr. Patterson used to meet at the office, or at Captain Claxton's or Mr. Guppy's house, and often sat far into the night discussing the details of the design of the ship.[3]

One of the most important questions which occupied Mr. Brunel's attention was the selection of the builders of the engines. Tenders were invited; and on receiving them, he addressed the following report to Captain Claxton, the Managing Director:—

June 18, 1836.

In considering the three tenders for the supply of marine engines for your first vessel, which you have submitted to me for my opinion, I have assumed that the interests of the company are paramount, and that all feelings of partiality towards any particular manufacturer or any local interest must yield to the absolute necessity, in this the first and the boldest attempt of the kind yet made, of not merely satisfying yourselves that you will obtain a good engine, but also of taking all those means of securing the best which in the eyes of the public may be unquestionable. In this view of the case, if you agree with me, I think you will consider that, provided the prices are fair individually, the relative amount of the tenders is a secondary consideration.

I assume, also, that the high respectability of all these parties would ensure equally from either the best materials and workmanship, and I shall confine myself simply to pointing out a few of the conditions peculiar to the engines which you require, and the means which the different parties have of complying with these conditions.

I need hardly remind you that, owing to the lateness of the season, you will require that the vessel should be prepared to run her first voyage almost immediately after the engines are fixed. You will remember, also, that it will be the longest voyage yet run; that in the event of unfavourable weather a total failure might be the result of the engine not working to its full power, or consuming too great a quantity of coals—a very common occurrence with engines apparently well made, after six or eight days' constant work; and, lastly, that the future success of the boat as a passenger ship—nay, even of the company's boats generally, and, to a great extent, and for some time, the reputation of Bristol as an American steamboat station, may depend upon the success of this first voyage. It is indispensable, therefore, to secure as far as possible a machine which shall be perfect in all its details from the moment of its completion. There may be time for a few trials for ascertaining the fact of its completion, but there will be none for effecting any alterations should they be found necessary, or for making any experiments. The machinery which you require to be so perfect is by no means an ordinary steam-engine.

Marine engines of 80 or 90, and even some of 100 horse-power, are mere models on a large scale of the ordinary-sized engines. Engines of 160 or 180 horse-power each would be unmanageable without many material modifications in the details; the arrangement must be different, and, as the strength of materials remains the same, the proportionate dimensions of the parts must be modified. Many contrivances of this description have been introduced into the large engines of 110 horse-power each made for the Navy. From 110 to 160 is still another and a very great stride. Those who have led the way in the first step are certainly the most likely to be aware of the difficulties of the second, and to be able to appreciate them better, and be more prepared to overcome them than those who have as yet only manufactured, however successfully, engines of the ordinary class. Of three parties tendering . . . Messrs. Maudslay have made by far the largest number, and have for some years led the way in the introduction of the largest armed steamboats; and there can be no question as to the fact that they are the oldest manufacturers of marine engines, that they are themselves the originators of the greatest number of the improvements of the day, that they have made the largest engines yet made, and the greatest number of large engines of all sizes; and, lastly, that they have the principal supply of engines for the large war ships now used for the Navy, and have had hitherto the sole supply of all above 70 horse-power. With these facts before you, it remains only for you to consider how far you agree with me in the conclusion I have come to, and which I have no hesitation in expressing—that I think you will be safest, in the peculiar case of the first ship, in the hands of the parties who have had most experience, and that Messrs. Maudslay are those persons. Their price is, I think, moderate.

This report was read by Mr. Brunel at a meeting of the Board, summoned at his request; the Directors adopted his advice, and accepted the offer of Messrs. Maudslay & Field, of Lambeth.[4]

The 'Great Western,' for so the ship was called, had not been long commenced when a somewhat celebrated controversy arose, in which the correctness of Mr. Brunel's views was questioned by the late Dr. Dionysius Lardner.

The circumstances which led to this discussion were as follows:—

The British Association for the Advancement of Science held its sixth meeting at Bristol in August 1836; and, as Dr. Lardner was announced to lecture on Transatlantic Steam Navigation, great interest was felt in Bristol on the occasion.

After some postponement, he delivered his lecture on August 25, to a crowded meeting of the Mechanical Section. The proceedings of the Association unfortunately do not give any report of Dr. Lardner's observations; but, as in his latest work on the subject[5] he speaks in commendatory terms of the report given in the 'Times' newspaper, that account may be relied on as correct.

In the 'Times' of August 27, 1836, it is stated that in the course of his lecture Dr. Lardner said,—

> Let them take a vessel of 1,600 tons, provided with 400 horse-power engines. They must take 2⅓ tons for each horse-power, the vessel must have 1,348 tons of coal, and to that add 400 tons, and the vessel must

carry a burden of 1,748 tons. He thought it would be a waste of time, under all the circumstances, to say much more to convince them of the inexpediency of attempting a direct voyage to New York, for in this case 2,080 miles was the longest run a steamer could encounter: at the end of that distance she would require a relay of coals.

There is no detailed report remaining of the animated discussion which followed the lecture, and in which Mr. Brunel took part. He exposed several errors in Dr. Lardner's calculations, but failed to produce any effect upon the majority of those present, who were powerfully impressed by the lecturer's dogmatic assertions.

Those assertions seem to have had a wide circulation beyond the walls of the lecture-room; and if Dr. Lardner's arguments were sound, and if transatlantic steamers ought to have taken their departure from 'the most western shore of the British Isles,' the enthusiastic advocate of a railway scheme in Ireland might well exclaim,—

> The promoters of this vast object stand forewarned of defeat. Dr. Lardner, who has bestowed a great deal of pains in arguing the bearings of this undertaking, has pronounced it impracticable; and I entirely agree with him in his conclusions. The effort, nevertheless, will be made; the genius of English enterprise will hazard the consequences; and every honest spirit that shall hear of the brave British crew which will embark upon that perilous expedition will feel his heart beating high for the merchant-sailor, whom nothing can deter. He will sail; but, though dangers will encompass him, and destruction appear, there is yet a hope for his ultimate success. Let us cheer ourselves with the expectation that, as the exhausted mariner returns, he will fall in with the western shores of Ireland; that, worn out and hopeless of home and comfort upon earth, the Shannon will win him to her bosom; that, invited by the graceful sinuosities of that noble stream, and the rich and fertile lands around, he will advance to this convenient and improving city; and as he rests within its walls that he will exclaim, 'This is the place from which I ought to have set out, for here have I returned with ease and safety!'[6]

Dr. Lardner's views are repeated in an article in the 'Edinburgh Review' for April 1837 (vol. lxv.); and in the report of the proceedings of the British Association for 1836 (p.130 of the proceedings of the sections), the reader is referred to this article, apparently as a substitute for an abstract of the lecture.

The following *résumé* is there given of the lecture:—

> The conclusions at which he arrived were briefly these: that, in the present state of the steam-engine as applied to nautical purposes, he regarded a permanent and profitable communication between Great

Britain and New York by steam-vessels making the voyage in one trip as in a high degree improbable; that since the length of the voyage exceeds the present limits of steam-power, it would be advisable to resolve it into the shortest practicable stages; and that, therefore, the most eligible point of departure would be the most western shores of the British Isles, and the first point of arrival the most eastern available parts of the western continent; and that, under such circumstances, the length of the trip, though it would come fully up to the present limit of this application of steam-power, would, nevertheless, not exceed it, and that we might reasonably look for such a degree of improvement in the efficiency of marine engines as would render such an enterprise permanent and profitable. (p.119.)

Among other objections to long voyages the reviewer enumerates the incrustation of boilers, and the choking of smoke flues; and then, with reference to the quantity of fuel required, he proceeds:—

In proportion as the capacity of the vessel is increased, in the same ratio or nearly so must the mechanical power of the engines be enlarged, and the consumption of fuel augmented . . . It is therefore demonstrable that, in the present state of steam navigation, if this voyage shall be accomplished in one uninterrupted trip, the vessel which performs it must, whatever may be her power and tonnage, be capable of extracting from coals a greater mechanical virtue, in the proportion of three to two, than can be obtained from them by the combined nautical and mechanical skill of Mr. Lang, the builder of the 'Medea,' and Messrs. Maudslay and Field . . . That the passage from Liverpool to New York cannot on any occasion be made in one run by a steam-ship we do not maintain . . . The average time of the outward voyage to New York is thirty-six days, and we say that when the circumstances of wind and water are such that a sailing vessel would require that time to make the passage, a steamer cannot make it without an intermediate supply of fuel. (pp.127, 139, 143.)

To sum up Dr. Lardner's views in his own words written at about this time,[7] 'We have as an extreme limit of a steamer's practicable voyage, without receiving a relay of coals, a run of about 2,000 miles.'[8]

It will be seen from these extracts that the proposition Dr. Lardner laid down as the basis of his 'demonstration' was, that the power of the engines must be increased as the size of the vessel. Were this true his conclusion would also be true—namely, that the capacity of a given vessel regularly to accomplish a given voyage does not increase with the increase of size, since the consumption of fuel is augmented in about the same ratio.

This assumption is directly opposed to the opinion held by Mr. Brunel, and acted on by him in his recommendations to the Steam-Ship Company—namely, that while the tonnage of a ship is increased as the cube of her dimensions, the resistance is increased only about as the square.

This question was the main point at issue between Dr. Lardner and Mr. Brunel; and the proposition which Mr. Brunel then asserted is at the present time the basis of the calculations which determine the proportion between the tonnage of a steam-ship and the length of voyage she has to perform without a relay of fuel.

The history of the 'Great Western' steam-ship has been interrupted by this examination of Dr. Lardner's propositions. The weight at one time attached to his opinions, the sinister influence they exercised over the early efforts of those who differed from him, and the great and enduring importance of the points at issue, have made it necessary to refer to them at length.

The ship had been steadily proceeded with, notwithstanding the adverse criticism of philosophers, and she was launched on July 19, 1837. On August 18 she left with a tug-boat for London to take her engines on board, and arrived in the Thames after a passage of four days, four-fifths of the way under sail.

When anchored in the river she was crowded with visitors, who, according to the newspapers of the day, were astonished at 'her magnificent proportions and stupendous machinery.'

The engines were at length completed, and received in every detail Mr. Brunel's constant supervision.

Extraordinary efforts were made to get the ship back to Bristol and to start her on her voyage across the Atlantic before the departure of the 'Sirius'—a vessel of about 700 tons and 320 horse-power, bought by the St. George's Steam Packet Company in order to anticipate the 'Great Western.'

At length the 'Great Western' left Blackwall for Bristol, at 6.10 A.M. on Saturday, March 31, 1838, having on board Captain Claxton, Mr. Guppy, Mr. Brunel, and many other persons interested in her success. All went well at first, but at about half-past eight o'clock a very alarming fire broke out. The felt which covered the boilers had been carried up too high, and the red lead which fastened it became hot; oil gas was generated, and it burst into a fearful flame, setting fire to the beams and under part of the deck. The ship was immediately run ashore on a mud-bank not far from the Chapman Beacon, while Captain Claxton, Captain Hosken (the commander), and Mr. Pearne (the chief engineer) endeavoured to extinguish the fire.

Captain Claxton went below through the engine-rooms, and forward between the boilers to the fore-hatch, and in a stifling atmosphere of burning paint and felt he directed the nozzle of the fire-hose against the flames. While he was at work, something heavy fell on him from above. On recovering from the blow, he stooped down, and found the body of a man, who was lying insensible, with his head covered to the ears with the water which had collected on the floor. Captain Claxton called for a rope, and the almost lifeless body was hauled up. It was not till he went on deck some time afterwards that he learnt that the person who had fallen on him was Mr. Brunel, and that he had saved the life of his friend.

It appeared that Mr. Brunel was going down to Captain Claxton's assistance by the long ladder which reached from the fore-hatch to the keelson, and put his foot on a burnt rung. He fell about 18 feet, striking an iron bar in his descent. Had he not fallen on Captain Claxton he must have struck the keelson or floor and been killed, and had not his head been raised at once he would have been suffocated by the water into which he fell. He was so severely hurt that he could not move, and he was laid on a sail on deck until the fire was extinguished, and then lowered into a boat, and landed on Canvy Island, where he remained some weeks. Although his sufferings were very great, he was able, within three days of the accident, to dictate a long letter to Captain Claxton on the state of the ship and engines.

The fire was soon got under, the ship resumed her voyage to Bristol, and anchored at Kingroad in the afternoon of Monday, April 2, to the great surprise of the good people of Bristol, who had heard that she had been burnt in the Thames. Their astonishment was increased by finding no outward signs of the disaster; but, as a fact, the deck above the boiler was charred a fourth of its thickness, and so remained till the ship was broken up.

The 'Great Western' started on her first voyage to New York on Sunday, April 8, at 10 A.M.,[9] and struck soundings off Newfoundland on the ninth day. She arrived at New York at 2 P.M. on Monday, the 23rd, having consumed three-fourths of the coal she had taken on board.

She found that the 'Sirius' had arrived before her; but under all the circumstances the palm was due to the 'Great Western,' for the 'Sirius' had left Cork eight hours before the 'Great Western' left Bristol (which lies a whole day's run further from New York), and had only arrived at New York in the morning of the day in the afternoon of which the 'Great Western' came in; and, what is after all the most important point for comparison, the 'Great Western' had nearly 200 tons of coal left, while the 'Sirius,' when she dropped her anchor at Sandy Hook, had not only consumed all her coal, but also all the combustible articles which could possibly be thrown on the fire, including (to repeat the well-known anecdote) a child's doll!

The 'Great Western' was received at New York with well-deserved honour. According to the journal of one of her passengers, 'Myriads were collected, boats had gathered round us in countless confusion, flags were flying, guns were firing, and cheering rose from the shore, the boats, and all around loudly and gloriously, as though it would never have done. It was an exciting moment, a moment of triumph.'

The ship started on her return home on May 7, 1838, with sixty-eight passengers on board. She made the voyage in fourteen days, although twenty-four hours were lost by a stoppage at sea.

After this she ran regularly between Bristol and New York till the end of 1846. In April 1847 she was sold to the West India Mail Steam Packet Company, and became one of their best vessels.

At length in 1857 she was broken up by Messrs. Castle, of Vauxhall. Among those who went there to take a farewell of her before she finally disappeared was Mr. Brunel; thus he saw the last of his famous ship.

## NOTE
### Dimensions of the 'Great Western' Steam-ship

|                                                      | Feet  | Inches |
|------------------------------------------------------|-------|--------|
| Length from fore-part of figurehead to after-part of taffrail | 236   | 0      |
| Length between the perpendiculars                    | 212   | 0      |
| Length of keel                                       | 205   | 0      |
| Breadth                                              | 35    | 4      |
| Breadth over paddle-boxes                            | 59    | 8      |
| Depth of hold                                        | 23    | 2      |
| Draught of water                                     | 16    | 8      |
| Length of engine-room                                | 72    | 0      |
| Tonnage by measurement                               | 1,340 | tons   |
| Displacement at load draught                         | 2,300 | "      |

### Dimensions of Engines, &c.

| | |
|---|---|
| Diameter of cylinders | 73½ inches |
| Length of stroke | 7 feet |
| Weight of engines, wheels, &c. | 310 tons |
| Weight of boilers | 90 " |
| Water 20 tons to each boiler | 80 " |
| Diameter of wheel | 28 feet 9 inches |
| Width of floats | 10 feet |

1. It has been observed with much truth that full justice has not been done to Mr. Brunel's exertions in this department of practical science.—See 'Address of George Parker Bidder, Esq., on his election as President of the Institution of Civil Engineers, January 10, 1860.'

2. Captain Claxton died on March 27, 1868, in his 79th year. The manuscript of this and the three following chapters was fortunately completed in time to be submitted to him. He spared no trouble either in giving or procuring original documents and other materials for all parts of this book, in the preparation of which he took the liveliest interest.

3. To enable the ship to resist the action of the heavy Atlantic waves, especial pains were taken to give her great longitudinal strength. The ribs were of oak, of scantling equal to that of line-of-battle ships. They were placed close together, and caulked within and without before the planking was put on. They were dowelled and bolted in pairs; and there were also four rows of 1½ inch iron bolts, 24 feet long, and scarfing about 4 feet, which ran longitudinally through the whole length of the bottom frames of the ship. She was closely trussed with iron and wooden diagonals and shelf pieces, which, with the whole of her upper works, were fastened with bolts and nuts to a much greater extent than had hitherto been the practice.

The principal dimensions of the hull and engines are given in the note to this chapter (p.183).

4. The engines as designed by Messrs. Maudslay were beam engines, although Mr. Brunel had strongly urged them to adopt the most compact form of direct-acting engines. They, however, thought it better not to depart from what was then the usual form.

5. *Steam and its Uses,* by Dr. Lardner, 1856. Chapter on 'Steam Navigation,' section 10.

6. 'An exposition of the advantages of the proposed Railway from Limerick to Waterford.'

7. *The Steam Engine: its Application to Navigation and Railways, with Plain Maxims for Railway Speculators,* 5th edition, 1836, p.307.

8. It is right to add that, according to the report given in the *Athenæum* newspaper of the meetings of the British Association at Liverpool in September 1837, 'Dr. Lardner addressed the section (Mechanical) on his old subject, the application of steam to long voyages. His remarks and calculations were to a great extent identical with those brought forward by him last year at Bristol, and published long since in his work on the steam-engine, but the conclusions were somewhat varied. The Doctor did not now deny that the voyage might be practicable, but he did not believe that it would be profitable' (*Athenæum,* September 23, 1837). Dr. Lardner was answered by several speakers, and among them by Mr. Guppy, who pointed out in much detail the unreliable character of Dr. Lardner's data; while nothing was suggested about commercial profits or subsidies. It may therefore be inferred that Dr. Lardner's arguments as to the consumption of fuel remained the same, although he may have abandoned the conclusion which legitimately followed from them—namely, that the long voyages were practically impossible.

   The Report in the *Bristol Mirror* newspaper of the same date (copied from the *Liverpool Standard*) is as follows:—'Dr. Lardner's speech was little beyond a repetition of his discourse last year in Bristol, re-published by him in the *Edinburgh Review.* The voyage to America by steam he treated as practicable, but so uncertain as to render a profitable result hopeless . . . During nearly all the year there was an adverse west wind, and the Gulf Stream was to be avoided.'

9. She had only *seven* passengers on board; fifty, it is stated, were deterred from going in her by hearing of the fire.

# IX

# THE 'GREAT BRITAIN' STEAM-SHIP

THE Directors of the Great Western Steam-ship Company, encouraged by the success of the 'Great Western,' determined shortly after her first return to England to lay down a second ship of not less than 2,000 tons burden. As they did not at that time contemplate the use of iron, a portion of the timber was purchased, and drawings were put in hand for a wooden ship. The proposed vessel was intended to be in all respects a companion ship to the 'Great Western;' only she was to be of larger dimensions, as it was found that additional cargo space would be remunerative.

In October 1838, Mr. Guppy (one of the Directors) communicated to the Board the results of some calculations Mr. Brunel had made relative to the cost and efficiency of iron vessels as compared with wooden ones. Mr. Brunel then suggested that Captain Claxton and Mr. Patterson, accompanied by one of his assistants, should make a voyage to Antwerp and back in the 'Rainbow,' an iron steam-boat of 407 tons burden, and report on the subject. On receiving their report, which was revised by Mr. Brunel, and which was strongly in favour of the adoption of iron, the Directors resolved to build their ship of that material, and of not less than 2,000 tons measurement, the same size as that which they had intended for their wooden ship. They also determined to erect the shops, and provide the tools for building her themselves.

As in the case of the 'Great Western,' the details of construction were settled by the Building Committee—Captain Claxton, Mr. Guppy, and Mr. Brunel—who were assisted by Mr. Patterson.

The preparation of the design occupied some time. In each succeeding drawing an increased size was proposed; at length the fifth design, showing a ship of 3,443 tons burden, was finally approved of. On July 19, 1839, the flat keel plates were laid, and the construction of the hull was commenced.

It will be necessary to enter with some detail into the history of the construction of the engines of the 'Great 'Britain,' as it has often been stated that it was on Mr. Brunel's recommendation that the Company built their own engines. It appears, however, that Mr. Brunel repeatedly urged upon the

Directors the utmost caution and economy, and that they ultimately acted 'against his suggestion.'

When the Directors determined, in May 1838, to build a second ship, they did not entertain any idea of undertaking so great a responsibility as the manufacture of the engines; nor had they any intention of doing so, even when, towards the end of the same year, they resolved to build the ship themselves, and to construct her of iron.

The dimensions of the proposed paddle engines (for at this date the use of the screw propeller was not contemplated) were sent, in November 1838, to Messrs. Maudslay and Field, Messrs. Hall, and Messrs. Seaward.[1]

Messrs. Maudslay declined to tender, and the negotiations seem to have fallen through at the time; but they were renewed in April 1839, when estimates for engines (with cylinders of 100 inches diameter and seven feet stroke) were again invited from several makers.

The contest lay eventually between Messrs. Maudslay and Mr. Humphrys (whose patent for trunk engines was worked by Messrs. Hall). At Mr. Brunel's desire they prepared designs for engines with cylinders of 120 inches diameter. He twice induced the Directors to postpone coming to a decision on the subject, in order that Messrs. Maudslay might mature their new patent for double-cylinder engines.

When their tender was placed before the Board, the Directors were of opinion that it largely exceeded the estimate of Mr. Humphrys. Mr. Humphrys' estimate, however, had been more than once sent back to him for revision, at the suggestion of Mr. Brunel, who expressed doubts as to the possibility of Mr. Humphrys being able to construct his engines within the sum named by him.

Messrs. Hall stated that if they tendered for the supply of engines on Mr. Humphrys' plan, large tools would have to be purchased by them, and the cost charged on the one pair of engines; they therefore strongly recommended the Company to become their own engine makers.

Influenced by these considerations, the Directors determined to adopt the plan of Mr. Humphrys, and to construct their own engines; and they appointed him the superintending engineer of their works.

It appears from a report by the secretary, Captain Claxton, dated March 23, 1840, that 'previous to coming to this decision, Mr. Brunel succinctly laid before the Directors his views of the matter, and his opinion of the great responsibility they would incur if they made their own engines; and doubtless the Directors would have yielded to his suggestions, but for the report of Mr. Humphrys, showing the utter hopelessness of getting the engines made piecemeal in Bristol.'

The following is the report of Mr. Brunel on the subject:

June 12, 1839.

At the request of Mr. Maze and Mr. Scott, whom I had the pleasure of meeting on Saturday last, I send you the following observations on the two plans and the estimates of Messrs. Maudslay, and of our Mr. Humphrys.

I have a copy of Messrs. Maudslay's letter of the 29th ult. containing their tender, and a subsequent letter of the 11th inst. in reply to some enquiries of mine respecting their tender and Mr. Humphrys' estimates, according to which the total cost of a pair of engines of 110 inches diameter and 8 feet stroke, upon his plan, and I presume modified as last recommended by Mr. Guppy and myself, including boilers and fixing on board, would be £29,296, or, as stated by Captain Claxton in a letter to me of the 1st inst., £30,700.

First, as to the comparative merits of the plans, I consider them both excellently adapted to our particular case, and that the choice will depend upon other circumstances than the construction of the engines, and these circumstances, I consider, would be, the relative cost and the advantages of forming an establishment which will eventually become necessary for the repair and maintenance of our engines, contrasted with all the advantages to be derived from the responsibility and experience in all the details of a first-rate manufacturer, and to which I attach very great value, particularly in the early proceedings of a Company like ours. As regards the cost, I understand Messrs. Maudslay's tender to be for an engine of four 75-inch cylinders, which is equal to a pair of ordinary engines of about 106 inches.

| | £ |
|---|---|
| Engine, boiler, and paddle-wheels, fixed on board, supposing the vessel in London, and with reduced size of boiler | 41,400 |
| Deduct allowance for coal-boxes and combings for hatchways as proposed by Messrs. Maudslay | 500 |
| | 40,900 |
| Additional expense incurred by Messrs. Maudslay in consequence of the engines being fixed on board at Bristol instead of London, I estimate at | 250 |
| Total amount to be paid Messrs. Maudslay | 41,150 |
| In addition to this will be the freight and insurance, which we are to pay, and also the unloading at Bristol and placing in the vessel, which I take at Captain Claxton's estimate | 2,000 |
| Making a total of | 43,150 |

It is to be observed that this includes Mr. Field's apparatus for changing water, Kingston's cocks, casing the cylinders, and all those extras which were applied to the 'Great Western,' and also the paddle-beams and paddle-wheels. Without these latter the nett cost of the engine, fixed in place, and including all other extras, would appear to be about £40,000 or £40,500. If the cylinders be increased to 77¾, which would be equivalent to the pair of 110 inch, and supposing the cost of the engines to increase in the same ratio as the power resulting from this increase, but which ought not to be the case, the total cost, according to Messrs. Maudslay's estimate, will be £46,500 and deducting the paddle-beams or frame work for carrying the paddles, which do not, I think, form part of Mr. Humphrys' estimate, probably about £45,500 as compared with Mr. Humphrys' estimate of £30,700. With respect to this latter estimate, I cannot help expressing the fears I entertain that Mr. Humphrys is over-sanguine, and that the cost would greatly exceed the sum named. The items seem to me to be moderate prices only for each article named, and I see no allowance for those alterations, damages, and waste of parts, and a variety of other contingencies, which in a piece of machinery of this magnitude and novelty is certain to amount to a very large sum.

In his estimate of the fittings and smaller parts, I think also he has greatly underrated them.

The outlay for tools and tackle would, I think, also be greater than he seems to anticipate, and on the whole I cannot but come to the conclusion in my own mind, and I should not act rightly if I did not communicate that opinion to you, that the first outlay will be fully as large and probably larger by adopting the plan of making our own engines than by employing a manufacturer. It is true we shall have some valuable and costly tools and shops included in this outlay, and a fine establishment formed, which may be rendered fully competent in point of means to continue the manufacture of engines for others, and to keep up the repairs of any number of engines which the Company are likely to have at work. My only fear would be that of the risk of the undertaking being too great for a newly-formed establishment. The making of the vessel itself is no mean effort, and to superadd the construction of the largest pair of engines and boilers yet made, and upon a new plan, is calculating very much upon every effort being successful, and particularly upon the continued assistance of those who have hitherto attended to the subject; as it must be well known to the Directors that if Mr. Guppy, for instance, should be prevented from giving his time as he has hitherto done, or if Mr. Humphrys should, from illness or other causes, leave us, the manufactory would be brought to a stand, and the loss would be serious. I have no wish to deter the Company from becoming their own manufacturers—I think it a course which must ultimately be adopted if the Company thrive—but I should have much preferred that it had been adopted gradually, that we had commenced with a vessel, and then proceeded with boilers and repairs; and, as our establishment became formed and matured, and when we might no longer depend entirely upon the engineering talents and of one Director, who may be unable to attend to it, or upon the health of one superintendent who, as yet, is alone in possession of all our plans and ideas, and at present is alone capable of carrying them out, we might then have ventured upon making the engines perhaps for the third vessel. Circumstances may, however, render it necessary that we could proceed more expeditiously, and I am only anxious that the Directors should be aware of the difficulties that we may have to encounter, and that they should not form expectations as regards economy in which they may be disappointed. The result of the best consideration I have been able to give to it is, that the question does not seem to be one of cost. In that respect, according to my view, the two modes of proceeding would be nearly balanced, but it resolves itself into the following question:—Is it better in our present position to enter at once upon the manufacture of the engines and boilers, in doing which we shall in part repay the cost of tools and shops, which must eventually be required, and by which we shall be more independent, and more capable of expediting the works, should it become desirable to incur any additional expense for that purpose, or to throw all the responsibility and risk on another party or contractor—the vessel, for which we could not easily contract, being still made in the Company's yard?

I have thus reduced the question to that state in which I can offer no further opinion or advice; it is now for you to determine. The question is one which has frequently to be decided upon by the Directors of public works; it is very much a matter of feeling, but it is simplified in the present instance by the circumstance that the expense in either case will be, to my view at least, about the same, and the work, I have no doubt, equally good in either case.

Upon this point, as perhaps upon the subject of cost, I have no doubt there will be some difference of opinion. It will be said that the work done under our own superintendence can be more relied upon than the work of a manufacturer, and that even in the engines of the 'Great Western' steam-ship, coming from one of the most experienced manufacturers, many defects may be pointed out.

I should agree fully with both these arguments, but I think these advantages are fully counterbalanced by that of the experience in all the details which is brought into operation in an old-established manufactory, and the great relief from responsibility and risk obtained by contracting for the whole work.

The Directors having determined to make the engines, erected shops and fitted them up with proper tools. The services rendered to them by Mr. Brunel at this period were fully acknowledged at the next meeting of the shareholders.[2]

Mr. Brunel's attention was now anxiously devoted to the consideration of the numerous questions involved in the construction of the ship and her engines; and, in order to obtain reliable information on many points, he sent one of his assistants, Mr. Berkeley Claxton, in the 'Great Western.' His sole occupation during six voyages was to note the amount of rolling and pitching, and the exact performance of her engines, with the effect of the use of the expansion valves on her speed, and on the consumption of fuel. The reports furnished Mr. Brunel with information which was of great value, especially when, shortly afterwards, he advised the Directors to adopt the screw propeller instead of paddlewheels.

The circumstances which led to the adoption of the screw propeller in the 'Great Britain' instead of paddle-wheels were as follows:—

In the early part of the year 1840, the performances of the 'Archimedes' steamer began to attract the attention of scientific men. This vessel, which was fitted with the screw propeller patented by Mr. Francis Pettitt Smith, arrived at Bristol in May. A few trips were made up and down the Float, but the advantages of the screw propeller were not fully appreciated by those to whom they were explained.

But Mr. Guppy, who had attended some of these trials, went round in the ship to Liverpool. On his return he made a report to the Building Committee, and the Directors, on Mr. Brunel's advice, passed a resolution delaying the progress of the engines of the 'Great Britain,' and of those parts of the frame which would be affected by any change of plans. Mr. Brunel was also requested by them to give his attention to the question of the adoption of the screw, and to report thereon.

During the next three months experiments were made by Mr. Brunel, assisted by Mr. Guppy and Captain Claxton, on the screw propeller in the 'Archimedes.'[3] These experiments afforded ample opportunity of trying the performances of several forms of screws.[4]

On October 1, Mr. Brunel attended a special meeting of the Board, and read and explained a report he had drawn up, in which he laid before the Directors at great length the results of the different experiments he had made, and the advantages which he believed would attend the use of the screw propeller.[5] A resolution was passed adopting it for the 'Great Britain.'

Mr. Brunel at first thought that he would be able to retain the form of engines which had been originally determined on for working the paddlewheels; but, on consideration, this was found impracticable. As the Company had by this time erected complete engine works, there could now be no question as to their undertaking the construction of the new description of engines required for working the screw propeller.[6]

Mr. Humphrys resigned the post of superintendent of the works, and Mr. Harman was appointed assistant engineer under Mr. Guppy, to whom the Directors, on the advice of Mr. Brunel, entrusted the supreme control of their manufacturing establishment.

The duties and responsibilities which devolved on the Building Committee—Captain Claxton, Mr. Guppy, and Mr. Brunel—were most arduous. To design and construct a steam-ship larger than any that had, up to that time, been launched, to make this ship of a material which had but lately been introduced into shipbuilding, and which had never before been employed on a large scale, to adapt to this ship a novel form of propeller which had not previously been used save in a merely experimental steamer, and to build in a newly opened manufactory marine engines of a much greater size than any that had hitherto been contemplated, and of a totally different character, was indeed a bold enterprise. Mr. Brunel had, as has been shown, recommended the Company not to undertake one part of the work, that, namely, of the manufacture of the engines, which he thought would have been better entrusted to the most experienced engine builders. But although the Directors had acted contrary to his advice, this circumstance in no way diminished the zeal with which he and his coadjutors entered upon their task.

A short statement of the principal dimensions of the vessel and engines is given in a note to this chapter; but some of the more remarkable features in the design may be mentioned here.

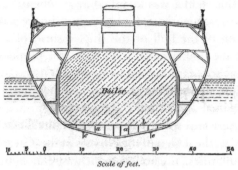

Scale of feet.

Fig. 13. 'Great Britain' Steam-Ship.
*Transverse Section.*

In the construction of the 'Great Britain,' the same care which had been spent in securing longitudinal strength in the wooden hull of the 'Great Western,' was now given to the suitable distribution of the metal. Over the transverse angle iron ribs at the bottom of the ship were laid ten deep longitudinal beams (see woodcut, fig. 13, a), which, over the greater part of the bottom of the ship, were covered with an iron deck (b) riveted to their upper edges by angle irons, thus forming a cellular structure which added greatly to the strength of the ship. It does not appear that this deck was designed to be water-tight, so that it did not form the same security against accident as the inner skin of the cellular structure which Mr. Brunel afterwards adopted in the 'Great Eastern.'

The upper part of the sides of the ship, in the middle of her length, were carefully designed so as to give her longitudinal strength. The side plates were thickened, and were riveted to iron shelf-plates three feet broad (c); and two bands of iron, six inches wide and one inch thick, with the joints strengthened, ran along the top of the ship's side. There were bands of iron riveted to the shelf-plate, and iron deck beams crossed diagonally under the planking of the upper and main decks. Also at the junction of the ship's side with the shelf-plate there ran longitudinally a tie of Baltic pine timber, 340 square inches in section (d); this being well secured to the shelf-plate and ribs, added considerably to the strength of this portion of the hull.

The ship had five watertight bulkheads, and was thus separated into six compartments.

She had no keel, as there did not appear to be sufficient advantage gained by such an appendage to make up for the increase of the ship's draught by the amount of the depth of the keel. There were two side or bilge keels (e), reaching down to the level of the keel plate of the ship, so that when grounded in dock she might rest on three points in her width.

The 'Great Britain' had what is termed a balanced rudder, a portion of the rudder (in this case about one third) being in advance of the pivot on which it turned. The result of this arrangement was that, the pressures on either side of the pivot nearly balancing one another, there was no difficulty in putting the helm over rapidly. This rudder was knocked away when the ship ran ashore at Dundrum, and was subsequently replaced by an ordinary rudder.[7]

In the construction of the hull of the ship, instead of a mere imitation of the arrangements of the timber in wooden ships, the proper distribution of the material to receive the strains that would come upon it was carefully considered. In the result, the ship contained, in the structure of her bottom, bulkheads, deck shelves, and longitudinal kelsons, the longitudinal principle of construction which Mr. Brunel afterwards so fully developed in the 'Great Eastern.'

Apart from their size, the design of the engines of the 'Great Britain' necessarily presented many peculiarities. The boilers, which were six in number, were placed touching each other, so as to form one large boiler about thirty-three feet square, divided by one transverse and two longitudinal partitions. This boiler, which was fitted in between the longitudinal bulkheads of the ship, had a double set of furnaces, and therefore of stoke-holes, one at the fore end, and the other at the after end, next the engine room.

It would seem that the boiler was only worked with a pressure of about eight pounds on the square inch.

The feed water for the boiler was passed through a casing surrounding the funnel, in which it was heated before passing into the boiler. This casing was open at the top, and the water flowed thence into the boiler by gravitation. A similar arrangement was adopted by Mr. Brunel in the 'Great Eastern.'

The condensers were made of wrought iron, being in fact part of the frame of the ship. The main shaft of the engine had a crank at either end of it, and was made hollow, a stream of water being kept running through it so as to prevent heating in the bearings. An important joint in the design was the method by which the motion was transmitted from the engine-shaft to the screw-shaft, for the screw was arranged to go three revolutions to each revolution of the engines. Where the engines do not drive the screw directly, this is now universally effected by means of toothed gearing; but, when the engines of the 'Great Britain' were made, it was thought that this arrangement would be too jarring and noisy. After much consideration, chains were used; working round different-sized drums with notches in them, into which fitted projections on the chains. The greater part of the length of the screw-shaft consisted of a hollow wrought-iron boiler-plate tube, the metal being thus very advantageously placed for taking torsional strain, and the shaft was in this way made very light. The engines were designed to work expansively, the steam being cut off at one-sixth of the stroke.

The completion of the 'Great Britain' was delayed many months, owing to the financial difficulties in which the Great Western Steam-Ship Company had become involved; the profit on working the 'Great Western' having been seriously diminished in consequence of the competition of the Cunard steamers.

At length, however, the ship was finished; and she was floated out of dock into the Floating Harbour on July 19, 1843, in the presence of His Royal Highness Prince Albert.

This seems a fitting place to insert the following letter from Mr. Brunel to Mr. Guppy, written at the beginning of August 1843:—

I have been thinking a great deal of your plans for iron-ship building, and have come to a conclusion which I believe agrees with your ideas; but I will state mine without reference to yours. At bottom and at top I would give *longitudinal* strength and stiffness, gaining the latter by the former, so that all the metal used should add to the *longitudinal tie,* while in the neutral axis and along the sides, and to resist swells from seas, I would have vertical strength by ribs and shelf-pieces, thus: the black lines being sections of longitudinal pieces, the dotted lines vertical and transverse diagonal plates, throwing the metal as much as possible into the outside bottom plates, and getting the strength inside by form, that is, depth of beams, &c., the former being liable to injury from blows, &c., the latter being protected.

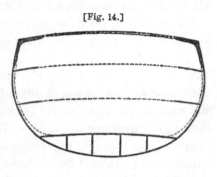

[Fig. 14.]

And now for the screw of which I am constantly thinking, and in the success of which for the 'Great Britain,' remember, I am even more deeply interested than you.

If all goes well we shall all gain credit, but *'quod scriptum est manet,'* if the result disappoint anybody, my written report will be remembered by everybody, and I shall have to bear the storm—and all that spite and revenge can do at the Admiralty will be done! The words 'better sailing qualities than could be given to the "Polyphemus,"' which I used in my first report to the Admiralty, I believe have never been forgotten.

Well, the result of all my anxious thoughts—for I assure you I feel more anxious about this than about most things I have had to do with—is first that we must adopt as *a principle not to be departed from,* that all mechanical difficulties of construction must give way, must in fact be lost sight of in determining the most perfect form—if we find that the screw determined upon cannot be made (but what cannot be done?), then it is quite time enough to try another form; though even then *my* rule would be to try *again* at making it . . .

The 'Great Britain' was built wider than the locks through which she would have to pass, as it was supposed that the Dock Company would allow them to be temporarily widened.[8]

After a good deal of discussion, negotiations were satisfactorily concluded, and the requisite alterations were made: the ship passed through into Cumberland Basin, and the upper lock was restored to its original state in a few days.

On December 10, everything was ready for her passing into the Avon through the lower lock. A steam-tug commenced towing her at high water, but, before she had moved half her length in the lock, it became evident to Captain Claxton, who was on board the tug, that there was not an inch to spare; she was touching the lock walls on either side—in fact, she had stuck between the copings. Upon this he gave orders to haul her back again as quickly as possible. This was hardly effected before the tide began to fall; a few minutes later, and the ship would have remained jammed in the entrance.

As the tides had passed their highest, it was necessary immediately to widen the lock, in order not to lose the spring tide; and this was accomplished under Mr. Brunel's superintendence, just in time to get the vessel through that night.

Mr. Brunel described this occurrence in the following letter, written to excuse himself from keeping an important engagement in Wales:—

December 11, 1844.

We have had an unexpected difficulty with the 'Great Britain' this morning. She stuck in the lock; we *did* get her back. I have been hard at work all day altering the masonry of the lock. Tonight, our last tide, we have succeeded in getting her through; but, being dark, we have been obliged to ground her outside, and I confess I cannot leave her till I see her afloat again, and all clear of her difficulties. I have, as you will admit, much at stake here, and I am too anxious about it to leave her.

The 'Great Britain,' after making several experimental trips, sailed for London on January 23, 1845, and, although she experienced very severe weather, made an average speed of 12⅓ knots an hour.

The excitement caused by her arrival at Blackwall was very great. Thousands of persons flocked to see her, and she was honoured by a visit from Her Majesty and His Royal Highness Prince Albert.

She left Liverpool on her first voyage on August 26, and arrived at New York on September 10, having made the passage out in fourteen days and twenty-one hours. She made her return passage in fifteen days and a half.

She started again in October, taking sixteen days and a half across. On her homeward passage, she broke her screw, and got home under canvas after eighteen days of rough weather which fully tested her sailing qualities.

The experience of these voyages showed that the supply of steam from the boilers was defective; the necessary alterations were carried out during the winter months, and the ship was fitted with a new screw.

In the beginning of 1846, everything seemed to promise well for the success of the 'Great Britain.' She started on May 9, with her full complement of passengers and cargo, but again an accident happened, which prevented this passage from affording a trial of her steaming power. On May 13, the guard of the after air-pump broke; but up to that time her speed had averaged eleven and three-quarters knots.

She returned from New York in thirteen days and six hours, against adverse winds for ten days, with a speed varying from eight and a half to twelve knots.

On one day of her voyage, June 13, she ran 330 knots in the twenty-four hours, or nearly sixteen statute miles an hour. This was said to have been the quickest passage which had, up to that time, been made under similar circumstances of wind and weather.

She left Liverpool again at the beginning of July, and arrived at New York in thirteen days and eight hours, or, deducting stoppages, in twelve days and eleven hours—the shortest passage then and for some time afterwards recorded. Her homeward passage was accomplished in thirteen days, including a stoppage of eighteen hours to repair the driving chains which had been damaged.

She started again from Liverpool on her outward voyage on the morning of September 22, 1846, having on board 180 passengers (a larger number than had ever before started to cross the Atlantic in a steamer), and a considerable quantity of freight. A few hours after her departure, and at a time when it was supposed that she was rounding the Isle of Man, the ship ran ashore, and all immediate efforts to get her off were unavailing. When daylight came, the captain found, to his surprise, that she was in Dundrum Bay, on the north-east coast of Ireland. The passengers were landed safely when the tide ebbed.

Captain Claxton, the managing Director of the Company, went at once to the ship. He found her lying at the bottom of a deep and extensive bay; the ground on which she rested had an upper surface of sand, but underneath this were numerous detached rocks. The ship had settled down upon two of them, and had knocked holes in her bottom. Her head lay NW., leaving her stern and port quarter exposed to a heavy sea, which, at Dundrum, always accompanies southerly gales.

When Captain Claxton got to the ship, he made arrangements for trying to get her off at the next spring tides, which were on the following Monday (September 28); but on the Sunday, a gale of wind from the south sprung up, and at the night flood-tide the water broke over her; nothing remained to be done but to drive the ship higher up the beach into a position of greater safety. Sails were therefore set, and she was driven forward a considerable distance.

Mr. Patterson was sent by the Directors to Dundrum with Mr. Alexander Bremner (who had had considerable experience in floating stranded ships), and they endeavoured to protect the vessel by breakwaters. These, however, were soon carried away; and, after this misfortune, the Directors seem for a time to have lost all hope of saving their ship.

On December 8, when the immediate pressure of Parliamentary work was over, Mr. Brunel went to Dundrum, having some time before been requested by the Directors and underwriters to examine and report on the ship. He was delighted, he said, in spite of all the discouraging accounts he had received, to find the 'Great Britain' 'almost as sound as the day she was launched, and ten times stronger and sounder in character,' though at the same time he was grieved to see her lying 'unprotected, deserted, and abandoned.'

Whatever may have been the misgivings of others, he felt no doubt as to the possibility of saving the ship, by at once protecting her by a breakwater made of fagots; and before he left Dundrum he set Captain Hosken at work at the new arrangements, and he guaranteed the immediate expense in the event of the Directors not sanctioning the measure.

Immediately on his return to town, he wrote the following somewhat vigorous letter to Captain Claxton:—

December 10, 1846.

I have returned from Dundrum with very mixed feelings of satisfaction and pain, almost amounting to anger, with whom I don't know. I was delighted to find our fine ship, in spite of all the discouraging accounts received, even from you, almost as sound as the day she was launched, and ten times stronger and sounder in character. I was grieved to see this fine ship lying unprotected, deserted and abandoned by all those who ought to know her value, and ought to have protected her, instead of being humbugged by schemers and underwriters. Don't let me be understood as wishing to read a lecture to our Directors; but the result, whoever is to blame, is, at least in my opinion, that the finest ship in the world, in excellent condition, such that £4,000 or £5,000 would repair all the damage done, has been left, and is lying, like a useless saucepan kicking about on the most exposed shore that you can imagine, with no more effort or skill applied to protect the property than the said saucepan would have received on the beach at Brighton. Does the ship belong to the Company? For protection, if not for removal, is the Company free to act without the underwriters? If we are in this position, and if we have ordinary luck from storms for the next three weeks, I have little or no anxiety about the ship; but if the Company is not free to act as they like in protecting her, and in preventing our property being thrown away by trusting to schemers, then please write off immediately to Hosken to stop his proceeding with my plans, because I took the pecuniary responsibility of the cost of what I ordered until he could hear from you, and of course I do not want to incur useless expense, but still more I do not wish any proceeding taken as from me to be afterwards stopped. I will now describe as nearly as I can what I have seen, and what I think.

As to the state of the ship, she is as straight and as sound as she ever was, as a whole. She is resting and working upon rocks, which have broken in at several places, and forced up perhaps 12 to 18 inches many parts of the bottom, from the fore stoke-hole to about the centre of the engines, lifting the boilers about 15 inches and the condenser of the fore engine about 6 or 8 inches; the after-condenser, perhaps, half an inch. The lifting of the fore-condenser has broken that air-pump, the connecting rod having been unwisely left in, and the crank being at the bottom of the stroke. Of course the air-pump could not help being broken; except this, the whole vessel, machinery, &c., are perfect. I told you that Hosken's drawing was a proof; to my eye, that the ship was not broken: the first glimpse of her satisfied me that all the part above her 5 or 6 feet water line is as true as ever. It is beautiful to look at, and really how she can be talked of in the way she has been, even by you, I cannot understand. It is positively cruel; it would be like talking away the character of a young woman without any grounds whatever.

The ship is perfect, except that at one part the bottom is much bruised, and knocked in holes in several places. But even within three feet of the damaged part there is no strain or injury whatever. I think it very likely that she may have started leaks where she has been pounding away upon the rocks, but nothing more; and as I said before, all above her 5 or 6 feet water line is uninjured, except her overhanging stern; there is some slight damage to this, not otherwise important than as pointing out the necessity of some precautions if she is to be saved. I say 'if,' for really when I saw a vessel still in perfect condition left to the tender mercies of an awfully exposed shore for weeks, while a parcel of quacks are amusing you with schemes for getting her off, she in the meantime being left to go to pieces, I could hardly help feeling as if her own parents and guardians meant her to die there.

Why, no man in his senses can dream of calculating upon less than three months for the execution of any rational scheme of getting her off; and no man in his senses, I should think, would dream of taking her across the channel in the winter months, even if he had got the camels or floats fast. Of this I don't feel so competent to form an opinion, though I think I can judge, and I should consider it a wanton throwing away of my shares if the Directors allowed her to be taken out, even if afloat; but at all events I am competent to judge of the probable time occupied in getting means to float her, and I maintain that it would be absurd to calculate upon less than two or three months. It is not therefore the mode of getting her off that we ought to have been all this time thinking of, but how to keep her where she is. I feel so strongly on this point that I feel quite angry. What are we doing? What are we wasting precious time about? The steed is being quietly stolen while we are discussing the relative merits of a Bramah or a Chubb's lock to be put on at some future time! It is really shocking.

Having expended a little of my feeling, I will tell you what I have done, and what I should recommend.

First, instantly to disconnect the engines and air-pumps, and remove all the working gear, so at least to leave the mischief to the lower part. By the bye, the cylinders are not disturbed or hurt in any way at present, but with the engines exactly at the bottom of the stroke, and the connecting rods on, it is a wonder they are not. If the air-pump had been disconnected it would not have been broken. I have taken upon myself to order this.

Secondly, I suggested to Hosken, in which he quite agreed with me, to take off all strain from her extreme stern. At present she has cables out from this prodigious mass overhanging nearly 30 feet from any part capable of bearing the strain.

I recommend his taking the chain cables through the ship's side, and making them fast with a spar or timber outside the starboard side, about as far forward as the capstan. At present she is canted seaward.

I thought she was better so than presenting the hollow lines of her quarter to the sea, and both Hosken and Bremner came round to my opinion.

Thirdly, there is a stream of water which now washes away the sand from her bottom. I think it essential this should be diverted, and kept so.

Fourthly, my plan for protecting her is totally different from any that have been proposed, and if we have not such excessively bad weather as would prevent anything being done, I believe it may easily be done. And if done will, I am convinced, be perfectly good; while any solid timbering, even if made, would, I think, be most likely the cause itself of tremendous damage, if once beat by the sea. I will only premise by saying that both Hosken and Bremner came to the conclusion that it was the best thing that could be done.

I should stack a mass of large strong fagots lashed together, skewered together with iron rods, weighted down with iron, sand bags, &c., wrapping the whole round with chains, just like a huge poultice under her quarters, round under her stern, and half way up her length on the sea side.

The detail of the mode, and the precautions of detail, I have not time now to describe. I am as certain as I can be of anything that, once made, such a mass of fagots would stand any sea for the next six months, and the chances of making it (after one or two failures, no doubt) are so good, that if properly taken in hand, I look upon it as certain. I will write more fully to-morrow—in the meantime I have ordered the fagots to be begun delivering. I went myself with Hosken to Lord Roden's agent about it, and I hope they are already beginning to deliver them. Write and stop them or not—if not, of course my responsibility ceases. I will write again to-morrow, but let me know by train how we stand with the underwriters.

This letter was a few days later supplemented by the following formal report to the Directors, which was printed and circulated amongst the proprietors.

December 14, 1846.

According to your request I have, as soon as my engagements would allow of my leaving London, paid a visit to the 'Great Britain,' and I now beg to report to you the state in which I found the vessel, and my opinion of the best means to be taken for recovering the largest possible amount of the property invested in her. If I state these opinions concisely, and without any qualifications, you will not suppose that I have the presumption to think them infallible, but merely that I am compelled, by the shortness of the time left me to write to you, to avoid all circumlocution, and to give you as simply and briefly as possible the opinions I have formed—at the same time I am bound to say that I have not formed them hastily, and that my convictions upon the several points upon which I may express my feelings are very strong.

First, as regards the present state of the vessel, I was agreeably disappointed, after the reports that had reached me, to find her as a whole, and, independently of the mere local damages of which I will speak presently, perfectly sound, and as strong and as perfect in form as on the day she was launched.

In receiving this statement you must bear in mind the great difference between an iron vessel and a timber-built ship. In the former, parts may be considerably damaged or even destroyed, and the remainder may not only be untouched, but may be left unstrained and uninjured. In a timber ship this can hardly be the case; if any considerable portion of a ship's bottom is stove in, the timbers or ribs completely across the ship and the planking longitudinally, cannot fail to be strained to a very considerable distance in both directions. You must therefore remove from your minds all impressions derived from your experience of damages sustained by timber-built ships in order to understand my statement, which is strictly correct:—that, except the parts actually damaged, the extent of which is comparatively small, the ship is perfectly sound, and as good as at the hour when she struck. This soundness and freedom from any damage extends from about the 5 feet water line to the top (with the exception of the injury sustained by the knocking away of the rudder post and a blow under the stern); nearly the whole of the vessel is therefore sound: the principal injury is in her bottom under the boilers and engines. The vessel has evidently been thumping upon the rocks, and almost entirely upon this part of the bottom from the first few days after she grounded; and at present in all probability her whole weight is resting upon this part; yet notwithstanding this she is perfectly straight, and has not broken nor even sprung an inch in the whole length. The boilers have been forced up about 15 inches, and one of the condensers has been lifted about 8 inches, breaking the air-pump. At present this is nearly the extent of the damage done; all of which could easily be repaired if the vessel were in dock.

I will now state my opinion of the best means of recovering the largest possible amount of the property which has been invested in her. In this view I can only imagine two alternatives—the one to break her up on the spot and make the most of the materials; the other to get her afloat and into port, and restore her into good condition, or sell her to those who would so restore her.

The first alternative may I think be discarded at once; the plates and ribs of an iron vessel are difficult enough to convert into useful materials for any other purposes, even in the midst of workshops and with tools and appliances at hand. In such a place as Dundrum Bay I do not believe the materials would pay the expense of cutting up; the masts, spars, chains, &c., in fact the stores and perhaps a few of the lighter part of the engines, might repay the cost of removal, but the whole would certainly not amount to many thousands; probably hundreds would be a safer estimate of the amount to be realised clear of all expenses. To remove the vessel and take her into port, and either restore her or sell her, is then the only means of recovering any part of the whole of the capital invested in this ship. If she is so brought into port she may be worth, unrepaired, £40,000, £50,000, or £60,000 according to the opportunities that may offer themselves of employing her usefully or selling her. The only question is, then, how at the least expense, and at the smallest risk, is the vessel to be got into port? But, as I will now endeavour to prove to you, the mode of getting the vessel off the shore and into port is again quite secondary to the consideration of how to preserve her where she is so that she may be in a condition to be removed, *and to be worth removing,* when the means of doing this are ready, and the proper time is arrived for attempting it.

In the first place I assert unhesitatingly, that no man in his senses and who thoroughly understands the circumstances of the case, the weight and position of the vessel, the amount of the rise and fall of the tide, and the draught of water around her, and the extraordinarily exposed situation, would dream of calculating upon completing the requisite means for floating her under three months.

*In the meantime the ship must be protected.* Even if it were practicable to construct the necessary apparatus, and to float the vessel to-morrow, it would be little short of madness to go to sea with her at this time of the year; but I am doing wrong to discuss a case which cannot arise. The vessel cannot, according to any rational calculation of chances, be got off under three months, and it is equally against all probability that, if left unprotected, there would be any thing worth taking off at that time. It is useless, therefore, at this moment discussing the best mode of floating the vessel; and I think, under such circumstances, it would be most unwise hastily to determine upon any plan. The first thing is to know whether there are any means of preserving the vessel, and whether any such plans can be carried into effect at some reasonable cost, which it may be worth incurring. I have looked at the vessel, and considered the very exposed situation in which she is placed (and a more exposed one could hardly be found), and I am convinced that no fixed breakwater of ordinary construction could be made at any reasonable expense, or in time to prevent mischief. There is no depth of sand into which to drive piles, and the rock is too uneven and broken to allow of any framing being constructed and secured to it; and any framework would be liable to be destroyed during the progress of its construction, as that already attempted has been, and the timber might be the cause of serious damage to the ship. The plan I should recommend would at least be free from these objections, would be comparatively inexpensive, and I am firmly convinced would be perfectly effectual as a protection. At the same time few persons who have not seen the effect of a sea beating against fagots will share in that conviction; what I recommend is, to form under the stern and along the exposed side of the vessel a mass of fagots made of strong and long sticks, and used in the manner which has been so successfully practised in Holland and elsewhere, for the repair and protection of banks against the sea. The fagots should be packed closely, and for a considerable thickness against the ship's side and up to the level of the decks, and secured with rods run vertically through the mass, and chains laid horizontally and binding the whole tightly to the ship. The heaviest sea has no effect upon such a mass, and I believe the vessel would remain as uninjured and indeed as unaffected by the sea as if in dock; 8,000 or 10,000 fagots, 300 or 400 fathoms of 1 inch or ¾ inch second-hand chain cable, none of which need be lost, 300 or 400 ¾ inch rods sharpened at the ends, 1,000 bags to fill with sand, with what stores you have on board, would suffice; and, if next coming springs and the gales which have hitherto accompanied them are safely passed, I cannot foresee any difficulty whatever in the way of completing the protection I propose. Of course, in all works dependent upon wind, weather, and tides, certainty cannot be obtained; but of one thing I am quite certain, that no other plan offers the same chance of success, or at so small a cost. I have communicated to Captain Claxton the steps which I took, in conjunction with Captain Hosken, for procuring the fagots before I left Dundrum; I have also communicated to him the directions which I gave with respect to the engines, &c., and it is unnecessary therefore that I should repeat them. I will only recapitulate in a few words the substance of the advice I have above given, and of my reasoning. You have a valuable piece of property lying on a most exposed shore; if preserved for a few months that property will in all probability be worth £40,000 or £50,000; if neglected for a few weeks longer it will probably be worth nothing. Can you, as men of business, under such circumstances, waste your time at this moment in discussing what you will do in three months hence, and what plan you will then adopt to take your property to market, but will you not rather first and immediately adopt decisive steps for preserving that property, and then consider what you had best do with it?

I have no wish to escape the responsibility of advising you, as you request me to do so, as to my opinion of the best plan to be adopted hereafter for removing the ship; but adhering to the principle that I have laid down, I should decline to do so at present, did I not see reason to fear that you might be losing time and money by relying upon expectations which I am convinced would be disappointed. I would strongly urge upon you not to place your reliance upon any plan which depends upon floating the vessel, by camels, and taking her to sea unrepaired, and therefore entirely dependent upon those camels. The immense breadth of these floating camels, and the risk of taking such an unmanageable floating ill-connected mass to sea, cannot have been correctly or sufficiently estimated; and the certainty of the whole going to the bottom, in the event of even a very moderate gale of wind or a slight swell, has been apparently quite lost sight of. I am also of opinion that the difficulty of lifting the vessel at all by auxiliary floats has been underrated. The vessel has worked herself about 5 or 6 feet into the solid rock and sand, and may very probably get a little deeper before the time for lifting arrives. She must therefore be lifted, say at least 4 feet to 4 feet 6 inches, before she could be got out of the dock she has made; and the floating power must therefore be calculated at least to raise her 5 feet 6 inches, so as to be quite sure of moving her. Now there is not more than 10 feet water around the vessel even at high water of ordinary springs, and it would be impossible to calculate upon more than 9 feet as a certainty. No floating power worth having can be got in the vessel, and the floating vessels must therefore be capable of sustaining the whole weight, with a draught when the vessel is lifted of only about 4 feet. The weight to be sustained is 2,000 tons, and to do this with vessels drawing only 4 feet would require that they should be upwards of 30 feet in breadth on each side, forming with the ship a total width of upwards of 100 to 120 feet, and upwards of 300 feet in length. I do not say that this operation is impracticable, but it is at least a very difficult one, and must be almost entirely dependent upon weather, and, unless in perfectly smooth water, a very hazardous one. My belief and conviction is that the safe mode of proceeding, and by far the cheapest, will be to lift the vessel by mechanical means, to lay ways under her, and to haul her up sufficiently far for her to be safe from the sea; to repair her just sufficiently to make her water-tight, then launch and bring her to Liverpool or Bristol. But, as I have before stated, there is time to consider these points, if in the meantime we take steps to preserve the ship. If the property is not even now worth protecting, it will indeed be waste of money to be preparing at some considerable expense to remove what will in all probability be only then a valueless carcase.

If the Directors should determine upon adopting the course I have recommended, I must remind them that the plan is one depending entirely on the skill, the vigour, the aptitude for expedients, and possibly, if bad weather should come on, and day after day the work be destroyed, on the unwearying perseverance and determined confidence in the plan of the person directing it, and the sufficiency of means at his command.

I cannot conclude without doing justice to Mr. Bremner, whom I met on board, and acknowledging the friendly and liberal manner in which he discussed the various means to be adopted, and assisted me with his valuable advice; and, although I may have somewhat differed with him as to the advisability of attempting to float the vessel away to sea without first repairing her, yet upon most points we were perfectly agreed; and I firmly believe that if any man could take her off (and if it would be prudent to let him do so) Mr. Bremner's great experience and sound practical knowledge and good sense in devising any plan, and his energy and skill in carrying it out, would ensure every chance of success which the circumstances admit of.[9]

The Directors adopted Mr. Brunel's suggestions; and, at his urgent request, they appointed Captain Claxton to superintend the execution of his plans.

Captain Claxton thereupon went to Dundrum, where he took sole charge of all the subsequent operations.

The following is selected from the many letters written by Mr. Brunel to Captain Claxton at this time.

December 29, 1846.

You have failed, I think, in sinking and keeping down the fagots from that which causes nine-tenths of all failures in this world, from not doing quite enough. Two and a half hundredweight of sandbags, weighing barely one hundredweight [*in water*], would not of course keep quiet a large fagot of five bundles, and two and a half hundred weight of fire-bars, I should think, would only barely do. The load must always be excessive, to make sure of a thing . . . I would only impress upon you one principle of action which I have always found very successful, which is to stick obstinately to one plan (until I believe it wrong), and to devote all my scheming to that one plan, and, on the same principle, to stick to one method, and push that to the utmost limits before I allow myself to wander into others; in fact, to use a simile, to stick to the one point of attack, however defended, and if the force first brought up is not sufficient, to bring ten times as much; but never to try back upon another point in the hope of finding it easier. So with the fagots—if a six-bundle fagot wont reach out of water, try a twenty-bundle one; if hundredweights wont keep it down, try tons.

The able manner in which Captain Claxton carried out Mr. Brunel's plans, and suggested important modifications of them, is acknowledged by Mr. Brunel, in a report to the Directors, written after the successful completion of the breakwater.

February 27, 1847.

I beg to enclose Captain Claxton's account of the proceedings at Dundrum Bay during the time that he has been engaged in forming the breakwater or protection to the ship in the manner recommended by me.

Notwithstanding the great difficulties he has had to contend with from almost incessant bad weather, with the wind blowing dead on shore nearly the whole of the month of January, and consequently preventing the tides from ebbing sufficiently out to allow of the work being properly proceeded with, and notwithstanding the occurrence of more than one storm at the most critical period of the work, he has, as I fully relied upon his doing, succeeded in so far protecting the ship, that she has been comparatively unaffected by violent seas, which, there is no doubt whatever, would otherwise have seriously damaged her. We may now calculate with tolerable certainty upon preserving her without further injury until the finer or at least more settled weather sets in.

In the work which Captain Claxton undertook, and has so successfully completed, he has been compelled to vary very materially the mode of proceeding first laid down; he has, in fact, been obliged to adapt his plans to his means of execution, and almost from day to day to devise modes of proceeding with only the experience of the past day to guide him. Numerous unforeseen difficulties have occurred, upon which he kept me daily informed; and simple as my plan might have appeared to others, it required much skill, contrivance, and unwearying perseverance to carry out so many alterations and improvements as it progressed. I had relied confidently on success when my friend Captain Claxton undertook the work, and the result has fully confirmed my expectations.

It is now necessary to turn our attention to the best mode of removing the ship. I hope in about a fortnight from the present time to be able to give you some opinion upon this point, but it is one requiring much consideration; and until I had the opportunity of conferring with Captain Claxton on the subject, and also had before me all the measurements and data which he has collected, it was useless to attempt it.

The construction of the breakwater will be understood by the following extracts from a report made by Captain Claxton to the Admiralty.

'Great Britain,' July 16, 1847.

. . . Mr. Brunel's instructions to me were principally by word of mouth; the difficulty to be got over, in his opinion, being the foundation upon sand, varying in depth according to the points or hollow of a substratum of rock, and according to the quarter from which the wind blew.

The foundation could only be made at low water, and as fast as a layer of fagots was laid, it was rapidly pinned down with iron rods, bent at the heads, varying from 9 to 6 feet in length, and driven to the rock under, loaded with stones quarried from the nearest reefs, and upon these, the last thing as the flood came in, chain cables, air-pump covers, fire-bars in large bundles, and the ship's guns were dropped, care being always taken to have the ends of the chains, and the slings of other heavy matters, fast to the ship. As the tide in smooth water began to recede, or in heavy seas began to lose effect in striking, these iron weights were lifted, and the fagots which were ready were . . . placed; and the same process followed tide after tide when the water ebbed sufficiently, which upon the neaps it never did at all, and upon the springs it only did for about an hour, unless there was either no wind at all, or unless the wind blew off the land, or from the West round by the North to ESE. It is necessary, to a proper understanding of the nature of this really large work, to describe a bundle of fagots, lest an idea should be formed that it is a small thing and easily handled. They averaged 11 feet in length, and 5 feet in circumference near the butts, which all pointed one way. When tied in Lord Roden's wood, many were 13 feet long, and none were taken under 10 feet. A cart of the country with one horse could carry about ten bundles when well lashed, and as they came down there was rarely more than the head of the horse to be seen. Sixpence per bundle, and sixpence for delivery, was the contract; the distance at high water nine miles, at low water six miles; and our sailors made them into large bundles of twos, threes, and fours; now and then we experimented with bundles of eight or even ten, in the middle, of which were bags of sand (old guano bags), varying in number, of 2 cwt. each, and sometimes as many as amounted to a ton in weight. These large bundles stood if the water remained smooth; but if, before we could build up to their height with smaller bundles, and, as it were, prop or shoulder them fore and aft, we were caught by a breeze and sea, we found them rolled up to high water mark, or on spring tides four hundred yards; while some are now being hove out of the sand entire, and with the sand bags and sand complete . . .

Having got the foundation, on which, I may mention, we placed one of the ship's iron life-boats, 30 feet long, 8 feet wide, and 5 feet deep, and loaded her with stones, and which also, although over bundles of fagots, went bodily down, until only the gunwale was above the level of the strand, we began to build the part which was to save the ship from the blows of the sea, and which I was instructed by Mr. Brunel to bring up to a point to the ship's gunwale in the form of a large poultice, occupying the whole space under her counter, the whole of the exposed quarter (the port quarter), and inclining inwards from the outside, and declining from the top to the same point forward, to the after end of the bilge keel. I was to be, and was, as careful as I could be to secure as well as to weight down as we built. Chains were secured in many places to the lower bundles of fire-bars sunk in the sand outside all, and these were brought into the arms of the screw, and to ring bolts let into the ship; and rarely were two layers placed without a repetition of the securing and weighting process all the while we had chains to use . . .

We were frequently beaten; whole masses were capsized, but it was found that even the foundation broke the ground swell, which, instead of having a fair run as it has over the strand, seemed stopped, and to break differently; and certainly the ship was daily eased of the blows of the sea . . . Finding the wind keeping on shore, the fagots we placed shrinking, sinking, or settling down, and, notwithstanding the weights and lashings, commonly breaking away, I proposed to try spars. Mr. Brunel acceded, and recommended my trying four long ones, to the heels of which were to be attached chains with a spread of eight feet, the spars being pointed at the heels, the slack of the chains, and the height at which they were stopped to the spars, being intended to be sufficient to embrace about a dozen bundles of fagots, the points of the spars sticking through the foundation below the level of the sand. After placing the first pair of spars at an angle from the gunwale of about 70 degrees, and before the second pair with its fagots could be got in place, and after heaving them down tight with a tackle to each from the head to the gunwale, a heavy sea came with the flood dead on, and although the spars were seven inches square, they stood the blows of the sea, bending in the middle full five feet when heavily struck. This gave me the idea of green trees: firs being the first that occurred to me; first because of their height, and next because in Lord Downshire's grounds at Dundrum the castle wood consisted of them only, and lastly, because they were cheap and at hand. The spars we fortunately placed, and which stood so well, were American elm, a large balk of which had been sawed in four lengths of 42 feet for framing, for Mr. Bremner's breakwater . . . The lengths required to allow the application of a tackle to the head to heave them tight down to the gunwale, or to the scuttle-holes of the ship, were from 45 to 50 feet. I found that the firs of this length were scarce, and too fine at the head; but on looking in Lord Roden's wood, a better substitute was offered in the form of beech trees of any size, and a contract was speedily made, and as speedily completed; the carts of the country bringing in one tree at a time, until about eighty were placed at the most exposed point, the quarter, in three rows, the outside row at an angle of 45 degrees. Beech-trees were found decidedly better than firs, or, I believe, than any other description of tree, as their weight was soon found sufficient to keep them in place without the heaving down tackles. They can be got of great length, without being so large and (carting, pointing, and handling considered) so unwieldy a butt as firs, and they are of greater diameter aloft,

consequently stronger, and altogether tougher than firs. Having got the whole
stem surrounded, and having continued them at from 4 to 5 feet apart up to the
bilge keel, or about 80 feet of the ship from the screw, I applied smaller spars
(easily and economically obtained from the Tyrella domain, or within half a
mile of the ship) laterally and diagonally, and about a foot apart in the former
case, and 3 or 4 feet in the latter; the number about 300, of all or any lengths
between 15 and 30 feet. While this was about, the fagot process was going on;
and before that was completed, I found the foundation with the boat, and its
40 tons of stones, and even the spars, although pointed at the angles, indicated
an inclination to move forward through the sheer force of the rollers. To check
this, tackles were made fast to three warps out to seaward, with two anchors to
each, and brought to the spars with four spans to the inner block. Thus twelve
of the largest spars were grappled, the falls taken on board, and all hove tight;
the whole of the spars having first been attached to one another by a round
turn of a half-inch chain with strong staples, and of course by the lateral spars
or spreaders, and their innumerable seizings. Suffice it to say, as regards the
spars, that when struck by violent seas from half-flood to half-ebb, they bend
in a body from 3 to 4 feet, and spring back after the blow; and this is, I believe,
the whole secret of the efficiency of the spar part of the breakwater, which has
stood the whole winter, only one having broken, and that because the head took
against the topsail yard, lashed along stanchions of the rails as a hold-down for
the tackles attached to their heads. As a proof that this is the probable cause
of its standing so well, I may refer to Mr. Bremner's breakwater, which went
with the first heavy gale and sea, although made of balks of from 17 inches to 13
inches thick: also to our having ourselves placed a pine spar 9 inches diameter
down by the side of the outside tree on the starboard quarter, and which broke
in two right in the middle with the first sea that struck it. The sea struck very
hard against the spars and framework, from which it was received by the fagots
without any shock to the ship worth speaking of. With respect to the fagots, I
could not find materials to go on weighting down as we went on piling up . . . I
therefore ran the remaining fagots up light. No sooner, however, was this done,
and a gale came on, than we were compelled to let go the lashings, and to help
some of the top ones to escape, as the whole of the unloaded body rose and
fell full three feet with the sea, and would have done mischief to the spars and
to the ship, by shouldering her quarter, and twisting her. About 200 bundles
broke away in this gale. They were, however, all recovered, and placed further
forward and low down, and then loaded with stones; about one third of the top
space having after this been left open, or one third down from the apex levelled.
Shrinkage did much towards this, and breaking in pieces a good deal; so it was
not deemed advisable to fill that space again.

About the 1st of May the progress made in lifting the ship by tightening her
from the inside, and by lightening her of everything moveable, led me to believe
that we should be about getting her off towards the end of this month. I therefore
felt that it was time to begin removing the fagot portion of the breakwater . . .

The process of levelling went on for about three weeks, when all above the loaded portion was taken away. We commenced on that portion from necessity, and found that by no contrivance of purchases and levers could our ten men get up or out more than four or five bundles per tide. This became, and even now is, a serious matter. We have got up all our chains and weights, and nothing remains but the fagots and stones, which are so embedded in sand as to form a mass which is more difficult to move than granite rock would be, as we cannot blast. Twenty labourers have been twenty-one tides at work, by contract, and certainly they have made an impression; but it is not lowered over two feet. The lower portion was made of furze bundles, here called whins, and they are great collectors of sand, and only come up at all by being cut to pieces. I mention this to show that furze and fagots loaded with stones on sand 500 yards from high-water mark, exposed to the sea, will form a foundation on which a building of any weight might be erected . . .

Mr. Brunel informed me, when I undertook to carry out his views, that there was nothing new in using fagots to stop breaches in sea-walls, as in Holland; and that he saw no reason why they should not stop the force of the sea in protecting ships as well, provided they could be secured, and a foundation got. They were to be made of alders, ash, holly, laurel, oak, or anything tough, to be cut down, and made up green, to be placed with all their leaves on, to be pointed to the sea at their butts; and inside the walls of them, if I may so speak, whin (furze) bundles were to be weighted down, the fagots placed in steps or rows of about 4 feet from the outside to the ship's gunwale.

As the summer came on, the mode of lifting and floating the ship had to be decided. On this point Mr. Brunel wrote to the Directors:—

May 4, 1847.

You have heard from time to time from Captain Claxton of the result of the means adopted by him for protecting the 'Great Britain' from the effects of the sea, and which, I am happy to say, have been quite successful; and you will have heard also generally from him of the steps which we have since taken, preparatory to getting off the ship.

I will now explain to you the object I have kept in view in these preparations, and the course I should now advise you to follow.

After completing the works for the protection of the ship, and before determining upon any mode to be recommended to you for getting her off, I thought it would be desirable to lighten her of all that could be easily removed, and to ascertain how much of the vessel could be made water-tight, and what extent of buoyancy could be obtained in the vessel herself, as, in my opinion, upon our success in this would depend altogether the practicability of lifting the ship by camels. If no great extent of buoyancy could have been obtained, I should certainly have recommended lifting the vessel by mechanical means, as I do not consider that, with the draft of water which could be calculated upon around her on ordinary tides, sufficient floating power by camels could be obtained in an easy practicable manner. It was also quite possible that we might succeed in making the vessel alone sufficiently buoyant to lift high enough on the spring tides; and by shifting her position, or by other means, to maintain the lift thus got, to allow of getting under her bottom to repair her.

By great exertion nearly the whole of the compartments forward and aft of the engine space, and part of the coal bunkers, have been made tight; and if the tides had ebbed as low as usual, the other bunkers and the boilers, fireplace and flues, would also have been made water-tight by these last springs.

Unfortunately, from the direction of the wind and other causes, these tides have neither ebbed nor flowed to their full extent; still, the vessel has been lifted, and some of this lift has been maintained, and if we were fortunate, it is evidently quite possible that our utmost expectations might be realized, and possibly the vessel might be lifted sufficiently to be made tight without any external assistance. This, however, would be too much to calculate upon, and the weight to be lifted having been reduced to one-half what it was, and being capable of still further reduction, the operation of lifting by camels becomes a much more practicable undertaking.

I should now therefore recommend that application be made to some parties who have had the most experience in such work, to lay their proposals before the Directors; and that in the meantime we should continue to make every effort to add to the buoyancy of the ship . . .

I believe that, in such a case, quiet, sober consideration, assisted by experience, and by careful examination of all the circumstances on the spot, will be infinitely more valuable than the most ingenious and brilliant schemes. And I believe we are most likely to obtain these conditions by calling in a man like Mr. Bremner, and leaving him to confer with Captain Claxton, and that he should have the benefit of our advice and assistance, and then lay his plans and proposals before the Directors.

Mr. Bremner and his son, assisted by Captain Claxton, made preparations for releasing the ship. The system followed was that sketched out by Mr. Brunel—namely, that of lifting the ship by mechanical arrangements, and then making good the leaks. The difficulties they had to contend with are graphically described in the reports which were sent almost daily by Captain Claxton to Mr. Brunel, and which were afterwards printed by the Directors at Mr. Brunel's suggestion.[10]

The principal leak was stopped, and the ship's head raised 8 feet 7 inches by ramming wedges and stones under her at high water. At last, on August 27, the ship was floated, and Captain Claxton wrote to Mr. Brunel:—

> Huzza! huzza! you know what that means . . . I made up my mind
> to stop her at the edge of low water, and then examine and secure all
> that might discover itself. The tide rose to 15 feet 8 inches. She rose

therefore easily over the rock, but was clear of it by only just five inches, which shows how near a squeak we had—it was a most anxious affair, but it is over. I marked 170 yards in the sand and on our warp, and at that extent I stopped her . . . I have no doubt that to-morrow we shall see her free.

The following day they started for Liverpool. One hundred and twenty labourers were hired to work at the pumps, but only thirty-six came on board, and their services were unavailable, as they spent their time in discussing how much they were to be paid. Consequently, when the ship was taken in tow at 4 A.M. on the 28th there was 6 feet of water in the engine room and 5 feet in the fore hold, and she was making 16 inches an hour. Men from Her Majesty's ships 'Birkenhead' and 'Victory,' which had been sent by the Admiralty to assist, were drafted on board, and the influx of water was reduced to four inches an hour. It was evident that Liverpool could not be attempted, so they made for Strangford Lough. A dense fog came on when they were off the entrance, and they pushed on to Belfast Lough, where the ship was grounded. During the night she was cleared of water, and the next day she started for Liverpool. The landsmen who had been hired the previous night to work the pumps were incapacitated by sea-sickness; and the ship was only kept afloat by the exertions of Captain Claxton and the dockyard hands who had been sent to assist in navigating her across. When she arrived at Liverpool she was placed over a gridiron, on which she sank when her pumps were stopped.

Notwithstanding the successful result of the efforts made for her rescue, the stranding of the 'Great Britain' in Dundrum Bay led to the ruin of the Company; and she was some time afterwards sold to Messrs. Gibbs, Bright, & Co., of Liverpool, by whom she was repaired, and fitted with auxiliary engines of 500 nominal horse power. On a general survey being made, it was found that she had not suffered any alteration of form, nor was she at all strained. She was taken out of dock in October 1851, and since that time she has made regular voyages between Liverpool and Australia.

She is known as one of the fastest vessels on that line; and remains to testify to the ability and wisdom of those who, more than thirty years ago, were daring enough to build so large a ship of iron, and to fit her with the screw propeller.

### NOTE
*Dimensions of the 'Great Britain' Steam-ship*

|  | feet | inches |
|---|---|---|
| Total length | 322 | 0 |
| The length of keel | 289 | 0 |
| Beam | 51 | 0 |
| Depth | 32 | 6 |
| Feet of water | 16 | 0 |

| Tonnage measurement | 3,443 | tons |
| Displacement | 2,984 | " |

### Dimensions of original Engines, &c.

| Number of cylinders | 4 | |
| Diameter of cylinders | 88 | inches |
| Length of stroke | 6 | feet |
| Weight of engines | 340 | tons |
| Weight of boilers | 200 | " |
| Water in boilers | 200 | " |
| Weight of screw-shaft | 38 | " |
| Diameter of screw | 15 | ft. 6 in. |
| Pitch of screw | 25 | ft. |
| Weight of screw | 4 | tons |
| Diameter of main drum | 18 | feet |
| Diameter of screw-shaft drum | 6 | " |
| Weight of coal | 1,200 | tons |

1.  These engines were to have had two cylinders of 88 inches diameter.
2.  It is interesting, in connection with this subject, to mention the following circumstance. At Mr. Brunel's recommendation, Mr. Humphrys consulted Mr. James Nasmyth as to the best means of forging the large paddle-shaft; as they could not get any manufacturer to undertake it. To accomplish this forging Mr. Nasmyth designed his steam-hammer, and though it was not then erected in Bristol, in consequence of the alteration of the form of the engines of the 'Great Britain,' it soon afterwards came into general use.
3.  In the Minute Book of this date it is mentioned as a reason for postponing any decision on the subject, that Mr. Brunel was making 'final,' and afterwards 'further,' experiments.
4.  See Mr. Guppy's paper, printed in the *Proceedings of the Institution of Civil Engineers for* 1845, p.151.
5.  This report is printed in Appendix II p.395. The paragraphs in which Mr. Brunel describes the advantages of the screw propeller will be found at page 406.
6.  A description of Mr. Humphrys' trunk engines is given in Tredgold's work on the steam engine (ed. 1838), p.390.
7.  The balanced rudder, which is peculiarly applicable to screw ships, has encountered much opposition; but it has lately been successfully introduced by Mr. E.J. Reed, C.B., into vessels designed by him for the Royal Navy.
8.  It may be convenient here to state that the dock in which the 'Great Britain' was built led into the Floating Harbour, which is a portion of the channel of the river Avon closed in. The Floating Harbour communicates through the Cumberland Basin with the river.

9.  One of the consequences of the publication of this report was that Mr. Brunel received so many letters containing suggestions for lifting and floating the ship, that he was obliged to have a circular letter printed declining assistance; and more than four hundred letters were also received by the Company's secretary.

10. 'The "Great Britain" Steam-Ship. Extracts from the letters of Captain Claxton, R.N., to I.K. Brunel, Esq., and the Directors. Bristol, 1847.'

Mr. Bremner's apparatus is also described in a paper read by him at the Institution of Civil Engineers, and printed in the twenty-first volume of the *Transactions*, p.160.

Accurate illustrations of the breakwater and floating apparatus will be found in the *Illustrated London News* of August 21, 1847.

In publishing the correspondence to which reference has been made, the Directors acknowledge their obligations to all concerned in the arduous task of saving the 'Great Britain'; and they add—'to Mr. Brunel above all their thanks are most due, for opening their eyes to what might be accomplished, and for taking upon himself the responsibility of her release, provided that Captain Claxton was employed to carry out his views.'

# X

# INTRODUCTION OF THE SCREW PROPELLER INTO THE ROYAL NAVY

SOON after Mr. Brunel had taken the bold step of recommending the adoption of the screw propeller in the 'Great Britain,' he was asked to send a copy of his report to the Admiralty. He did so; and in the course of a few months was invited to attend the Board on the subject of some experiments their Lordships proposed to make.

An interview took place on April 27, 1841; of which Mr. Brunel gives the following account:—

> I attended the Board: Lord Minto stated that he wished a complete experiment to be made on the applicability of the screw to Government boats, and he proposed to place the conduct of the experiments in my hands as a professional man. I stated that I should have great pleasure in doing it, and should take great interest in it, provided they intended to make a good experiment and would place it entirely in my hands, without the intervention of any Government officers, but that I should communicate direct with the Lords, and of course with Sir E. Parry.[1] He said he proposed to build a vessel and engines on purpose, and that he particularly wished it to be left entirely in my hands, and took me apart to the window to impress this last condition on me.

Within a fortnight of his appointment, Mr. Brunel invited Messrs. Maudslay and Field and Messrs. Seaward to send him designs for the engines: they were to have been of 200 horse-power, with a stroke of 4 feet. The engines were to make a smaller number of revolutions than the screw, the motion being communicated from the engine shaft to the screw shaft by drums and straps. This arrangement was adopted in preference to tooth-gearing, in order to facilitate the variation of the number of revolutions, in the experiments with different screws.

The designs ultimately sent to Mr. Brunel for his approval were those of Messrs. Maudslay and Messrs. Forrester. He reported to the Admiralty in favour of Messrs. Maudslay's engines.

Before drawing up the detailed specifications, Mr. Brunel was desirous of procuring data for estimating the surface of the screw required to obtain the same resistance as that offered by the paddles of a ship similar to the new vessel. With this object he applied to the Admiralty for permission to make an accurate trial of the performances of the paddle steamer 'Polyphemus' at various speeds. This permission was granted; the trial, however, was fixed for such an early day that Mr. Brunel had barely time to make the preliminary arrangements, with the assistance of his friend Captain Claxton.

When Captain Claxton arrived at Southampton, the day before the trial, he found that there was no measured mile set out. He immediately hired men, got chains, staffs, and flags, and set out both a nautical mile and a statute mile. They made half a dozen runs each way with the 'Polyphemus,' noting carefully all particulars of speed, revolutions, &c.; and the results obtained were considered very satisfactory.

On October 1, Mr. Brunel received information from the Admiralty (now under the administration of the Earl of Haddington) that Mr. Maudslay and Field's tender was accepted. The engines were immediately put in hand, under Mr. Brunel's supervision.

When they were approaching completion, he became anxious to learn something about the progress of the ship which was to have been built for them. Nowhere could she be found. The minutes were searched at the Admiralty, and it was ascertained that the ship was ordered, but that no ship had been laid down. This discovery, as might be supposed, excited considerable surprise. Mr. Brunel was sent for to the Admiralty to see Sir George Cockburn, the First Naval Lord. Almost the first words to him were: 'Do you mean to suppose that we shall cut up Her Majesty's ships after this fashion, sir?'—Sir George at the same time pointing to a model of the stern of an old-fashioned three-decker, in which large slices were taken off to give room for the screw, and the whole of the lower deck exposed to view, thus making the application of the screw look very ridiculous. On the model was written, 'Mr. Brunel's mode of applying the screw to Her Majesty's ships.' Mr. Brunel smiled, and denied its being his idea at all; he had never seen it before, and knew nothing about it. 'Why, sir, you sent it to the Admiralty.' This also Mr. Brunel denied having done. While an enquiry was being made as to where the model came from, Mr. Brunel employed himself in effacing the inscription with his knife. When the messenger returned, he reported that the model had come from the office of the Surveyor of the Navy. He was sent for, but did not appear. Mr. Brunel, to terminate this awkward interview, pleaded business, and bowed himself out.

Mr. Brunel often told this anecdote, and spoke of the adverse influence which had been exerted in some department of the Admiralty to prevent the successful issue of these experiments.

Soon afterwards Mr. Brunel was informed that the 'Acheron' would be prepared for the screw. He thereupon represented to their Lordships that the 'Acheron' could not, from her full after-body and other defects, be converted into such a ship as the Board had originally determined to construct for trying the screw; and, indeed, her unfitness was admitted by the authorities themselves.

To a letter on this subject, in which he stated that the attempt to apply the screw propeller to the 'Acheron' would not answer any of the objects which their Lordships had in view, Mr. Brunel received no reply, and for the next four months was kept quite in the dark as to what was going on. At length he made enquiries as to the cause of a treatment which, he said, he had never before experienced from any public body.

These enquiries proving fruitless, he at last wrote to the Admiralty, declining further interference, as it had appeared on investigation that the condition on which he had accepted his positions from Lord Minto—namely, that he should have the entire superintendence of the experiments—had not been observed.

On the receipt of his resignation, Mr. Brunel was summoned to the Admiralty. He thus described his interview in a letter to a friend:—

> Not a word was said about my complaint of the past, but they said they wished me to continue the experiments, and that my screw was to be tried first. I said that was not at all what would suit me; that I would, if they wished it, conduct an experiment as originally proposed; that I had no screw, that I was no competitor, but an arbitrator in whom the Admiralty had perfect confidence; that I was this or nothing. Then commenced a tedious fencing . . . However, it ended in all parties being written to, and told that they were to follow my directions, and that I was to proceed to give such instructions as should enable a full experiment to be made of all screws generally. I then requested that this time I might have my instructions in writing.

In a few days Mr. Brunel received an official intimation that the 'Rattler' was to be adapted for the screw, under his directions.

The 'Rattler' was a vessel of 888 tons burden, 176 feet 6 inches between the perpendiculars, and 32 feet 8½ inches beam.[2] She had been commenced for a paddle-wheel steamer shortly before, and was of nearly the same tonnage and midship section as the 'Polyphemus,' but she had not the fine lines aft which were so important for the use of the screw.

She was not launched until April 13, 1843, just two years after Mr. Brunel's first interview with Lord Minto; and she was then delivered to Messrs. Maudslay in the roughest possible state.

Except during the time that Mr. Brunel was prevented from attending to business by the half-sovereign accident, he was in constant communication with the dockyard authorities on matters relating to the construction of the ship, with Messrs. Maudslay as to the multiplying-gear and engines, and with Mr. F.P. Smith as to the forms of the various screws to be tried.

The engines and screw were fitted in the ship, after considerable delay; and, on October 24, 1843, Mr. Brunel reported on them in their completed state. On the 30th, the experiments were commenced.

More than twenty trials were made between that date and the following October, when the 'Rattler' went to sea, and Mr. Brunel could not of course any longer personally superintend the experiments; but, except on one or two occasions when his place was supplied by an assistant, all the trials that were made during the first year were conducted in his presence, and he transmitted the results from time to time to the Admiralty.[3]

The performance of the 'Rattler' was found to be satisfactory; and the position of the engines and screw being below the water line was so pre-eminent an advantage, that in 1845 the Lords of the Admiralty ordered more than twenty vessels to be fitted with the screw;[4] and since that time it has gradually superseded the paddlewheel for ships of war.

The services which Mr. Brunel rendered to the country during the whole of these proceedings were given entirely without pecuniary recompense, and in the face of opposition and discouragement; but he had the satisfaction of knowing that he had been mainly instrumental, not only in introducing the screw propeller into the mercantile navy, but also in securing its adoption in Her Majesty's fleet.

1. Sir E. Parry was at that time Controller of Steam Machinery, and a warm supporter of those who desired to make a fair trial of the screw propeller.

2. Bourne, on the *Screw Propeller,* ed. 1867, p.263.

3. The records of these trials, found among Mr. Brunel's papers, show results which coincide in all material points with those given by Mr. Bourne at p.284 of his work on the *Screw Propeller,* ed. 1867.

4. Bourne, Appendix, p. xxxiii.

# XI

# FROM THE COMMENCEMENT OF THE 'GREAT EASTERN' STEAM-SHIP TO THE LAUNCH

MR. BRUNEL'S earlier labours in connection with the progress of Ocean Steam Navigation have been described in the chapters on the 'Great Western' and 'Great Britain' steam-ships.[1] 'Great Eastern' is but the result of the application, under different circumstances, of the same principles which had guided him in his previous undertakings, the practical working out of the 'idea which he 'had frequently entertained, that, to make long voyages economically and speedily by steam, required the vessels to be large enough to carry the coal for the entire voyage at least outwards; and, unless the facility for obtaining coal was very great at the out port, then for the return voyage also, and that vessels much larger than had been previously built could be navigated with great advantage from the mere effect of size.'[2]

In 1851, four years after the release of the 'Great 'Britain' from Dundrum Bay, Mr. Brunel became again connected with the construction of steam-ships. In that year he was consulted by the Directors of the Australian Mail Company upon the class of vessels which it would be advantageous for them to purchase, in order to carry out their contract for the conveyance of the mails to Australia. He advised them to have ships of from 5,000 to 6,000 tons burden, in order that they might only have to touch for coal at the Cape.

Some of the Directors would not hear of so startling a proposition; but they nevertheless asked Mr. Brunel to become their Engineer; and he retained the post till February 1853. Two ships were built under his direction by Mr. J. Scott Russell—the 'Victoria' and the 'Adelaide.'

It was, no doubt, his connection with the Australian Mail Company that led Mr. Brunel to work out into practical shape the idea of 'a great ship' for the Indian or Australian service, which had so long occupied his mind; and it appears that in the latter part of 1851 and the beginning of 1852 he devoted much time and thought to the subject. He collected facts relating to the trade with India and Australia which demonstrated the advantages to be gained by a rapid and direct communication for the conveyance of passengers and troops, as well as of merchandise. It was with these enlarged views that Mr.

Brunel entered upon the construction of the 'Great Eastern.' He writes in February 1854, 'In February and March 1852 I matured my ideas of the large ship with nearly all my present details, and in March I made my first sketch of one with paddles and screw. The size I then proposed was 600 x 70, and in June and July I determined on the mode of construction now adopted of cellular bottom; intending then to make the outer skin of wood for the sake of coppering.'

In the spring of 1852 he communicated the results at which he had arrived to Mr. John Scott Russell, Captain Claxton, and other scientific friends, and also to several Directors of the Eastern Steam Navigation Company.

This Company had been formed in January 1851 for the purpose of establishing an additional line of steam communication by the overland route, for the conveyance of mails, passengers, &c., between England, India, and China, with a branch to Australia. However, in March 1852 the Government determined to grant the contract for the whole service to the Peninsular and Oriental Company. The Directors of the Eastern Steam Company were therefore obliged to report to their shareholders that the object for which the Company had been incorporated could not be carried out.

At about this time Mr. Brunel's scheme was brought before the Directors, and he submitted to them a detailed statement of his project.

After describing the size and capacities of the vessels then used on the route between England and the East, and the amount and cost of the coals they consumed, he continued:—

June 10, 1852.

The same amount of capital and the same expenditure in money for fuel now required for a line of ships of the present dimensions would build and work ships to carry in the year double the number of passengers, with far superior accommodation, and in about half the time, and two or three times the amount of cargo; the whole difference being produced simply by making the vessel *large enough to carry its own coal*, exactly as when the 'Great Western' was projected for the New York line, the passage had been considered an impossible one for steamboats, or, if possible, only at a total sacrifice of all return for the cost. Certainly, no steamboat then built could get across except by a chance fair weather passage, and then only by being completely filled with coals and leaving no room for passengers or cargo. Simply by building a ship of the size necessary to take the coal, over and above the accommodation required for a due number of passengers and a reasonable quantity of cargo, the passage was rendered perfectly easy and certain, and has since become a mere matter of course, and an ordinary and profitable trading voyage.

The increased size, instead of being a disadvantage, was found, as predicted by the projectors, to be a great benefit, and gave increased speed, even beyond that proportionate to the power; and this steamboat, built in 1836, is still as good as any of her size afloat.

Nothing more novel is proposed now, but again to build a vessel of *the size required to carry her own coals for the voyage.* The use of iron, which has since 1836 become common, removes all difficulty in the construction, and the experience of several years has proved, what was believed before by most unprejudiced persons, that size in a ship is an element of speed, and of strength, and of safety, and of great relative economy, instead of a disadvantage; and that it is limited only by the extent of demand for freight, and by the circumstances of the ports to be frequented.

A Committee was appointed to confer with Mr. Brunel and with Mr. Scott Russell, 'who was fully acquainted with all Mr. Brunel's plans, and had ably assisted him in maturing them.'[3]

The Committee reported to the Directors that they had met on the day after their appointment, when, Mr. Brunel being unavoidably absent, Mr. Russell had attended and entered into a very full explanation of Mr. Brunel's plans, and that a long investigation of his proposition had taken place; that a few days later they had met again, when Mr. Brunel attended, and that after a further and most satisfactory investigation, they had come to an unanimous decision in favour of the scheme. This resolution was adopted, and Mr. Brunel was appointed Engineer to the Company.

The following extracts from his reports and correspondence carry on the narrative till the date of the next meeting of the shareholders (December 1, 1852), when the details of the project were laid before them:—

*Extract from a Letter describing the Scheme*

July 1, 1852.

The principle is, as I explained to you a very simple one—that of building ships to carry their own coals, instead of incurring large expenses and great delay in coaling at numerous intermediate stations; and the result is a large vessel certainly, but one which, at the same cost of fuel as is now required for small ones, has, besides that room, for 4,000 or even 5,000 tons (measurement) of cargo, and as many passengers as offer. Thus the capital embarked in the one vessel is not so great in proportion to the tonnage space for cargo as the capital embarked in several smaller vessels carrying the same amount; while the current expenses are greatly less, and the speed, and economy of time by that speed and by avoiding tedious stoppages, greatly in favour of the large one. Practical men concur with me, not merely in the practicability of constructing the vessel, but in the great advantage as regards speed, seaworthiness, and safety resulting merely from the increased size; while all the mercantile men concur in the opinion that if goods can be carried direct in thirty to thirty-five days, the certainty of freight ensures a return far beyond all present proportion of return to cost . . . On these points, of course, I quote only the opinions of the Directors. On the mechanical part I offer my own opinion, and may quote those of the first practical men of the day—Messrs. Maudslay, Messrs. Watt and Co., and J. Scott Russell, all of whom have assisted me in the project, and are prepared to join in it.

*Letter to J. Scott Russell, Esq., on the Form and Dimensions of the Great Ship*
July 13, 1852.

The adoption of this plan being now determined upon, we must proceed to determine the details, and the first step unquestionably is the determination of the size and form of the ship. Now, in preparing the general design, I think the following conditions should be strictly complied with. If any of them appear to involve any great sacrifice in cost, or to involve any other peculiar difficulties, these difficulties can be considered afterwards; but the wisest and safest plan in striking out a new path is to go straight in the direction which we believe to be right, disregarding the small impediments which may appear to be in our way—to design everything in the first instance for the best possible results strictly according to the principles which theory, so far as it is supported by practice, teaches us, and without yielding in the least to any prejudices now existing unsupported by theory and practice, or any fear of the consequences; we can then afterwards weigh and balance deliberately the advantages of adhering to or giving up this or that particular part, or modifying dimensions, either from motives of economy, or as yielding to public opinion from motives of policy.

In determining the lines of the ship, for instance, I should adopt that which we have reason to believe the best possible without any concession, or any compromise or regard to any assumed difficulties of construction, or regard to assumed opinions; these difficulties will very likely vanish afterwards if disregarded in the first instance, as in the case of the continuous curve, in which my fresh ideas had the advantage even of your much greater knowledge, hampered by a little preconceived idea. With respect to the size, to arrive at it by constructive calculations from the fixed conditions that we can lay down is perhaps possible, but rather difficult, and I think we know sufficiently nearly now what the minimum size must be to work upon; that, and a trifling alteration afterwards in the scale, will suffice to bring it to the exact required capacity.

The positive conditions, then, are a maximum draught of water of 24 feet, when leaving the Hooghly with the coals for the voyage home; and the capacity must be at least 21,000 tons of displacement at this draught of 24 feet.

I think you will find that to effect this comfortably you must give a length of 650 feet at least, and an extreme breadth of 80 feet, but this beam of 80 feet requiring no fuller entrance than you would make with a beam of 70 feet, the 80 feet being obtained entirely by continuing a gentle curvature throughout the whole length, instead of having any parallel lines.

If the experiments upon the friction of surfaces turn out as I hope, and give us reason to expect a very much less resistance from a copper surface than that now created by painted iron, I suspect we may be led rather to increase our length and diminish the proportion of beam; but this is a very serious question, not entirely dependent on the consideration of the form of least resistance including friction, but also materially affected by the consideration of the advantages of the extreme steadiness of motion which length seems to give. It is a subject which must be well discussed and well considered, with the assistance of all those whose opinions and expertise are likely to be of use to us. My own impressions, I confess, derived from considering the cases which we have, even after the striking result of the 'Ocean Queen,' are that positive length, independently of relative length, has much to do with it. When I see that the 'Great Britain,' although with a beam of about one-sixth of her length at the water line, and a midship section favourable to rolling, is nevertheless steady, I must conclude that positive length may compensate very greatly for a relatively wide beam. Now, we shall unquestionably have abundance of positive length. We must then be careful not to sacrifice much to keep a small beam, without being very sure that there are very great advantages; and, except for the assumed advantages of the long parallel or equal bearings, the form of least resistance, including friction, with a draught limited to 24 feet, and a required displacement of 21,000 tons, would, I apprehend, give us a beam nearer 90 feet than 70 feet. I should like to know exactly what the proportion would be without regard to the theory of the long narrow parallel forms; and then let us consider how much, if anything, should be sacrificed to attain the advantage assumed to be attained by relative length.

Let us therefore have at once the draft of a vessel of 21,000 tons displacement at the 24 feet water line, and of such form as will in your opinion give the greatest speed in smooth water, without seeking to make it narrow.

We must, of course, also bear in mind the comparative weakness of form caused by length, and the consequent increased thickness of material required, besides an actual increase of surface, involving a very considerably greater quantity and weight of material in the ship, which last consideration is very greatly in favour of breadth of beam; for I think you will find that the quantity of iron in two ships of 600 and 700 feet in length respectively, with the same displacement and the same ultimate strength to resist strains, will be fully in the ratio of their length.

### Report to the Directors on Mode of Proceeding

July 21, 1852.

Since the adoption by the general meeting of the plan recommended by the Directors, I have been engaged very constantly in maturing the details of that plan, and considering the course which it would be necessary to follow in order to carry them out in the surest, safest, and most efficient manner.

The steps which are about to be taken are unquestionably in the right direction, but they are considerable ones, and must be taken with deliberation and certainty, and without leaving anything doubtful; and, when determined upon, they must be followed up with decision.

Although you will probably determine upon constructing not less than two vessels in the first instance, yet they must both be proceeded with at once, and must in fact be exact duplicates of each other. The success of the two, therefore, depends upon that of each; there can be no average struck in such a case, but the two ships must be designed and executed on such principles and with such perfection that no doubt can exist of the result.

By well considering all that has been done, by selecting all that has been most successful, and by a judicious application of such results to the peculiar circumstances of our case, all this certainly can, I think, be assured, but it can be assured only by proceeding with the caution and the decision which the circumstances demand.

In the first place, as to the designing of the whole, the principle being determined upon, much may be ascertained by mere calculation, but for these calculations data are required, which nothing but experience can furnish. I have, therefore, availed myself of the assistance of those most competent to afford the required information. I have called in to my assistance the gentlemen whom I had already named to you as best able to give strength to our position by the value of their opinions, and best able to execute the various parts of the work with that experience and perfection which are essential to our success.

With respect to the form and construction of the vessel itself, nobody can, in my opinion, bring more scientific and practical knowledge to bear than Mr. Scott Russell. As to the proportion of power to be adopted, the form and construction of the engines, screw, and paddles, besides Mr. Scott Russell, I have had the benefit of the deliberate consideration and advice of Mr. Field, of the firm of Maudslay and Field, and of Mr. Blake, of the firm of Watt and Co. I have written also to my friend Mr. F.P. Smith, to whom the public are indebted for the success of the screw, for his advice on the subject. With such assistance I think we may rely upon the certainty of being able to design and to execute all that is best in the mechanical and ship-building department. In the naval department I have had the opportunity also of consulting two gentlemen, Captain Claxton and Captain Robert Ford, who possess special knowledge and experience on the subject. I have had several conferences with all these gentlemen, I have explained fully my views, and, with their assistance, settled preliminarily some of the principal points of detail. What I should propose to the Directors now is, that with that assistance I should proceed to prepare in detail the design of the ship, and the exact dimensions and form of the engines; that, in the meantime, I should obtain information upon certain points which will govern you as to the mode of contracting for the construction of the ships, and also that I should be authorised to adopt some means of determining one or two most important points which must govern some of the principal dimensions.[4]

*Report to the Directors on Enquiries relating to the Draught*
*and Form of the Vessel*

October 6, 1852.

Since the date of my last letter to you, recommending that certain enquiries and investigations should be set on foot to determine several points which would materially influence the plans I should have to submit to you, many circumstances have occurred to delay these investigations. Not having sent any competent person expressly to Calcutta to ascertain with certainty the draught of water that might be adopted, I have endeavoured to obtain as much information as possible upon this point from persons capable of affording it, who might be in England.

Several very competent men, captains of long experience in that particular navigation, and even local pilots of the first standing, happened to be within reach, and I have had personal communication with these gentlemen. Notwithstanding, however, these fortunate opportunities of obtaining information from the best existing authorities, we are left in pretty nearly the same state of doubt as to the maximum depth as we should be by a mere inspection of the charts, the opinion of very competent men varying so much as to fix this maximum as low as 21 and as high as 23½ and even 24 feet. They all concur, however, in fixing Diamond Harbour as the point in the Hooghly which may easily be reached, but beyond which it would be almost impossible to go.

A question as to the extent of swell which in so large a ship might be given to the sides, increasing the capacity without materially increasing the resistance, involved one of the experiments to which I referred in my former letter; these experiments have been made, and the result, such as it was, of the enquiries before referred to as to the navigation of the Hooghly, led me to direct the preparation of draughts of three different models of ships, and upon further consideration of these three, and under the circumstances, I have come to the conclusion of recommending one which will have the following dimensions:—namely, 670 feet in length, 85 feet beam, and a deep water draught of 30 feet.

Such a vessel would be able to carry her own coal for the voyage home out of the Hooghly with about 23 feet draught; but if between now and the period when the exact arrangement must be determined, it is found expedient not to attempt so great a draught in the Hooghly, the same vessel will, by coaling at Trincomalee on the return voyage, be exactly adapted to work out of the Hooghly with a good cargo of goods and coals for Trincomalee with only 20 to 21 feet draught.

I have been in communication with the eminent engine builders whose names I have mentioned on a former occasion, and with Mr. F.P. Smith, the inventor of the screw propeller. Some trials and investigations are still in progress to determine the relative advantages of a copper and iron bottom, on which question may depend the arrangements which may be requisite to provide for docking or rather laying up for cleaning, and when these points are determined I shall be prepared to lay before you a complete design of ship and engines for your consideration.

Efforts were made to induce the public to assist in carrying out the project. In February 1853, the Chairman (the late Mr. Henry Thomas Hope) and several of his colleagues formed themselves into a committee for the purpose of communicating with Mr. Brunel on the subject of his plans, and reporting to the Board thereon.

The results of this conference were embodied in the following report which Mr. Brunel addressed to the Directors:—

### Report on the Proceedings of the Committee

March 21, 1853.

To enable the Committee to arrive at a correct conclusion on this difficult question, I had gone through rather lengthy calculations of the minimum dimensions and the comparative estimated cost of the ships which would in each case answer the purpose under the several different circumstances which might be assumed; and the results of these calculations were laid before the Committee, and the relative advantages and disadvantages of each assumed case discussed . . .[5]

After much discussion, and comparing the probable receipts with the estimated expenditure, and allowing fully for interest of capital embarked, I think it appeared to the Committee that as a mere mercantile transaction, and with reference only to the Australian trade, the larger vessel would be the most economical, showing not merely the means of securing the largest return, but involving an actual diminution of annual expenditure.

The Committee were unanimously of opinion that the largest size of the first class would be the best, and would in every way answer the objects of the Company.

The dimensions arrived at by calculation for this ship would be in round numbers, 670 feet long, 80 feet beam.

This sized vessel would combine most of the advantages which we seek to obtain. It would carry coal to Diamond Harbour and back to Trincomalee; it would afford room for about 800 separate cabins larger than those now fitted up in packet ships, with large saloons, capable of accommodating 1,000 or 1,500 first and second-class passengers; and would carry 3,000 tons weight of cargo, without making any allowance for that increase of speed proportionate to the mere increase of size of which we see every day fresh proofs; the average speed of the ship, with the proposed power of engine and calculated consumption of coal, would be 14 knots at the average, making the passage out in 34½ days, say 36; but with that increased speed which has been shown to take place with increased dimensions, we may speculate upon the voyage being performed in 30 days.

This same vessel, fitted up for the Australian voyage, and loaded deeper, would carry coals to Australia and back, would take out 3,000 passengers easily, and a small amount of cargo only, but could bring back any amount that could be

1. Lithographed portrait of Brunel from a painting by J.C. Horsley and Brunel's signature

*Images supplied by STEAM Museum of the Great Western Railway, Swindon*

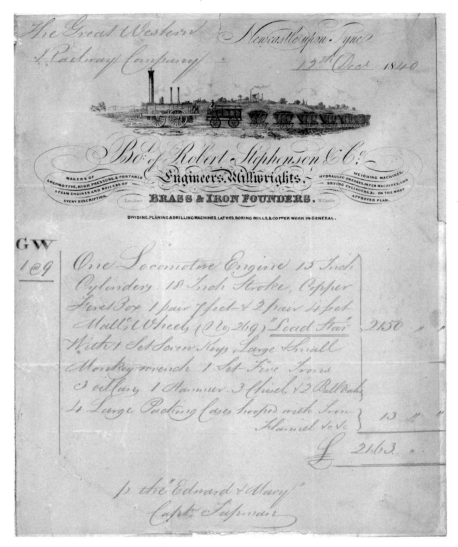

2. (above) Invoice from Robert Stephenson & Co. for the purchase of the locomotive *Lode Star* for the Great Western Railway, 1840

3. (below) Newton Abbot Station, 1846, part of the South Devon Atmospheric Railway

*Images supplied by STEAM Museum of the Great Western Railway, Swindon*

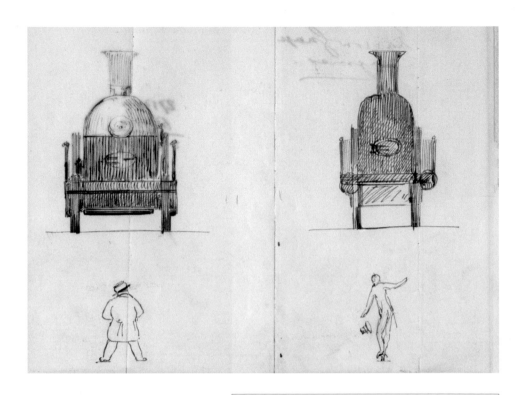

4. (above) Cartoon sketch depicting the 'Battle of the Gauges': Broad Gauge vs Narrow Gauge, 1840s

5. (right) During the 'Battle of the Gauges' publications like this one, opposing Brunel's broad gauge, were common

*Images supplied by STEAM Museum of the Great Western Railway, Swindon*

THE

# BROAD GAUGE

𝔗𝔥𝔢 𝔅𝔞𝔫𝔢 𝔬𝔣 𝔱𝔥𝔢

## GREAT WESTERN RAILWAY

### COMPANY.

WITH

AN ACCOUNT OF THE PRESENT & PROSPECTIVE LIABILITIES
SADDLED ON THE PROPRIETORS BY THE PROMOTERS
OF THAT PECULIAR CROTCHET.

**By £. s. d.**

" A barbe de fol, on apprend à raire."
[Which, being translated for the benefit of Country Gentlemen, means]
" Mr. Brunel has learnt to shave on the chin of the
Great Western Proprietors."

*SECOND EDITION.*

LONDON:
JOHN OLLIVIER, 59, PALL MALL.
1846.

6. (above) *North Star*, the first broad-gauge locomotive to run on Brunel's Great Western Railway in 1837

7. (below) A map of the Great Western Railway from the 1834 Prospectus

*Images supplied by STEAM Museum of the Great Western Railway, Swindon*

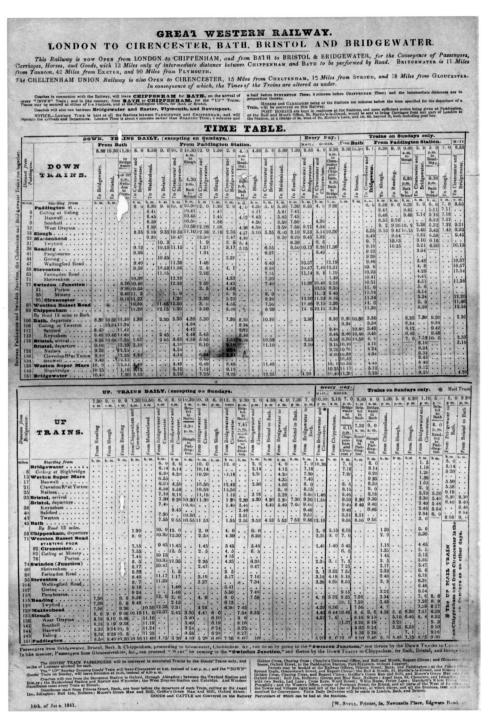

8. GWR Timetable dated 1841. At this time Box Tunnel had yet to be completed and the thirteen miles between Chippenham and Bath was taken by road.

*Image supplied by STEAM Museum of the Great Western Railway, Swindon*

9. (left) Paddington Station: an Edwardian view along Platform 1. The impressive wrought-iron arched roof-spans, designed by Brunel, can clearly be seen here.

10. (below) The west portal of Box Tunnel from a lithograph by J.C. Bourne

*Images supplied by STEAM Museum of the Great Western Railway, Swindon*

11. (right) The Thames Tunnel as a pedestrian walkway. From a print by Tombleson, *c*.1830.

*Image courtesy of The Brunel Museum, London, www.brunelenginehouse. org.uk*

12. (below) The Thames Tunnel workings, 1876, by George Yates

*Image courtesy of Southwark Art Collection*

## BRISTOL & EXETER RAILWAY.

### VISIT
OF HIS ROYAL HIGHNESS

## THE PRINCE CONSORT,
TO THE

### OPENING
OF THE

# ROYAL ALBERT BRIDGE,
AT

## SALTASH,
ON

## MONDAY, 2nd May, 1859.

### ROYAL TRAIN TIME BILL.

| DOWN. | | | UP. | | |
|---|---|---|---|---|---|
| | DEP. a.m. | ARR. a.m. | | DEP. p.m. | ARR. p.m. |
| WINDSOR ... ... | 6 0 | — | SALTASH ... ... | — | — |
| Bristol ... ... ... | | 8 35 | Cornwall Junction | | — |
| " ... ... ... | 8 45 | | " | 6 50 | |
| Taunton... ... ... | | 9 35 | Newton ... ... | | — |
| " ... ... ... | 9 38 | | " | | |
| Exeter ... ... ... | | 10 25 | Exeter ... ... | | 8 15 |
| " ... ... ... | 10 35 | | " ... ... | 8 25 | |
| Newton ... ... ... | | 11 5 | Taunton... ... ... | | 9 12 |
| " ... ... ... | 11 10 | | " ... ... ... | 9 15 | |
| Cornwall Junction | | 12 0 | Bristol ... ... ... | | 10 5 |
| " | 12 5 | | " ... ... ... | 10 15 | |
| SALTASH ... ... | | 12 15 | WINDSOR ... ... | | 12 50 |

*The following arrangements will be necessary for the proper working of this Train, which must be strictly attended to:—*
The 7.50 a.m. Down Passenger Train is to Shunt at Tiverton Junction.
The 8.0 a.m. Goods Train Down will not start from Bristol until after the Royal Train.
The 8.0 p.m. Up Train is to Shunt at Tiverton Junction.
The 9.20 p.m. Short Train from Weston is to Shunt at Yatton.

Bristol, 29th April, 1859.

13. (above) The Royal Albert Bridge, Saltash, during construction

14. (left) Poster advertising the opening of the Royal Albert Bridge just a few months before Brunel's death

*Images supplied by STEAM Museum of the Great Western Railway, Swindon*

15. (above)  The Royal Albert Bridge, Saltash, today

*Photograph © John Christopher*

16. (below) A view across the Thames to Maidenhead Bridge, 1895

*Image supplied by STEAM Museum of the Great Western Railway, Swindon*

17. (previous page) The Clifton Suspension Bridge

18. (above) The piers of the old Moorswater Viaduct beside the modern viaduct

*Photographs © John Christopher*

19. (above) The SS *Great Britain*

20. (below)  The screw propeller of the SS *Great Britain*

*Photographs © John Christopher*

21. The *Great Eastern* at Neyland, 1872. The date written on the photograph is the date she was launched.

*Image courtesy of the J. & C. McCutcheon Collection*

*Great Eastern 1858*

THE GREAT EASTERN LAYING THE ATLANTIC CABLE.

22. (above) The *Great Eastern* laying the cable for the Atlantic Telegraph
*Image courtesy of the J. & C. McCutcheon collection*

23. (below) Brunel's office at 18 Duke Street London, *c.*1840s
*Image supplied by STEAM Museum of the Great Western Railway, Swindon*

conveniently collected, or if provision were made for taking in 3,000 or 4,000 tons of coal in Australia, that additional amount of cargo might be taken in the passage out. The passage out to Port Philip should be made easily in 36 days, and home by Cape Horn in the same time.

The Committee having come to the conclusion that this class of vessel would best fulfil the several conditions which the circumstances imposed, I have been engaged in determining the several details consequent upon this selection of size, and have put in hand drawings of the ship, which will enable me to arrive more correctly at the cost, and will enable us to obtain tenders for the construction. These details involve a great deal of study and consideration, and the making of the drawings alone requires some considerable time, so that I do not think much advance can be made under three weeks from the present date, but I will endeavour to expedite the work as much as possible.

Mr. Brunel was authorized to continue his communications with engine-makers and ship-builders, and to invite tenders.

After very detailed and careful calculations of the smallest capacity that would secure the attainment of the objects sought for, the dimensions of the vessel and the power of the engines were finally determined; tenders were received from Mr. Scott Russell, Messrs. Watt and Co., and Messrs. Humphrys and Co. for both sets of engines, and from Mr. Russell for the construction of the hull of the ship and for placing her afloat.

A meeting of the Directors was held on May 18, when Mr. Brunel submitted the various tenders he had received. The Board agreed to adopt his recommendations, and to accept the tenders of Mr. Scott Russell for the ship and paddle engines, and that of Messrs. James Watt and Co. for the screw engines.

The following report, which Mr. Brunel placed before the Directors, gives a detailed account of the steps he had taken to procure the tenders, and the grounds on which he had formed his judgment of them:—

*Report on Tenders*[6]

May 18, 1853.

According to your instructions, I applied to the several parties with whom I had previously been in communication on the subject of the engines and ship, for tenders for their supply. As regards the engines, I drew up a short specification, defining generally what was required, and leaving the parties to make their own designs and propose to me the form of engine they would adopt.

With respect to the ship, where no such variation could be permitted, I have had very detailed drawings and specifications prepared.

Copies of the specifications are annexed.

In defining the power of the engines and boilers, I have, in conformity with my own views and of those of several of the Directors, who have expressed themselves strongly on the subject, required a very full amount of power, without naming the nominal horse-power, which is a very vague mode of defining anything but the cost which by custom is made dependent upon that nominal power; but I have defined dimensions of parts and surface of boilers which will ensure the means of exerting a very large amount of power.

As regards the ship, I have not spared strength of materials, and have required the best workmanship.

The result of this application for designs and tenders is, upon the whole, very satisfactory, although two of the parties from whom I had hoped to have received proposals have not been able to send any . . .

Mr. Blake, of the firm of Watt and Co., and Mr. J.S. Russell and Mr. Humphrys, have, as I had before reported, devoted much attention to the subject: from these gentlemen I have received distinct well-considered designs of the screw and paddle-engines.

I have been in frequent communication with these gentlemen, and have seen their plans while in progress, and have made my suggestions upon them, and assisted more or less in maturing them, and at all events in preventing the adoption of any principle or arrangement that I should afterwards object to. Notwithstanding this, the three designs, particularly for the screw engines, are totally dissimilar, and I am placed in the difficulty of having to choose between three totally different plans, each designed by skilful and experienced men, and each possessing many known and acknowledged advantages.

I should also observe that, although I have known the general arrangement which each would adopt, I did not receive the plans or the tenders which I now forward until last evening, and that of Mr. Humphrys as late as 10 o'clock this morning, and that I have not therefore had time to draw up any very detailed report upon them.

The following are, however, the principal considerations which influence me in the selection which I am disposed to recommend.

It must be borne in mind that the screw engines will be the largest engines that have yet been made. The principal part of the propelling power of the ship will be thrown upon the screw; and upon these engines therefore will mainly depend the performance of the ship, and particularly upon their constant never-failing working, probably for thirty or forty days and nights, must depend the certainty of the ship's performance.[7] Under all these circumstances, the compactness and stiffness of framing, the greatest possible simplicity of construction, and the fewest possible number of parts, and, finally, the absence of any novelty, however promising, that can introduce any unforeseen difficulties, are conditions which would outweigh in my mind many advantages that might, and I think would, be attained by several arrangements which have been suggested.

All the designs now submitted comply with these conditions to a very considerable extent . . . but the extreme simplicity of Mr. Blake's engine leads me to prefer it.

As regards the paddle engine, I unhesitatingly give the preference to that proposed by Mr. Russell. I believe it to be as simple an arrangement as can be adopted for engines with such a slow motion and so long a stroke as these must have, and the single crank in the centre I consider a great advantage . . .

As regards a contract for the ship, I have found it more difficult to proceed in the ordinary course. The conditions which we must ensure of quality of workmanship and execution, under close inspection and within reach of one's own supervision, are not easily attained; and, though as a matter of course readily promised and undertaken by all ship-builders, they are rarely secured. It is essential also that the ship should be built where the engines can be readily fixed on board before launching, and in a yard which can be devoted to the purposes of the ship, and whence the launching can be effected with the engines and boilers on board.

All these are conditions not easily secured. I have been in communication with one or two parties, and the result is a tender from Mr. J. Scott Russell, which I enclose, and which has in fact been framed upon my calculations. Owing to the recent rise in iron it is somewhat, but not materially, above the amount at which I had originally estimated the vessels; but the tender, founded upon the supposition of two such vessels being ultimately ordered, is as nearly as may be the same as my original estimate, and upon the whole I consider the tenders both for engines and ships very satisfactory and confirming fully our previous calculations.

These tenders do not include, in the case of the engines, either the screw itself or the paddle wheels, nor, in the case of the vessel, the cabin fittings, masts, and rigging, boats, or stores. I should estimate these roughly at £50,000 more,—at least such an allowance ought to be ample; but these details will require a great deal of consideration, and could not be included in the original contracts; while at the same time they can mostly be better and more economically supplied by competition or by arrangement with the special makers of the respective articles . . .

During the next six months Mr. Brunel was engaged in preparing the formal contracts and specifications. These documents were settled with much care, and after frequent communications with the contractors, who consented to the insertion of clauses which gave the full control and supervision over every part of the work to the Engineer, with very large powers of interpretation. They required, however, that, should Mr. Brunel cease to act as Engineer, any disputed point should be settled by arbitration, and not by his successor.

Besides the delay occasioned by the magnitude and novelty of the undertaking, there were other difficulties which helped to postpone the commencement of the works. The Directors were unable, under their charter, to enter into any contracts until a certain amount of their capital was actually paid up; and, as several shareholders had retired when the change of plans was determined upon, it was no easy work to get the shares taken. That this was eventually accomplished was due mainly to the exertions of Mr. Brunel and Mr. Charles Geach, one of the Directors.[8] 'Could I have foreseen,' Mr. Brunel writes, 'the work I have had to go through, I would never have entered upon it; but I never flinch when I have once begun, and do it we will.'

Several times they nearly broke down, but at length the contracts were signed, and on the same day, December 22, Mr. Brunel gave the formal notice to the contractors to proceed with the works. 'After two years' exertions (he wrote), 'we are set going, the contracts entered into, and the work commenced.'

### Extracts from Mr. Brunel's Memoranda, A.D. 1852–1853

*July* 11, 1852.—The dimensions I commenced with in March last, of 650 x 70 x 30 appear after all to be not far wrong, according to present views. I make them now 700 x 70 x 24 about; but much depends upon the last dimension, the draught. If another foot or two can be safely taken it will be of great advantage . . . With this size of vessel, having a midship section of about 1,800, and a length of 700, I assume a nominal horse-power of about 2,500. The first question of importance is, in what proportion shall this be divided between the screw and paddles? . . . My present impression is to halve the power between the two.

In both the engines every known means must be adopted to secure efficiency:— 1. An excess of boiler power; 2. expansion permanently, say at ⅓; 3. steam of not less pressure than 20 lbs., and I should prefer 25 lbs.; 4. that cylinders, particularly top and bottom, slide chest, and steam pipes, be all jacketed, and the jacket supplied with steam from an auxiliary boiler of at least 10 lbs. more pressure than that of main boilers; and it would be very desirable to make some experiments to determine whether it is not worth having a heating apparatus to heat the steam immediately before it enters the cylinders.[9]

*July* 17.—After a long conference with Mr. Field, I continue of the opinion that it would be well to apply about three-fifths of the power to the screw and two-fifths to the paddles, and probably, as the vessel gets light, diminishing a little the expenditure of power on the paddles, and keeping up the full power on the screw. Mr. Field is not in favour of increasing the pressure of steam beyond 12 lbs. or 15 lbs., on the ground that all the mechanical difficulties increase rapidly without a corresponding advantage, particularly where size and weight are not so important. There seems much truth in this . . . The possible advantages of a slight increase are not sufficient to justify the risk of the possible new difficulties in a work on so large a scale. Nothing uncertain must be risked. These arguments do not apply to the jacketing and heating, which Mr. Field also deprecated, or rather discouraged, simply on the ground of the trouble and difficulty of effecting it, but he admitted that all experience went to show the advantages of it; and as to the difficulties, which I could not see, they involve no other risk than that of being useless: they cannot do mischief. The heating of the top and bottom of the cylinders, I think, must be particularly important in a short-stroked engine working expansively. In a cylinder of 80 inches diameter and 40 inches stroke, having regard to the *time* of contact, the area of the bottom will be nearly equal in effect to the surface of the cylinder.

*July* 19.—After much consideration, I think I feel satisfied that the best construction will be to have strong bulkheads every 30 feet or thereabouts, this distance being dependent on what is required for one set of boilers and its stock of coals; these bulkheads being carried right up wherever practicable—I think every alternate one may be—and then place the main ribs of the ship, and even at least two main deck beams, *longitudinal* instead of *transverse*.

*February* 2, 1853.—Several drafts of ships have been made and much consideration given to the subject, and frequent discussions with various parties. The result of all is that my present views are as follows:—

The ship, all iron, double bottom, and sides up to water line, with ribs longitudinal like the Britannia tube. I have not been able to devise any good mode of determining the relative amount of friction of a copper and an iron surface; and, although I believe in copper, it would not do to act on mere belief. I therefore at present settle iron, the surface being carefully made smooth. Doubts have come across me also as to whether with a very long surface the difference between the smoothness will so much affect the total resistance. Is not a film of water, after a certain distance, carried with the body? and, if so, its greater or less roughness, if not producing currents, is almost unimportant. Would there be any difference in the resistance of a fine file or a rough one drawn through tallow, if they both covered themselves with grease? Is there any similarity? As to size, if we are to go round the world,[10] I do not think we can do with less than—length, 730; beam, 85; draught deep, 34; and I assume a nominal horse-power of engines equal to 1¼ of the sectional area at 30 feet; but, taking consumption as a better measure, and assuming that every possible economy is practised, and every refinement introduced that can produce economy, I shall assume 7½ lbs. per hour per nominal horse-power, or say 0.08 ton per day per horse-power; and as I assume the horse-power to be 1¼ sectional area, it makes the consumption = 0.1 ton per day per foot of sectional area. And this is a very large allowance and ought to ensure a very high speed. In order to effect the utmost economy, I should work up to 20 lbs. steam (calling it 16 lbs.), cutting off certainly at ⅓ the stroke, and adopting every precaution to keep the steam hot and the condenser cool. The latter depends, I believe, solely upon the perfect dispersion of the injection water, so that the condensation of the steam may take place suddenly, otherwise, the same amount of water may condense the steam *in time,* the same amount of heat be given off, the same quantity of injection water used, and yet the condenser be always full of steam at a good pressure. It might be well worth the experiment to try the effect of a large injection at the moment of the exhaust port being opened; but above all things I believe the heating of the steam to be important; and for this purpose I should jacket the steam pipes and cylinders top and bottom, and heat with high pressure steam, say at 60 lbs.—I have increased this pressure the more I think of it; 60 lbs. would be above 300 degrees, and 20 lbs. not quite 260 degrees; therefore there would be a full 40 degrees of surplus to ensure the temperature. I have a great tendency to believe in the advantage of further heating even, which might be done by a Perkins' arrangement of hot water; but possibly the new conditions, as regards oiling, &c., might involve difficulties not desirable to introduce in this case. In the boilers it will also be necessary to adopt every refinement which has been found really to answer, although not always adopted; above all, every means of keeping them clean—scum pans, and Field's exchanging apparatus. But what would be even more effectual would be some easy means of removing a whole bundle of tubes and replacing them by clean ones; and surely this would not be difficult, the tubes being large and with plenty of space, so that a man could pass his arm between. A rather important addition to boilers would also be a means of blowing off without noise. Several modes would seem to be possible, but whatever plan is adopted, it should be one which is completely self-acting, and perfectly effectual when used suddenly and without any preparation, and at a moment of confusion and alarm. Blowing through a wire gauze pipe would probably be as likely a way as any.[11]

The more consideration I give to the subject the more disposed I am to adopt oscillating engines for both screw and paddles. The extreme simplicity and small number of parts, and compactness, and the direct action of every resistance to the force which it is wanted to resist, seem to leave nothing to be desired, and would seem to make it a better and more mechanical arrangement of a cylinder and crank than any other, quite independently of the object for which it was originally designed, which was simply 'stumpiness.'

February 21.—The original line (to Calcutta) seems likely, after all, as usual with most original ideas, to be the best; at all events, so good that the vessel must be built to be able to go there. The dimensions best fitted for this would seem to be—length, 700 feet; beam, 85 feet; depth of hold, 58 feet; screw, 24 feet; paddle 60 feet. If arranged for Calcutta, we must arrive there on an even keel, and therefore, to maintain the most equal level for the paddles, they must be kept well forward, and the change principally at the stern. Engines indicated horse-power 8,000; steam at 25 lbs.; auxiliary steam at 60 lbs.

The ship to be lighted with gas, to be thoroughly ventilated by mechanical means, having large air trunks, with small pipes and valves to each cabin, with the means of warming this air in cold latitudes and seasons, and cooling it in the more frequent cases of hot climates. The ship must be steered from the forecastle, whence a perfect look-out must be kept with fixed telescope, &c., and speaking pipes and bells to the engine rooms.

March 14.—At a meeting of the Committee, held this evening here, the several costs and qualities of four different sizes of ships, of which all the calculations had been made by me, namely:—

| No. | Length | Breadth | Mid. sec. | Draught |
|---|---|---|---|---|
| 1 | 663 | 79.9 | 1,646 | 24 |
| 2 | 634 | 76.39 | 1,640 | 25 |
| 3 | 609 | 73.5 | 1,639 | 26 |
| 4 | 730 | 87 | 2,090 | 28 |

were discussed, and the No. 1 determined upon as the best under all the circumstances. I should propose, therefore, to make the dimensions of No. 1:— length, 680 feet; beam, 81 feet, to be swelled to 83 feet; extreme draught, 30 feet; mean, 24 feet; daily consumption, say 200 tons.

This ship can carry her coal to Calcutta, and arrive and leave with only 21 feet 6 inches draught, having 9 days' coal and 3,000 tons cargo; or she could first go to Australia and back, without or with very little cargo out, and consequently would take out as much cargo as you might choose to send coal for her to Australia . . .

These dimensions are worked out in the design No. 5 (April 9, 1853), but they would be better for a slight increase, if the 83 feet were made 85 feet, and the 680 feet were made 700. We should have an increase in capacity of 83 x 680 = 56,440 to 85 x 700 = 59,500, or 6 per cent of displacement. This would bring the displacement at 32 feet draught up to 31,250 tons.

*March* 22.—Settled the various dimensions of scantlings with S. Russell to enable him to direct drawings of all details to be got out.

*April* 28.—We are now seeking tenders for engines and ship of the following dimensions:—Length, 680 feet; beam, 83 feet; mean draught, about 25 feet; screw engine, indicated horse-power, 4,000; nominal horse-power, 1,600; paddle, indicated horse-power, 2,600; nominal horse-power 1,000; to work with steam 15 lbs. to 25 lbs.; speed of screw, 45 to 55 revolutions; paddle, 10 to 12.

Among the details of improvements still to be considered are the receiving through measures the coal from the bunkers, and running it on tramways and waggons to the front of the fires, thus at the same time measuring out accurately the hourly consumption, and saving labour; but a still more important object, the use of clean water—that is using the same water over again—is well worth considering and it is well worth the experiment, whether cooling down the water of condensation to use again is not in fact the easiest way. With an unlimited supply of cooling water this ought to be easy.

*August* 7.—Memoranda for engines.—Very sensitive governors to be applied to both engines to prevent running away.

*November* 18.—It is curious that the above should be the last memorandum, as I now open the book to make the same in consequence of the accident to the 'Agamemnon.'[12]

There can be no reason why a sensitive governor should not act in less than one revolution of the crank, and act upon a tumbler which should shut off instantly the expansion valve. There should be two such governors, one to each end of the crank shaft, and they should work direct from a spur wheel from the shaft without any intermediate shafting, to give elasticity, or to risk breaking. (Query, hydraulic governors?)

The auxiliary engine and boiler to be at least 20 feet from bottom, and, better still, above load water line, or so boxed as to be out of reach of water; so that if the ship grounded and filled, this engine would remain serviceable for pumping or anything else.

The history of the 'Great Eastern' has now been traced up to the date of the contracts for the construction of the ship and her engines.

The following selections from Mr. Brunel's memoranda illustrate the progress of the design during the early months of the year 1854:—

*February* 26, 1854.—The details of construction, both of engines and ship, involve an immense amount of thought and labour. I have devoted a great deal of time to it already, and yet even the preliminary details either of engines or ship are far from being satisfactorily settled. I have no record of the many consultations hitherto held on the subject, but shall hope to keep one hereafter. On the 6th inst., some of the drawings of the ship and of parts of the engines, having been several times revised and altered, being ready, I spent the greater part of the day at Millwall[13] in going again into them, and settled some parts, such as the dimensions, &c., of cranks, and bearings, general form of engine frame, and some of the general principles of framing and plating of the ship. Some other consultations have been held, and again to-day (February 26) I have spent some hours at Millwall . . . Discussed the details of a midship section, the drawings of which were in a forward state; directed that the cabins should positively be made 6 feet 6 inches each in the clear, and the bulkheads made subordinate to this; found that it could be done without difficulty, and without causing any mechanical objections in construction. I am anxious to have some approximate estimates of weights.

It is evident that large weights may most easily be wasted or saved by a careless or close consideration of the best application of iron in every single detail. I found, for instance, an unnecessary introduction of a filling piece or strip, such as is frequently used in ship-building to avoid bending to angle irons; made a slight alteration in the disposition of the plates that rendered this unnecessary; found that we thus saved 40 tons weight of iron, or say £1,200 of money in first cost, and 40 tons of cargo freight—at least £3,000 a year. The principle of construction of the ship is in fact entirely new, if merely from the rule which I have laid down, and shall rigidly preserve, that no materials shall be employed on any part except at the place, and in the direction, and in the proportion, in which it is required, and can be usefully employed for the strength of the ship, and none merely for the purpose of facilitating the framing and first construction.

In the present construction of iron ships the plates are not proportioned to the strength required at different parts, and nearly 20 per cent of the total weight is expended in angle irons or frames, which may be useful or convenient in the mere putting together of the whole as a great box, but is almost useless, or very much misapplied in affecting the strength of the structure as a ship.

All this misconstruction I forbid, and the consequence is that every part has to be considered and designed as if an iron ship had never before been built; indeed I believe we should get on much quicker if we had no previous habits and prejudices on the subject.

*March* 3.—Mr. Blake [*of the firm Messrs. James Watt and Co.*], called, and went fully into the general drawings which he brought. On the necessity of large surfaces he quite concurred with me. The extent to which such a general principle should be carried is of course very difficult to determine; my idea is that it has never yet been approached . . .

*March* 10.—Engaged all the afternoon at Millwall . . . Settled and signed the drawings of crank and piston rods. Went into many details of ship . . .

The extracts from Mr. Brunel's correspondence which follow, have been selected as containing a definition in his own words of the position he held as Engineer to the Company by which the great ship was built.

### *Letter to the Secretary of the Eastern Steam Navigation Company*

[This letter was written in consequence of a resolution of the Directors, asking Mr. Brunel to recommend them a resident engineer, in order that constant supervision might be exercised over the works, and frequent reports made to the Board.

The Directors rescinded their resolution; but this letter is inserted as showing, in clear and forcible language, Mr. Brunel's view of the nature of his duties and responsibilities, and as laying down what in his opinion ought to be the relations between the Directors of a Company and their Engineer.]

August 16, 1854.

. . . It surely cannot be necessary to remind the Directors that the very unusual stake which as a professional man I have been willing, perhaps imprudently, to risk on the success of this project—I mean stake of professional character, not mere pecuniary risk—must secure a much greater amount of attention to any step, and supervision of any detail on my part, than any ordinary professional engagement would obtain; the heavy responsibility of having induced more than half of the present Directors of the Company to join, and the equally heavy responsibility towards the holders of nearly half of the capital, must ensure on my part an amount of anxious and constant attention to the whole business of the Company which is rarely given by a professional man to any one subject, and, as it seems to me, ought to command a proportionate degree of confidence, or rather command entire confidence, in me, if any at all, for in such a case there can hardly be any medium. The fact is, that I never embarked in any one thing to which I have so entirely devoted myself, and to which I have devoted so much time, thought, and labour, on the success of which I have staked so much reputation, and to which I have so largely committed myself and those who were disposed to place faith in me; nor was I ever engaged in a work which from its nature required for its conduct and success that it should be entrusted so entirely to my individual management and control . . .

The Directors have a right to expect, and will ever receive, from me the fullest information and the most unreserved communication upon all points as they arise, as from one who feels the responsibility of being their sole professional adviser in a very important and serious business, in which we are all embarked, and all deeply interested; but I cannot act under any supervision, or form part of any system which recognises any other adviser than myself; or any other source of information than mine, on any question connected with the construction or mode of carrying out practically this great project, on which I have staked my character; nor could I continue to act if it could be assumed for a moment that the work required to be looked after by a Director, or by anybody but myself or those employed directly by me and for me personally for that purpose.

If any doubt ever arises on these points I must cease to be responsible, and cease to act.

### Letter to the Secretary of the Eastern Steam Navigation Company

[In explanation of the following letter it need only be stated that an elaborate article on the great ship appeared in one of the London newspapers of November 1854. Mr. Brunel's name was only once mentioned throughout the whole of it, and in these words: 'Mr. Brunel, the Engineer of the Eastern Steam Navigation Company, approved of the project, and Mr. Scott Russell undertook to carry out the design.']

November 16, 1854

Since I wrote to you I have taken the trouble to read through the long article in the ——, and am much annoyed by it. I have always made it a rule, which I have found by some years' experience a safe and profitable one, to have nothing to do with newspaper articles; but then, if on the one hand the works I have been connected with have rarely been puffed (never by me), they have also been rarely affected by misstatements; as such notices, when not inserted by interested parties, are always slight and unauthentic, and drop without producing any effect. This article in the ——, however, bears rather evidently a stamp of authority, or at least it professes to give an amount of detail which could only be obtained from ourselves; and if, as I think is the case, copies of it have been circulated by us, it may acquire the character of being an authorized statement; and, as such, I am individually much annoyed by a great deal that is in it, and by the omission of much that might with propriety have been introduced.

What is constantly repeated or implied, and remains uncontradicted, is at last received almost unconsciously as fact even by those who have the means of knowing it to be incorrect, if they thought about it; and, although from system I have never interfered with newspaper statements, it has not been from any affected or real indifference to public opinion, perhaps it was more from pride than modesty, and therefore I am by no means indifferent to a statement which would lead the public, and perhaps by degrees our own friends, to forget the origin of our present scheme, and to believe that I, happening at the time to be the consulting Engineer of this Company, which I was not, and having had no peculiar connection with previous successful improvements in steam navigation, allowed them to adopt some plan suggested by others, who I suspect, if even such were the case, would never appear to share with me the responsibility if any failure should result. Of this certainly I have no fear, but at the same time I am desirous of something more than mere immunity from blame.

I not only read this article once, but I was so struck by the marked care shown in depreciating those efforts which I had successfully made in advancing steam navigation, and mainly on the strength of which no doubt I originally obtained the confidence of the Directors, which induced them to enter upon our present bold undertaking, that I read the paper a second time, and for the very reason that I have for so many years shunned public writings, namely, to escape misstatements, I feel compelled on the present occasion to take some steps publicly to correct those erroneous impressions, which must be created by a document having the appearance of emanating from ourselves . . .

The objectionable points that I refer to evidently did not strike you, and that is a strong proof how easily incorrect impressions insinuate themselves unawares; but I feel strongly that a judicious friend would not have failed to do justice to the spirited merchants of Bristol, who, in spite of the strongest condemnation of the plan by the highest authorities, and the ridicule of others, persevered in building and starting the first transatlantic steamer. The circumstances as regards the 'Sirius' are coloured so as to be quite incorrect; and the same friendly hand would not have thrown ridicule, and that by a positive false statement, upon that which he at the same time admits to have been the means of almost introducing two of the greatest improvements in steam navigation. A writer wishing success to our enterprise would not have omitted to mention that I had a claim to public confidence on this occasion, for the reason that I was at least the principal adviser in those previously successful attempts.

And lastly, I cannot allow it to be stated, apparently on authority, while I have the whole heavy responsibility of its success resting on my shoulders, that I am a mere passive approver of the project of another, which in fact originated solely with me, and has been worked out by me at great cost of labour and thought devoted to it now for not less than three years . . .

The works had been commenced in the spring of the year 1854, amid the progress of the ship towards completion was eagerly watched, both by scientific men and by the general public.

The newspapers and periodicals of the day frequently contained descriptions of the work, and statements of the anticipated performances of the ship, often very much exaggerated. The writers seem to have been quite at a loss how to convey to their readers any idea of her size, and they generally attempted to do so by comparing her dimensions with those of some of the well-known streets and squares in London.[14]

In the beginning of the year 1855, the longitudinal and transverse bulkheads, which formed the main frame-work of the ship, were completed for nearly 400 feet of the centre portion and the plating was being fixed in place.

As the general design was now settled, it was thought that the time had arrived when it would be desirable for Mr. Brunel to describe at some length the nature of the undertaking, and the manner in which it was being carried out.

*Report to the Directors of the Eastern Steam Navigation Company*
[This Report was published at the time and excited much attention.[15] The paragraphs which describe the arrangements proposed for launching the ship have been omitted, as they will more conveniently be inserted in the following chapter.]

February 5, 1855

. . . . .

Although the simple description of the present state of the works of the ship and engines, and of what has been done during the last six months, may be summed up in a few words, I shall, in compliance with the request of the Directors, embody in this the substance of the several other reports which I have from time to time made to the Court of Directors during this half year, and take this opportunity of laying before the proprietors the fullest information upon our plans and proceedings. In doing this it maybe difficult to avoid some appearance of repetition of statements previously made; but I have thought it better, even at the risk of this, to refer to the objects we have had in view, and explain fully the nature of the works we have undertaken, and the manner in which we are carrying them out.

The construction of the vessel is the portion of our work which, without being actually novel, involved in all its details the greatest amount of special consideration and contrivance.

The unusual dimensions, the general form and the mode of construction of all the parts involved by these dimensions, the necessity of studying each part in detail, so as to obtain, by judicious mode of construction alone, the greatest amount of strength with the minimum amount of material; all these circumstances, and particularly the last, have rendered necessary a very large, though unseen, amount of labour in the preliminary plans and stages of the work; and, although I had for nearly two years before the contracts were entered into, devoted a great deal of time and thought to the subject, yet of course until the exact size of the vessel, and the general plans of the Company, had been finally determined upon none of these matters could be entered into in detail. Much time has consequently been required to mature and prepare these plans; and as I have made it a rule from the first that no part of the work should be commenced until it had been specially considered and determined upon, and working drawings in full detail prepared, and, after due deliberation, formally settled and signed, the work did not make at the onset that display of progress which might have been made, if less regard had been paid to establishing a good system which would prevent delays hereafter, and ensure a more perfect and satisfactory result. I am not prepared to say that the work is in that state of progress which will ensure its completion within the period fixed in the contract; but I am quite certain that if we had proceeded with less system we should have considerably delayed the final completion.

I shall now refer to a few of the principal peculiarities in the construction of the ship.

In the preparation of the detailed plans, I have carried out fully those principles which I originally described as leading features of the construction.

The whole of the vessel is divided transversely into ten separate perfectly water-tight compartments by bulkheads carried up to the upper-deck, and consequently far above the deepest water lines, even if the ship were water-logged, so far as such a ship could be; and these are not nominal divisions, but complete substantial bulkheads, water-tight, and of strength sufficient to bear the pressure of the water, should a compartment be even filled with water; so that if the ship were supposed to be cut in two, the separate portions would float; and no damage, however great, to the ship's bottom, in one or even two of these compartments would endanger the floating of the whole, or even damage the cargo in the rest of the ship, or above the main-decks of the compartment in question, and all damageable cargo would be stowed above that deck. Besides these principal bulkheads there is in each compartment a second intermediate bulkhead, forming a coal bunker, and carried up to the main-deck, which can, on an emergency, also be closed. There are no openings under the deep-water line through the principal bulkheads, except one continuous gallery or pipe near the water line, through which the steam pipes pass, and which will be so constructed as to remain closed, the opening being the exception, and the closing again being easy, and the height being such that, under the most improbable circumstances of damage to the ship, ample time would be afforded to close it leisurely, and to make it perfectly water-tight. I have also adopted the system, to be followed rigidly and without exception, of making no openings whatever—even by pipes and cocks—through the ship's bottom, or through the inner skin below the load water line, and I attach much importance to this system.

In the majority of cases in which steamboats are compelled to put into port from failure of bilge-pumps and other really trifling defects, no such serious consequence would have resulted, but from the difficulty and almost impossibility of remedying at sea any defects in the numerous pipes and openings now carried through the ship's bottom, wherever convenient, and without much regard to the danger of doing so.

I have found no great difficulty in carrying out this system completely, and the advantages, both as regards safety and the facility of remedying defects without delaying the ship on her voyage, must be obvious.

Independently of the security attained by the perfect division of the ship into really water-tight compartments of a sufficient number, so that the entire filling of one or even two of them will not endanger the buoyancy of the whole, the chances of any such damage as can cause the filling of one of them are greatly diminished by the mode adopted in the construction of the ship's bottom. The whole of the vessel (except the extreme stem and stern, the whole buoyancy of which is comparatively unimportant from the fineness of the lines), up to a height considerably above the deepest water line, is formed with a double skin, with an intervening space of about three feet. This arrangement resulted originally from the system of construction I adopted, in which the bulkheads, placed at intervals of twenty feet, form the main transverse frames or ribs 'of the ship, and in the intermediate space the material is disposed longitudinally in webs connecting the two skins, giving to the whole much greater strength with the same amount of material; but one of the most important results has been the great increased security attained, as the outer skin may be torn or rent against a rock without causing the ship to leak.

The space between these two skins is thus divided, by the longitudinal beams or webs and the principal bulkheads, into some fifty separate water-tight compartments, any one or more of which may be allowed to fill without materially affecting the immersion of the ship.

Besides the main transverse bulkheads, at about 60 feet intervals, there are two longitudinal bulkheads of iron running fore and aft, at about 40 feet in width, adding greatly to the strength of the whole, and forming, with the transverse bulkheads, being all carried up to the upper deck, fire-proof party walls, cutting up the whole into so many separate parts, that any danger from fire may be almost entirely prevented.

The transverse bulkheads being perfect, there being only one door—and that of iron—in each, at one of the upper decks, all currents of air or means of communicating fire may be completely cut off; and with an additional precaution, which I will refer to afterwards, besides the most ample means of supplying water, I believe that all possibility of danger from fire may be completely prevented.

All these principles of construction being kept in view, the details of construction—that is, the arrangement and due apportionment of the strength and sizes of all the plates, and the mode of fastening them—having been determined separately, the plates have been made at once of the required dimensions, and the work has proceeded systematically. This system is the most important, as securing not only good work, but affecting, to a much greater extent than might at first be supposed, the total weight of the ship; which, although the terms of the contract protect the Company against any excess of expenditure beyond a certain fixed sum, is yet of the greatest importance, as will be easily understood when I mention the fact that several merely trifling alterations in the modes of arranging the plates and other details have caused an economy of 20 to 50 tons each, and that the vessel may thus be made capable of carrying 200 to 300 tons more of coal, cargo, or provisions; or iron to the same amount may be usefully applied to strengthen other parts or effect useful additions.

The details of the engines have all been settled; and the principal parts, as already stated, are in an advanced state of completion.

In considering the plans of those engines, the largest that have yet been manufactured, I have endeavoured to ascertain what may be termed the weak points of the best engines hitherto constructed by the same or by other makers—those points in which experience has pointed out deficiencies—and to provide fully against similar defects in our case.

Before commencing the boilers, I have taken every means in my power of profiting by the experience of others, and have collected all the evidence and opinions as to the precise form and proportions which have been found most efficient; and particularly such as have been found best suited to the combustion of anthracite coal. A very great difference is found to exist in the useful and economical results of boilers, even of good manufacture. Some are noted for the power of producing rapidly abundance of steam, at the cost of great consumption of fuel; others have the opposite qualities, and some combine successfully both those qualities which are desirable. It might have been supposed that all points of such a simple subject would have been long since settled, and that no boiler would be made inferior to the best. Such is not, however, the case; and although the differences of construction are in themselves slight, the difference of result is often considerable.

I have taken some pains to satisfy myself on these points, and have endeavoured to select and to copy the most successful boilers; and in order to remove all doubts as to their fitness for the use of anthracite, I have made an experimental boiler, and, after numerous trials, determined upon the form and dimensions to be adopted.

In the consideration of these details, as indeed on all other points affecting the success of this undertaking, I have not hesitated to consult everybody whose opinions I considered valuable, and to bring the result of their opinions in aid of my own and the manufacturer's experience.

I have only to add, that after giving much consideration to the question of the diameter of the paddle-wheels and screw, I have determined them sufficiently for fixing the position of the shafts, and am now engaged in considering the best form and construction of the propeller itself, and also the construction of the stern-frame and rudder of the ship.

The position of the paddle-shaft, and the diameter of the paddles, have been questions of some difficulty. It being necessary to provide for a considerable variation in the draught of water, though not so great as with many existing large steamers, and to balance well the relative advantages of securing the highest average speed, at all the various draughts, or the highest speed at a light draught, and to combine as far as possible the two, so that the vessel may be as well adapted to perform comparatively short and very quick passages to ports not affording a great draught of water, as long voyages, heavily laden, at a more moderate maximum, but still a large average rate of speed. Although the full advantage of the great capacity of the vessel for carrying coal for long voyages would not be felt in a voyage, for instance, to New York, or in other short voyages, yet, unquestionably, she would exceed all other vessels in speed and extent of accommodation; and if it should be found desirable to make such voyages, your vessels ought to be able to command almost a monopoly by their superior capabilities, and I have therefore endeavoured so to place the paddle-shaft, and so to construct the wheels, that they may be adapted to the convenient application of the full power of the engine at a light draught of water and at very high speed.

As regards the screw, the same points have to be considered, and a choice made amongst the various forms and proportions more or less successfully adopted at the present time. I have always found the reports made upon the results of various forms of screws and propellers, and the performance of different vessels, so little to be depended upon, even when apparently made in good faith, and the results obtained from good authority, that I have been long since compelled to adopt no conclusion, unless from results witnessed by myself or by persons observing for me. I have for some time past availed myself of every opportunity that offered of observing and obtaining something like accurate results upon the various points affecting immersion of paddles or screw, and I am engaged in considering those results.

I have referred to the subject of protection from fire; it is one of considerable importance, and I have some hopes that a process, which has been recently patented by Lieutenant Jackson, may be successfully applied to rendering wood uninflammable. Some door panels have been already experimented upon with results which have induced me to pursue the experiments, and I am about to try the comparative inflammability of various qualities of wood, both prepared and unprepared; and if we can succeed in preventing the wood producing a flame, and thus communicating the fire; with the numerous metallic subdivisions we shall have in the ship, the spreading of fire, even from cargo or furniture, would become impossible. I am also engaged in determining the character and extent of mast and sail to be carried, as provision must now be made in the construction of the ship for receiving the masts.

The Directors are aware that I have been in communication with Professor Airy as to the instruments which may be used in such a ship, to ensure more accurate and frequent observations, and as to the nature of these observations; an enquiry into which he has entered with that liberality and desire to assist all improvements, in navigation especially, for which he is so well known. Several new instruments are now making for trials.

Sir W. Snow Harris has promised to turn his attention to the subject of the lightning conductors; and as soon as the iron work is a little more advanced, and while the form and position of all the principal masses are visible, the subject of local attraction and the adjustment of the compasses will be considered by those most competent to advise; and I am not without hope that the means of correction may be rendered much more certain and perfect than usual. I mention these, as some of the numerous points which require and are receiving attention.

In the last paragraph but one of this report Mr. Brunel mentions that he proposed to adopt a system for obtaining continuous observations in order to determine the position of the ship.[16]

His first letter to the Astronomer Royal, which is dated as early as October 1852, explains the object with which he began his investigations:—

October 5, 1852.

You may possibly have heard of a project in which I am engaged, of building some *very large* steamboats . . . Among the several requirements, which it appears to me are involved in such a large project, is one in which I hope for your advice and assistance, and I trust you will consider the subject worthy of your attention. In such a voyage, where so much depends upon perfect navigation, and with such a capital at stake, no means can be too perfect and no expense or trouble must be spared to ensure the constant determination, with the greatest attainable accuracy, of the ship's position and course. The determination of her speed, the effect of winds and currents, and the variation of her compasses, together with meteorological observations, are all involved in this; and I propose to have an observatory and establishment of observers, whose duty it will be to be constantly engaged, day and night, at any moment when anything can be visible in the heavens, in taking such observations as will determine more or less accurately, according to the nature of the observations that the state of the weather or other circumstances may admit of, the several points to be ascertained. Now, the questions to be determined are, what is the nature of the various observations that can best be made under all the different, and the favourable and unfavourable circumstances that may daily arise? what the instruments required, and what the character and extent of the staff of observers required; and even whether any, and if any what, new tables might be useful for such a purpose? The primary object being to be *constantly* determining either correctly or approximately the ship's true position, and in like manner *constantly* checking the compasses and giving her true course. I will not unnecessarily expose my ignorance in such matters by stating what I have assumed to be practicable. I will only remind you that experience proves that in a vessel of such size there will be great steadiness of motion, and therefore unusual facilities for accurate observation.

Mr. Airy entered very warmly into the subject, as did also Professor Piazzi Smyth, the Astronomer Royal for Scotland; and a long and interesting correspondence passed between them and Mr. Brunel. Experiments were conducted, principally with the view of obtaining a stand for astronomical instruments by means of the contrivance known as the gyroscope, the principle of which had been already adopted by Mr. Brunel in a level designed by him in 1829.

In connection with the observers' department, Mr. Brunel paid much attention to designs of sounding apparatus and means for accurately measuring the speed of the ship. He also intended to have a stream of surface water constantly pumped up through the observers' cabin, which should, by its change of temperature, immediately indicate the presence of icebergs, instead of the plan of an occasional bucketful being hauled up on to the deck according to the humour of the officer of the watch.

Mr. Brunel made a very curious contrivance for enabling the man on the look-out to keep his eyes open in a gale of wind. This consisted of two sets of vertical plates of tin placed one behind another, diverging from the direction of the wind, with a clear wide passage between the two sets of plates. The wind, entering at the end of the apparatus, became separated by the first two inclined plates, and the residue that passed on in the direct line was again subdivided, so that at the end of the last set of plates there was no rush of air between them, and a man looking through the aperture, with his face to the wind, was in a perfect calm. This was a useful arrangement, the look-out man's eyes being as well protected as though behind a glass. A glass would not answer the purpose, as it would become obscured with spray.

The unfortunate circumstances which attended the completion of the ship prevented the introduction of these and many other arrangements which Mr. Brunel had originally proposed.[17]

In November 1855, the Directors proceeded to appoint a commander to the ship; and their choice fell upon Mr. William Harrison, one of the most distinguished of the captains of the Cunard Company's steamers.

This appointment gave Mr. Brunel great satisfaction; he found in Captain Harrison a warm and faithful friend, and an able adviser on all matters connected with the completion and equipment of the ship.[18]

Before they came to a decision, the Directors asked Mr. Brunel to communicate to them his views on the considerations which ought to influence them in their choice.

The memorandum he drew up on the management of the great ship, and a letter on the duties of the chief engineer, are inserted at length, as forming a complete record of the principles which, in his opinion, should be followed 'in the use of this new machine,' while employed on the voyages which she was specially designed to make.

### Memorandum on the Management of the Great Ship

October, 1855.

The question of the principles to be followed in the use of this new machine, for such it must be considered, and the character and the qualifications of the man to whom it is to be entrusted, and the points to which his attention must be particularly devoted, have long been subjects of deep and serious consideration with me.[19]

And the difficulties are complicated by the consideration that public opinion, and those established and universally received opinions which, being the result of experience, cannot be called prejudices, have to be considered, and must be yielded to to a great extent, even when somewhat opposed to sound reasoning, or to conclusions deliberately formed upon a due consideration of the novel circumstances under which we have to act; but, while conceding much to past experience, and to preconceived opinions, the real requirements of the case must have the first consideration.

Thus, as it is a ship that we have to navigate, public opinion, if nothing else, requires that we should have a sailor of undoubted experience and skill to command her; but although seamanship may be one of the essential conditions or qualifications for the commander, I think we shall find, upon examination, that the work we have to do, and the duties to be performed, and the qualifications we consequently require in the man who is to conduct this work, are so peculiar that mere seamanship, in the ordinary sense of the term, if it were to be solely relied upon, would be even a disadvantage rather than otherwise, and that we require much that is not necessarily found in the most perfect specimen of a seaman, and much that is far more difficult to find than the ordinary qualifications.

I wish particularly to impress this upon you: the subject has occupied long, frequent, and serious deliberations. I have come to the conclusion which you will no doubt readily acquiesce in, that our first practical mechanical success, upon which so much of our financial success hangs (although others might reap the benefit even of our failure in the first instance), will depend mainly upon the skilful management in our earliest voyages of the machine we are about to set afloat; but I have also come to the conclusion, the correctness of which may not be so immediately apparent to you, that this machine, though nominally a ship, not only admits of, but requires, a totally different management from that which may have successfully navigated ordinary ships, and that most of the habits, feelings, and sensations, amounting to instinct, possessed by a good sailor, and the peculiar power of skilful adaptation of expedients to emergencies, which constitute the merits of a first-rate sailor, have to be put aside for a time rather than applied in the first instance to make a successful commander of the vessel.

In navigating a small vessel, a man has to study the appearance of the weather, the direction of the wind, and that of the seas; to consider the probabilities of change; to vary his course more or less with deference to any of these causes, and according not only to the then state of all these operating causes, but also with reference to the probabilities of change; and with their consideration his mind is principally occupied, and it is in the exercise of judgment and foresight in these points that he shows his skill.

Steam navigation, and the gradual increase in the size of vessels, have no doubt materially modified this, but much still remains to occupy the mind of the commander of a steamer, with which not only we shall have nothing to do, but too much attention to which, from previous habits, would divert his attention from that which, under the different circumstances, will become much more important.

In the same manner, in manœuvring such a vessel in harbour, the ordinary modes would be totally inapplicable; whereas, by entirely discarding all previous practices, and keeping in his mind and making use of the peculiar powers which are at command in this vessel, and judiciously attending to the immense effect of the *vis inertiæ,* or momentum of such a mass, so far beyond that of anything now afloat that it becomes a totally *new* influence, this machine may be managed by a skilful man with a facility which is not attainable with small craft; while such a mass would be destruction to itself or to anything that comes in contact with it, if treated at all in the same manner as even the largest steamers are now handled.

Another peculiarity to be attended to in the management of this mighty mass is that by no possibility must she be allowed to touch the ground.

Our ship will be found, I believe, if tried, to possess unusual strength: no combination of circumstances within the range of ordinary probabilities can cause such damage to her as to sink her, even if she were to be run on sharp rocks with deep water around; and I believe she might remain for months aground, exposed to the heaviest seas without serious damage. The lives of the passengers, and even the cargo generally, will be safe, almost perfectly safe, and so far she will be very different from and very superior to any vessel now afloat; but there will also be another great difference, although she may remain safely aground, she will probably remain so for a long time.

If she were to ground at high water or where there is a little tide, and when she happens to be rather light of coal, and should take in any water so as to deepen her, no ordinary appliances in the power of the crew will get her off, or, at all events, not without great loss of time, and at very heavy sacrifices. Such an event in the first year of her career would probably ruin our Company, as the grounding of the 'Great Britain' did the Great Western Steam-ship Company, although it proved the good qualities of that ship, and thus advanced iron ship-building. A sailor will no doubt receive with disgust and as an insult the suggestion of the possibility of his running his ship ashore. We cannot afford, however, to rest contented with this expression of feeling, natural and praiseworthy as it may be: our circumstances are totally different; and that which is considered certainly as a serious evil, but still only an evil more or less serious, would be to us fatal; it is no longer a question of degree of injury—it is death. I have had some costly experience; I have had to do with many steamers—several remarkable for their size and their value, according to the ideas of the time, and two of them each in their day considered Leviathans, more wondered at than even ours, and exciting much more anxiety on the part of their promoters, and there was as much care taken in the selection of commanders as I ask you to take now; nevertheless both of these, to the ruin of the Company in one case, and almost every other steamer I have had anything to do with, has not merely touched, but been aground. Not a season hardly passes without a case of even a Queen's ship going ashore, although navigated by men educated for the purpose, with the advantage of every appliance hitherto thought necessary and sufficient, and with fearful responsibility attached to them. All these are published cases known to the world; and if the log books of all the steamers were searched, I believe the result would be a list in which the ships that had not been aground would form the exceptions, and, in the majority of cases, the blunder is so gross, that the most far-fetched excuses of errors of compass and unknown currents, although in well-known channels, are obliged to be invented.

The real explanation is a simple one. Ships are navigated with far too reckless a confidence in the mere personal instinct and skill of those in command, and in their ability to get out of a scrape in time. Methodical systems and mechanical means of ensuring accuracy are far too much neglected, or rather have not kept pace at all with the great improvements in speed and the power of locomotion which science has introduced, whether in the construction of steam vessels or even of sailing ships, and which the advance of the day now calls for. There are no means at present used of taking soundings worth having while a ship is going fast through the water, or in time to be of any use if the soundings shoal rapidly; the trouble is so great and the operation so slow that it is not resorted to sufficiently often. The steering by compass is so rough and so coarse, that no real accuracy is attained. Good observations to determine position are taken at too long intervals of distance run, and consequently are subject to be interrupted for too long a period by what is called and supposed to be continual bad weather; although it would rarely happen, if persons were continually on the watch, who had nothing else to do or to think of, that twenty-four hours would pass without some glimpses of a star or of the sun. Thus with a speed of ten or fifteen knots, and cloudy weather, and, above all, with that unfortunate confidence of seamen to which I have referred, there is never any certainty just at the time when it is really required. No doubt, skilful seamen do generally arrive at an astonishing approximation in their estimates, and the results on the average are most successful, and remarkable proofs of the skill brought into action. But the instrumental means of attaining accuracy are lamentably in rear of the improvements that have been made in the means of locomotion, and have not at all kept pace with the vast increase of capital embarked in each individual case.

All these deficiencies are, however, easily remedied; the means do exist, or can be devised, and might be applied if we insisted on their application. We have seen numerous instances where the difficulties are admitted and grappled with systematically, and with what patience and skill a fleet has been piloted up unknown rivers full of shoals in China, and through intricate channels with covered rocks in the Baltic. In our case, the object to be attained is a vital one to us; and what I most dread is the confidence of our commander resulting from his previous experience preventing his appreciating the peculiarities of the case, and applying that greatly increased amount of method and system which is essential to change that, which is now rendered only highly probable by the skill of man, into mechanical certainty. The man who takes charge of such a machine, in which is embarked so large a capital, must have a mind capable of setting aside, without forgetting, all his previous experience and habits, and must be prepared to commence as an observer of new facts, and seize rapidly the results. A man of sense and observation, with a good mechanical head, and with decision and courage, would succeed without much previous nautical knowledge; but, unquestionably, a man familiar with all that is going on around would be much more competent, provided he does not allow his habits derived from former experience to induce him to neglect any of the new means of information in his power, to all which his former knowledge should be made subservient.

If much has to be unlearnt, or rather carefully set aside for the time, as generally inapplicable in its present shape, there is also much to be learnt in the navigation of such a vessel; and one of the most essential qualifications of the commander would be a belief in his want of experience, and a readiness to see the novelty of his position, and a cautious and sound but quick perception of the new and powerful influences and *new* effects of those same causes with which he may have been familiar under very different circumstances.

The great mass and size of this vessel must necessarily render it so much less affected by the ordinary disturbing causes of wind and sea, that, practically, little attention ought to be paid to these, at least not as expecting the ordinary results; but other effects may be produced which must be carefully studied and learnt in the early voyages, so that they may be met and counteracted as they have been by experienced men under the present system. Thus, as an example of this class of effects to which I would refer, there is reason to believe that this vessel, although apparently unaffected by a wind on the beam, or a cross sea, which would be very noticeable even with the largest of the present steamers, would nevertheless have a steady and rather strong tendency to come up to the wind. A certain direction of a cross sea is not unlikely to produce a contrary effect. It is quite possible also that the apparent lee-way, or the deviation of the axis (or keel) of the ship from the line of the course, will be greater with the wind on the beam than with smaller vessels, although the actual drift or lee-way may be less; and all these effects will result from her great length (unless counteracted by causes I do not now foresee), and will probably be totally different in degree from any similar effects now felt. These effects must be carefully studied, and with a mind prepared to consider them as new, for they will be new and distinct from anything now experienced. In the same manner, and to a much greater degree, the effects of the speed which we must hope to attain will upset all the usual methods of determining accurately the position and the course of the vessel, and still more the precautions to be taken in approaching land. A 24 hours' run, at 20 miles per hour, without a good observation, and with a possible error of estimate, or a doubt, at all events, of the exact effect of the set of currents, or of the speed of the ship through the water, or of the precise amount of lee-way, may easily make an error of 20 or 30 miles in position at the end of the day's run.

On the other hand, by almost abandoning the present modes, and adopting measures which I shall point out presently, adapted to the new circumstances, a much greater degree of certainty and accuracy may be attained than is now even sought for. I have had the best advice on these points, both astronomical and nautical: the Astronomer Royal, Mr. Airy, who both as a man of science and a practical man, and by his official position, is the first authority on such matters, and Professor Smyth, of the Edinburgh Observatory, Captain Beecher of the

Admiralty, and several other scientific naval men, have assisted me; and, under their advice, instruments are being constructed, and a system devised, which will admit of continuous observations being made with great accuracy at all times of the day, and particularly at night, and by means of which also a continuous correction of the ship's compasses with the most minute accuracy can be kept up, and her position be ascertained with almost the same accuracy as a point on land; and I have their authority for saying that such improved means are desirable, and are attainable.

The importance and advantage of continuous observation throughout the twenty-four hours, whenever a glimpse can be obtained of any object which will answer the purpose, and even when the horizon is invisible, may not at first be evident, but it cannot be over-estimated. An exact knowledge of the ship's movements will be soon acquired that would almost replace observation, and the average of numerous observations secures an accuracy which cannot be approached by any other method.

By a careful and continuous comparison of the exact distances run, with accurate records of the speed of the engine and of the ship through the water, registered by good instruments (the performance of which has been tested, and which are being made), and having proper reference to the varying draught of water of the vessel, the most precise measure of the speed of the vessel through the water can be learnt; and by careful observation of the force and direction of the wind, the effect of any given wind, and an accurate measure of her drift or lee-way, may be obtained; and then, by adhering strictly, without regard to wind or weather, to a course previously laid down on good charts, having upon them the sets of currents, as already observed, the true position of the ship may be determined at any moment, and there need never be the slightest hesitation, or any time or distance lost by doubts.

In addition to the instruments above named, I hope to be able to provide the means of sounding at moderate depths while the ship is going her full speed.

Having the means, then, of determining the ship's course and position with much greater accuracy than is now practised, we must seek to make that course, or the distance to be run, the shortest possible, that is to say, the course which will occupy the least time.

The exact course to be taken by such a vessel must be determined upon and laid down upon the charts after a due consideration of all the circumstances which can possibly affect the time occupied in the voyage between any two points, and particularly by examining and considering well all the information which scientific observers have collected and recorded as to the currents, probabilities of fogs, or of ice, and other impediments, the chances of meeting vessels, which must be avoided as much as possible, the average direction of prevailing winds, varying such course, perhaps, in certain latitudes with the season of the year, but certainly not with the temporary direction or force of the winds or state of weather at the moment; for the speed which we shall attain renders these causes of secondary consideration, and would interfere with all ordinary calculations derived from present experience of the probable state of the wind in the new position which the ship would attain after a few hours' run. The business of the commander would be, therefore, to adhere rigidly to the exact course previously deliberately determined upon, and not to be tempted to deviate except slightly, and even then only according to rules which he shall have previously laid down for himself.

The importance of adhering strictly to this rule cannot be over-estimated. The period occupied in a voyage will be materially influenced by the exact course followed, and as such course must be determined upon only by a calm and somewhat laborious study of documents, it cannot be safely determined upon except in the closet, that is, *before the voyage*, and uninfluenced by the excitement of hopes or by disappointments caused by difficulties.

The safety of the ship and of the enterprise may also mainly depend upon the rigid adherence to this rule.

With the means at his disposal for checking that course, and with the method and regularity involved by a rigid adherence to it, and a correction at once of any departure from it, a degree of certainty will be attained as to the ship's position which will almost preclude the possibility of an accident, and which I consider practically invaluable as affecting the safety of our vessel. This rigid adherence to an exactly prescribed course will prevent any risk of that which is the cause of the greater number of losses. No inducement of fine weather, or a happy state of mind and body of the captain, or a desire to save time, or to show the beauties of a coast, or any other temporary cause, will lead to a nearer approach to dangerous points than had been previously determined upon as safe. The shortest and best route, the safest from all dangers, and giving the widest berth to all shores or shoals, having been laid down, must be kept to against all temptations.

Economy of fuel is another consideration of the highest importance. The engines are of power sufficient, if fully worked, to consume considerably more than the cargo provided. Large as may be the supply which the ship is capable of carrying, it is no more than that which is calculated as necessary for the voyage intended, with a moderate allowance for contingencies; and it must be borne in mind that the ordinary means of making up any deficiencies would be totally inapplicable in our case, since no supply capable of being drawn in an emergency from other sources would be sufficient.

The precise quantity calculated and provided must be made sufficient, or the consequences would be serious. The usual practice, therefore, of going ahead as fast as the engines will take the ship must be entirely abandoned. Careful observations, systematically pursued, will show the speed which, under different states of immersion, can be attained without a disproportionate expenditure of power, and to this speed the engines must be limited. A little experience will probably show that, under certain circumstances of immersion of the ship, or of the state of the sea, it may be economical to force the screw or the paddle engine, the one more than the other. This can only be determined by somewhat delicate and very precise and accurate observations. These must be made. It is not merely a question of degree, but our whole success depends upon the application, not of one, but of all these refinements.

A close study of the relative speed of the ship, as ascertained by the self-registering logs, and by continuous astronomical observations, compared with the expenditure of power in the engines as indicated by the number of revolutions constantly registered, and the power expended as measured and recorded by the chief engineer, will be necessary for some time to come, and until indeed the management of such large ships becomes as much a matter of rule as it is now, or perhaps ought to be more than it is, with smaller ones.

As to the use of the sails, while the engines are nevertheless in full action, it must be entirely a matter of experiment, and an experiment in which, again, all previous habits and prejudices must be set aside.

Whether a sail steadies such a ship usefully or not can be, and must therefore be, positively determined by measurement with proper instruments, and not by the sensations; and the result upon the speed of the vessel as compared with the power expended must, in like manner, be ascertained by positive observations and measurements. No past experience can do other than mislead. It is quite possible that the same means which improve the rate of the present large steamers may be prejudicial to our performance.

The commander must appreciate the necessity of all this study and attention to what is rather mechanical than nautical, or our voyage will be a failure. The difference between the peculiar qualities of such a floating mass as compared with those of the largest steamer now afloat, is likely to be as great as between the last and a 100-ton cutter; and the most prejudiced believer in the acquired skill of an old sailor who had learnt to manage the small vessel in the most perfect and masterly manner, would not expect him to be able to handle one of the present large steamers, still less to elicit the best performance out of her. In the same degree the man who takes the command of our ship must, if he is to succeed, enter upon his duties with a belief that he has nearly all to learn, at the same time that this feeling is perfectly consistent with a proper confidence in his own powers to master the new circumstances, and to succeed with this as he may have done with other vessels.

Finally, the commander's attention must be devoted exclusively to the general management of the whole system under his control, and his attention must not be diverted by frivolous pursuits and unimportant occupations. I believe that even in the present large steamers much advantage would result from relieving the captain from all care of the passengers and cargo; but in our case, where we may have to provide for thousands instead of hundreds, and arranged in different classes, and living in completely separate saloons and compartments, the present system of a captain dining at the table and associating with the passengers would be impracticable, even if it were desirable. But for much more important reasons, and on general grounds, I think that while the commander is of course supreme over every department, he should not be embarrassed by undertaking any one, still less should he have his mind occupied with the troublesome and frivolous concerns of a vast hotel, nor should he be hampered by the necessity of attending to the hours and the forms of a large society. Moreover, I consider it essential that he should by his presence and control keep up the position and the sense of responsibility of the chief officers under him, by living and messing with them; the commander and those acting immediately under him must occupy a more dignified position than they now do.

The result of all these general views is, that the command of this ship must be considered to consist mainly in the superintending and keeping up in a high state of order the perfect working of a highly methodical pre-arranged system, by means of which the ship is to be made to go like a piece of very accurate machinery, precisely in the course which has been pre-arranged, and precisely at the speed, and with the consumption of power, which has been ascertained to be the highest attainable with the requisite economy; and there must be a proper establishment of assistants, competent to control each department of this system. As regards the constitution of this establishment, I consider the commander should have a staff of chief officers or captains. I believe three will be necessary, with a fourth performing general duties and ready to take the place of any one of the three; that one of them should always be in command of the ship (under the commander); that, beside these chief officers, there should be a master, corresponding to a master in a Queen's ship, who would have assistants and calculators or clerks, whose duties would be to keep the ship's reckoning, to keep up perfect and continuous observations, to calculate with precision and set from hour to hour the exact course by compass which has to be followed, to keep in the course determined upon by the captain, to keep a series of accurate observations and records of all matters that can affect the ship's movements—duties involving an amount of science and practical astronomical and mathematical knowledge which requires a superior education, and which is found only in this class of men. The duty of the master would be therefore to supply the science necessary for the conduct of the ship, and to be the commander's cyclopædia and book of reference, to be able at any moment to report to the commander the exact position of the ship and her course, and the variation of her compasses, and take the soundings, if any, to note the fact of a change in the temperature of the water, indicating approach of ice, and any other symptom or fact which can affect the ship's movements—all which should be determined by continuous observations methodically and mechanically made, and not be dependent upon the chance of the commander's anxiety or greater or less forethought. The chief engineer should also be a superior man, selected more for his general qualifications as a good director of men and machinery, than as a mere marine engineer. These should form a staff, and be of a standing to live and mess with the commander; so that each department should thus be furnished with a chief competent for the special duties of his department, and reporting to and acting under the general control of the commander.

That the principles thus laid down as to following exactly a prescribed and predetermined course, and as to regulating exactly the consumption of power and consequently of fuel, and the keeping up a system of what may be termed scientific observation for the purpose of ensuring this regularity, I submit should be rigidly enforced; and the commander should be required to adopt these principles as the guide of his conduct, and to use the measures that are placed at his disposal for working this machine in the manner and with the precision pointed out.

### Letter on the Duties of the Chief Engineer

March 19, 1857

The duties which I apprehend will devolve upon the chief engineer of our ship will be, firstly, the supreme direction and management of both the principal engines, and all the auxiliary engines and machinery worked by them in the ship; and as the construction of such a ship, and of many of its adjuncts, such as the iron masts and yards, steering apparatus, and other parts, are strictly of an engineering character, and such as in the event of repairs, particularly at sea or in foreign ports, would require a mechanical engineer rather than a shipwright, I think it must be made part of your duty, as the most competent officer, to make yourself thoroughly acquainted with the construction of the ship and all its parts, and all mechanism within it, so as to be prepared to take such share of responsibility as to the state of the structure of the ship as it may be found desirable to throw upon you as chief engineer, and at all events to be prepared to be the captain's chief authority and responsible adviser on all matters of mechanical engineering.

The principal duty, however, will of course be the management of the engines, including the care of the paddle, screw, and other machinery immediately connected with the engines, or worked by them; and as the success of the ship as a steamboat will depend entirely upon the amount of power developed by the engines, in proportion to the fuel consumed, there is no limit to the degree of attention, of judgment, and of skill, that not only may be usefully applied, but that must be applied to ensure success.

I have no wish to alarm you as to the amount of work or responsibility that will devolve upon you; my object is rather to show you the opportunity afforded of displaying judgment, skill, and assiduous attention, and thus, as I hope, to excite your ambition, when I seek to draw your attention very strongly to the peculiarities of this case.

In ordinary steam navigation, whatever perfection has as yet been sought for or attained, the business of the engineer has been mainly to keep the engines in perfect order, and to develop the greatest amount of power possible, and, secondly, to effect as great economy as possible in the consumption of them; but the latter has been merely a question of economy in a pecuniary point of view, and not of necessity, and has been entirely secondary to the first condition, so much so that the most successful ships have not been by any means the most economical, on the contrary, they have been rather extravagant consumers.

In the present case, the circumstances are totally different, and it will be essential that you should change altogether your accustomed views on this subject.

This ship is built to go round the world with a defined and limited amount of fuel, which you have no power to exceed, or rather, if you exceed it, at any part of the voyage, the whole is a failure. The circumstances are therefore reversed, with this additional condition, that there is no medium or partial success. In the ordinary cases you have a limited power of engine and an excess of fuel, in which it is desirable but not essential that you should effect economy; in the present case you have a limited and defined quantity of fuel to consume, with an excess of engine power, and the art will be to obtain the largest total amount of power from this fuel, expending it progressively, and in such a manner as to reach a given point. To effect this, and obtain the best possible results, will require of course that the engines should be kept in the best possible order; but this, although a preliminary condition, is an ordinary one, requiring no peculiar duties or exercise of judgment, and must be assumed, as on all occasions, a matter of course. The peculiar duties in our case will be the continuous study in every trifling detail that can effect the result of the means of obtaining the largest amount of steam from the defined expenditure of fuel and the use of this steam, so as to obtain the largest amount of power, and the largest amount of result.

The mere study of this question must necessarily occupy some time, and for several voyages it will be a subject of experiment; but the more rapidly positive information can be obtained, the more prompt and certain will be our success. To attain success will require a degree of attention to every minute detail, which it has never yet been necessary or profitable to devote to this branch.

The continuous weighing out of coals and measure of the products of each boiler (for which means will be provided), the continual observation of the extent to which blowing-off is desirable, the continuous measure by indicator of the performance of the engines under different pressures and degrees of expansion, so that you can at all times furnish the captain with the exact performances of the two engines, and the cost of fuel required to produce given results in each; so that he may have the means of comparing your expenditure with the results he obtained in the speed of the ship, and of learning the relative beneficial effects of employing more or less the paddle or the screw in different states of weather, or different immersions of the ship, will be required; and every method of increasing the performance of each gang of stokers, and of stimulating their skill and care, and every refinement in each separate branch of the work, to effect economy of fuel, or rather, development of power with a given amount of fuel, will be necessary. We all know full well how, if every effort is continuously made and every possible care is continuously bestowed in each department, 4 or 5 per cent can be saved or gained in many points, and at many times in the 24 hours, between the drawing of the coals from the bunkers and the development of power at the paddle board or the screw blades; and if only 1 per cent, can be thus gained in a few points, the aggregate will soon amount to 10 per cent, which with us may make the difference between success and failure. All these things will require judgment, thought and attention, rather than labour, and, above all, close watching and method, and good management of men.

Besides these more than ordinary duties during the voyage, the only peculiarity in the service will be that, with such a costly machine, the mere interest of money and fixed expenditure upon which will not be short of £200 a day, and the perfect state of which is so essential, you will be required to give more attention to the machinery when in port than is usually required.

I trust that this strong but not exaggerated statement of what would be expected of the chief engineer, will excite your desire to undertake the duties, rather than deter you from seeking the post; and that, if the Directors should accept the offer of your services, you will enter upon the duties with confidence, though with a sense of their serious importance. In the event of your appointment, it will be a necessary condition that you should be able to commence at once the supervision of the erection of the engines. This work is already much farther advanced than I should have wished it to be before the chief engineer had taken charge of it. I attach great importance to his having that familiar knowledge of all the parts and their condition, which no study of drawings can give so well as actual inspection during erection, and I wish also that he should satisfy himself of the perfect truth of every adjustment.

It will be seen from the documents which have been printed in this chapter how 'deeply and seriously' Mr. Brunel had considered all the conditions which were, in his opinion, necessary for the economical construction, and the successful employment, of the great ship; but it is hardly possible, by means of extracts from his correspondence, to convey an adequate impression of the amount of labour he expended—from the year 1852 to the last days of his life—on the supervision of every detail of the work. 'The fact is,' he said, 'I never embarked in any one thing to which I have so entirely devoted myself, and to which I have devoted so much time, thought, and labour, and on the success of which I have staked so much reputation.'

Heavy as Mr. Brunel's duties were in October 1854, when he wrote these words, a far greater amount of labour was subsequently imposed upon him.

During the year 1855 financial difficulties arose which interfered with the progress of the ship; and at last, in February 1856, although Mr. Brunel had done everything in his power to prevent the necessity of such a step, the works were suspended; and they were not resumed till the end of May, after which date they were carried on by the Company under the supervision of the existing staff. It was greatly against Mr. Brunel's wishes that this was attempted, except as a temporary measure, as he considered it impossible for a company to carry on such a work efficiently and economically.

Notwithstanding all these difficulties, that which seemed at first only a confused mass of iron assumed by slow degrees the graceful proportions of a 'great ship'; and the hull of the vessel was completed by the end of the summer of 1857, so far as it was desirable to proceed before the commencement of the launching operations.

1. The letter to Mr. Guppy of August 1843 (given above, p.193) contains a reference to some of Mr. Brunel's ideas upon iron shipbuilding, subsequent to the date of the design of the 'Great Britain.'

2. Extract from a memorandum by Mr. Brunel (dated February 25, 1854) on the early history of the great ship.

3. So Mr. Brunel wrote to the secretary when excusing his own absence from the meeting on account of illness.

4. Mr. Brunel concludes this report by indicating the best mode of contracting for the construction of the ship and engines, and suggests that an examination should be made of the river Hooghly.

5. From the great length of this report it has been necessary to omit the less important paragraphs.

6. The passages which are omitted refer to the relative cost and merits of the different tenders.

7. Two years later Mr. Brunel spoke of the success of the screw engines as 'a most grave question upon which hangs I fear to think how much.'

8. Mr. Geach was one of the most zealous friends of the great ship. His death, within a year after the commencement of the work, was the first of her many misfortunes.

9. Stringent conditions as to jacketing were introduced into the contracts; but they were not enforced by Mr. Brunel, in deference to the strong objections urged by the makers. He much regretted this concession.

10. The destination of the ship as proposed at this time was Australia.

11. The power of blowing off the steam without noise is of great importance, and Mr. Brunel made many experiments on the point. The noise of the steam blowing off when the engines of a ship are stopped for fear of an impending collision or other accident, is so great, that it is almost impossible to hear any orders that may be given.

12. The screw propeller of the 'Agamemnon' was broken in a trial trip outside Plymouth Sound, on November 9, 1853; and the result was that, from not having efficient governors, the engines flew round at an alarming speed.

13. Mr. Russell's works. The ship was built in a yard immediately adjoining them.

14. See, for example, an article in the *Quarterly Review* of March 1856, entitled 'The Triton and the Minnows.'

15. It must be remembered that this report describes the ship as she was at that time designed by Mr. Brunel.

16. Extract from memoranda of October 5, 1855:—

    'By constant observation, to lay down position and course of ship, and correct compasses.

    To record speed of ship through water by logs.

    Revolution of engines.

    Force and direction of wind. Draught and trim of ship, sails carried, &c.

    Temperature and peculiarities of sea water.

    As the result of these observations, to plot down hour by hour the position of the ship, to compute the speed, variations of compass, the direction and force of current, and true direction and force of the wind.'

17. Mr. Brunel also proposed to have charts prepared for the route of the great ship to Australia and Calcutta, similar in character to some which were made under his directions for the route of the 'Great Western' to New York, which he described as follows:—

'When we started the "Great Western" to New York, I had a chart drawn and engraved of the sea (that is, the lines of latitude and longitude, and the bearings of the compass, and the coast and soundings) on a cylindrical projection of a great circle from Bristol to New York; and we found it very useful for the captain to see his great circle sailing, and to see how much he was deviating from it.'

18. Captain Harrison's command of the 'Great Eastern' after she left the river was unhappily of but short duration, as he was drowned by the accidental upsetting of his boat in Southampton Water on January 21, 1860, four months after Mr. Brunel's death.

19. With this paper Mr. Brunel enclosed a letter, in which he points out with great earnestness the responsibility of his own position.

After calling the Directors' attention to the important step they were about to take, he proceeds:—

'I am not pointing out a danger without being prepared to propose a remedy. The same man who, after he has been selected and appointed on account of his previous character as a sailor, and as an experienced naval man, would probably feel disposed to reject advice coming from those who do not profess to be sailors, and to resist directions which might appear to him as trenching upon his authority, or as implying doubts of his ability, could have no such feeling (if he is a sensible man and fit for the position) if his attention had been drawn to these views before his appointment, and if he accepts the trust reposed in him on the understanding that he is expected to pay attention to these opinions, and if, as I shall urge upon the Directors to ascertain, he entertains no objection to the adoption of them, and agrees to follow the principles of action which I hope to induce the Directors to adopt as rules in the navigation of the ship.

'I propose therefore to lay before the Directors the result of my anxious consideration of this subject, to urge upon them the adoption of my views, and, if they adopt them, to urge that they should make it a condition in the selection and appointment of a commander that such views are approved of and adopted by him.

'This is a strong and plainly stated request, but not more strongly put than I feel that the occasion requires.

'I have an immense stake in the success of this enterprise. I do not refer merely to my pecuniary investment; but, as affecting my professional reputation, my stake is much deeper; as, although I was accidentally led by circumstances into proposing the plan we have adopted, and the Company was not originally formed to carry it out, and although the plan when proposed was well weighed and considered by men competent to judge, at all events, upon the prospects as a commercial speculation, and although it was adopted by them, and therefore they must share in the responsibility, and although many may share with me in the credit of our success;

yet there is no doubt that I should have to bear solely and very heavily the blame of a failure. On this ground alone, therefore, the Directors, I am sure, will willingly allow me to urge my views strongly, and will excuse the length at which I do so. But I shall rely upon satisfying them that my views and opinion should command their concurrence on their own merits; and with this preface, which has already reached an undue length, I will lay before them a paper on the subject, most of which has been written for some time in anticipation of the present circumstances, and having been thus written at different periods is, I fear, somewhat disjointed, and wanting in arrangement, and therefore much longer than it might have been.'

# XII

# THE 'GREAT EASTERN': THE LAUNCH

THE mode in which the great ship was to be launched had necessarily to be determined before she was commenced. In May 1853, when the contract for her construction was entered into, the question was left open, and the contractor was either to launch her, or to build her in a dock 'if it be found preferable.' With Mr. Brunel's full concurrence, Mr. Russell determined to build the ship on the river-bank, broadside to the river.

The reasons which led to this determination were fully described by Mr. Brunel in his report of February 5, 1855. This report has, with the exception of the parts relating to the launching operations, been printed above, p.236. The passages there omitted are as follows:—

One of the first points to be decided was the mode of launching the vessel, which of course would determine the position in which it was to be built; and I wish to take this opportunity of explaining my reason for adopting the plan I have decided upon, which, being unusual, might be supposed to be unnecessary.

Vessels are generally built above the level of high water, and then allowed to slide down an inclined plane into the water; occasionally, as in the case of the 'Great Britain,' they are built in a dry dock, into which the water is afterwards admitted, and they are floated out.

Both plans were well considered in the present case; but the size of the dock required, the difficulty of finding a proper site for such a dock, the depth required for floating a ship with her engines and boilers, which it was most desirable to introduce while building the hull, and the depth of channel required to communicate between such a dock and the deep water of the river, all combined to render the dock plan a very expensive, and, considering the nature of the soil in which it would have to be formed, a somewhat hazardous proceeding. Launching seems to offer the fewest difficulties and the greatest certainty; but the dimensions of the vessel required some modifications of the usual modes of proceeding.

Launching is generally effected by building the ship on an inclined plane, which experience has determined should be at an inclination of about 1 in 12, to

1 in 15, the keel of the ship being laid at that angle, and the head consequently raised above the stern, say one fifteenth of the whole length of the ship. In the present case this would have involved raising the fore part of the keel, or the forefoot, about 40 feet in the air, and the forecastle would have been nearly 100 feet from the ground; the whole vessel would have been on an average 22 feet higher than if built on an even keel.

The inconvenience and cost of building at such a great height above ground may be easily imagined; but another difficulty presented itself which almost amounted to an impossibility, and which has been sensibly felt with the larger vessels hitherto launched, and will probably, ere long, prevent launching longitudinally vessels of great length. The angle required for the inclined plane to ensure the vessel moving by gravity being, say 1 in 14, or even if diminished by improved construction in ways to 1 in 25, is such that the end first immersed would become water-borne, or would require a very great depth of water before the forepart of the ship would even reach the water's edge. Vessels of 450 or 500 feet in length would be difficult to launch in the Thames unless kept as light as possible; but our ship could not be so launched, the heel of the sternpost being required to be, as I before said, about 40 feet below the level of the fore-foot. Some mitigation of the difficulty might be obtained by an improved construction of the ways; but the great length of ways to be carried out into the river would, under any circumstances, be a serious difficulty.

These considerations led me to examine into the practicability of launching or lowering the vessel sideways; and I found that such a mode would be attended with every advantage, and, so far as I can see, it involves no countervailing disadvantages. This plan has been accordingly determined upon, and the vessel is building parallel to the river, and in such a position as to admit of the easy construction of an inclined plane at the proper angle down to low-water mark.

In constructing the foundation of the floor on which the ship is being built, provision is made at two points to ensure sufficient strength to bear the whole weight of the ship when completed. At these two points, when the launching has to be effected, two cradles will be introduced, and the whole will probably be lowered down gradually to low-water mark; whence, on the ensuing tide, the vessel will be floated off. The operation may thus be performed as slowly as may be found convenient; or, if upon further consideration more rapid launching should be thought preferable, it may be adopted.

I have entered at some length into an explanation of all the reasons which led to the adoption of this plan; as I am anxious that they should be known, and particularly that it should be well understood by the proprietors and those interested in our success, that I am not adopting any novelties; unless, so far as those modifications of the more usual practices which experience points out as necessary to meet the peculiarities of a particular case may be deemed such.

I should add that the necessity, arising from the same causes, of launching transversely has been felt with long vessels of another description, namely, pontoons, or floating piers; one of 300 feet in length, which I have built at

Plymouth, was so launched, and previously to this, one of 400 feet in length by Mr. Fowler on the Humber.[1]

I hope to be able to arrange that the machinery, which is to be provided by the contractor, for lowering the vessel down the ways will be also fitted to form a 'patent slip' arrangement for hauling the ship up for repairs; so that, if it should be found desirable to do so, such apparatus may be purchased for that purpose, and fitted up at the port which the ship will frequent. With the view of facilitating such an operation, or the grounding of the ship on a gridiron for examination at low water, a sufficient extent of the floor of the ship is formed perfectly flat, and is so strengthened as to allow the ship when loaded to be grounded without being unduly strained.

After it was determined that the ship should be built on the river-bank instead of in a dock, and parallel to the river instead of at right angles to it, the next point for consideration was, whether the ship should be lowered gradually to low-water mark, or whether a free launch should be attempted.

In a free launch the ship is allowed by the action of the force of gravity to run down the ways at a considerable velocity. In the case of the 'Great Eastern' there were insurmountable objections to this plan. Some of them might have been overcome by mechanical appliances; but these would have introduced complication and additional elements of risk.

In accordance with the opinion which he had from the first entertained, Mr. Brunel determined to move the ship slowly down the ways.

Subsequently to his determination to launch slowly, Mr. Brunel decided to employ sliding-surfaces of iron instead of greased wood.

In ordinary launches the ways are thickly greased, so that there is between the ways and the cradles a thick stratum of grease, which renders the friction very small. The conditions, however, do not remain the same throughout the passage of a ship down the ways; for, when she has moved some distance, the cradle has been rubbing away and squeezing out the grease; and therefore the part of the cradle which supports the middle and bows of the ship meets with increasing resistance from friction. Another and more serious cause of the destruction of the lubrication arises from unevenness in the ways.

The result of the action of the friction between the wooden surfaces after the destruction of the grease is sometimes so great that they become mutually imbedded, the fibres of the wood being rolled up together to such an extent that it has been found difficult afterwards to separate the timbers. The increased friction due to the deterioration of the sliding-surfaces of grease does not often produce failure in ordinary ship launches, because the vessel, while still on the fresh grease, acquires a momentum sufficient to carry it over the lower part of the ways, notwithstanding the retardation resulting from increased friction.

It was from a legitimate fear of the development of a retarding force due to the destruction of the grease, that Mr. Brunel hesitated to employ wooden sliding-surfaces. The ground was far from solid; and the use of piles as a foundation for the ways would not have prevented the possibility of excessive local pressure being brought on parts of the surfaces. The heat produced by undue pressure at any point under the great area covered by the cradles would tend to spread and aggravate the evil; and, had any considerable portion of the sliding-surfaces become wood-bound, the difficulty would have been far less remediable than in the case of an ordinary launch, where the cradles and ways are throughout accessible. But in the case of the 'Great Eastern' the space between the ship and the ways, over a considerable portion of the area covered by the cradles, was very confined, and it would have been a most tedious, if not a hopeless, task to get at the injured part so as to repair it properly.

At the end of the year 1856, when the construction of the ways had to be commenced, Mr. Brunel acted upon his views as to the dangers attendant on the use of wooden sliding-surfaces, and adopted iron. By this step, although there might be some fresh difficulties to be encountered, the disastrous consequences were avoided which might have followed from employing wooden surfaces.

Under two places in the length of the ship the ground had been prepared for the reception of the launching ways. These ways or inclined planes were two in number, and reached to low-water mark. They were placed at such positions as best to carry the weight of the ship without straining her. The ways, as originally designed by Mr. Brunel, were each 80 feet wide; but, with the desire of spreading the weight of the ship over a still larger area, he decided to add 20 feet to each side of each way, thus increasing their breadth to 120 feet. The ship's head pointed down the river; 180 feet of the bow projected beyond the forward way, 110 feet were unsupported between the two ways, and 150 feet of the stern projected beyond the after way. The distance from the starboard side, the side next to the river, down to low-water mark, was about 240 feet; and the actual length of the ways, including the portion under the ship, was about 330 feet.

At the same time that he decided to use iron as the sliding-surface, Mr. Brunel adopted means for ensuring, as far as possible, the even distribution of the weight upon the ways. With this object he did not attempt to make them unyielding, but allowed them to yield slightly, so that, like a cushion, they might adapt themselves to the under surface of all parts of the cradles with a sufficient upward pressure. The ways rested on the river-bank, and piles were used to prevent the earth under the edges of the ways from swelling out at the sides, and yielding more than the ground under the middle portion.

The ground having been prepared to the slope of 1 in 12, a layer of concrete of about two feet in thickness was laid over the area of the ways. On the concrete were placed timbers running at right angles to the ship. These timbers, which were imbedded in the concrete, were 1 foot square, with a space of 2 feet 6 inches between them. Across these timbers, and parallel to the ship, were placed other timbers, with intervals of 2 feet between them; and upon these again were laid rails 18 inches apart, parallel to the ways, and at right angles to the ship. The rails were of the ordinary kind used on the Great Western Railway.

Thus the ways consisted of a network of timber resting on a thin bed of concrete; and on the top of the timber network were placed the rails which formed the actual sliding-surface.

The under side of the cradles consisted of iron bars, which were laid parallel to the ship, and therefore across the rails of the ways. These bars were each 1 inch thick and 7 inches broad, with an interval of 11 inches between the bars. Upon these bars was fixed 6 inches of hard wood planking (see fig. 15, *a*), and on this again came the framing of the cradles. Tapered timbers (*b*) were driven in, so as to fill up the wedge-shaped space between the hard wood over the bars and the flat bottom of the ship. On the side next the river, between these timbers and the rounded part of the under side of the ship, were driven in separate wedge-shaped pieces (*c*), which were secured to the timbers below by long bolts, arranged so as to allow the removal of the wedge-pieces when required. The means of unbolting the wedge-pieces was an essential provision for floating off the ship, as they had to be removed before she could move sideways off the cradles.

Resting on the lower timbers of the cradle were stout props (*d*) which pressed against the ship's side higher up than the wedge-pieces, and took part of the weight, and spread it over the outer part of the cradle. There were similar props (*e*) on the landward side of the cradles.

There were 80 rails on each of the ways, and nearly 60 transverse bars under each cradle so that there were 9,000 intersections of the bars and rails. As the ship and the cradles weighed 12,000 tons, each intersection carried on the average a weight of 1⅓ tons.

After the construction of the ways was settled, the amount of power required to move the ship down had to be determined.

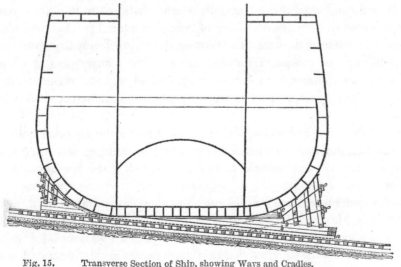

Fig. 15.    Transverse Section of Ship, showing Ways and Cradles.

*Scale of feet.*

The motive power was not simply the chains, tackle, presses, &c.; but there was also the action of gravity. One motive power, then, was not only available, but was inevitably present; and, as the ways were at an inclination of 1 in 12, the motive power of gravity upon the weight of 12,000 tons was 1,000 tons. The question to be decided was, whether the 1,000 tons of motive force was sufficient to overcome the friction; and, if not, then what additional force would be required to do so.

In January 1857, immediately upon the adoption of iron sliding-surfaces, an experiment was arranged on a considerable scale, in order to form some idea on this important point. Two rails were laid at an inclination of 1 in 12, and upon them an experimental cradle was placed, weighing some 8 tons, and representing a small portion of the actual cradle.

The effect of the friction of iron sliding-surfaces may be summed up very simply. It appeared that the motive power need not, at most, be more than would have been given by placing the ways at an inclination of 1 in 8, and that restraining power could not have been safely dispensed with if the ways had been placed at a greater inclination than 1 in 16; as it was observed that, contrary to received notions, the friction became less as the velocity increased, and that, in case any considerable velocity were attained, a great force would be required merely to overcome the motive power of gravity down the incline, independently of that required to destroy the velocity.[2]

The task of getting the ship from the place where she was built to her moorings in the river divided itself naturally into two parts—the moving of the ship down the ways, and the floating her from off her cradles.

This subdivision of the whole undertaking of the launch into two almost distinct operations is of great importance in considering the manner in which Mr. Brunel conducted them; especially when it is borne in mind that one, the moving down the ways, was capable of being, by careful precautions, rendered almost safe; whereas the other, the floating the ship off, was dependent on the successful issue of various minor operations, in the management of which the fallible human element had a greater share, and where small accidents, though, in their primary effects, productive only of delay, might cause irretrievable disaster.

In the operation of lowering the ship, there had to be provided both power to move her and power to check her motion. In floating, but one force was necessary, namely, that required to pull the ship off if she got jammed on the cradles.

With a desire to provide for the possibility of an extreme amount of resistance on the ways, Mr. Brunel designed a complete hydraulic apparatus, which would have been sufficiently powerful to move the ship down without interruption or delay. It is much to be regretted that he did not persist in carrying out his original intention.

In the operation of floating, chains and tackle were the best means of supplying the tractive force that might be required; and Mr. Brunel decided to have a very large amount of available power. If the weather were fine, and the tide at its calculated height, if no part of the cradles got disarranged, if the calculations as to the ship's draught of water were correct—if everything went right, there would be no necessity for any great hauling power; a few tug-boats would suffice to take the ship to her moorings. But Mr. Brunel determined that in this critical operation of floating he would not trust to good fortune, when the absence of it might produce grave injury. The power which he thought it desirable to provide in chain purchases for the floating was very considerable, being equal to a pull of 500 tons.

As it seemed probable that the ship would not require much force to move her down the ways, it seemed also probable that the river tackle (as the chains and appliances for hauling the ship off were called) would be sufficient for both purposes. This being the case, it at the time appeared right, in the embarrassed state of the Company's finances, to dispense with the more powerful and costly apparatus which Mr. Brunel had proposed for moving the ship down; there being no fatal consequences to be apprehended from a defect of power.

Influenced by these considerations, Mr. Brunel resolved to trust to the river tackle alone.

He referred to this decision in a letter to the Secretary of the Company written during the launch:—

November 26, 1857.

My original intention, the right one, was to fit up properly such an hydraulic apparatus as should be fitted to move the ship the whole length of the ways, and to depend upon the whole river tackle only in the event of her moving very easily, and for getting her off the ways at the end. From an unwise attempt to economise I determined to dispense with the immediate costly apparatus for pushing, and by sufficient power merely to move the ship at starting or in the event of sticking,[3] and to depend upon the same river tackle to keep her moving down the ways.

The experiments made with the trial cradle had shown the necessity of providing a certain amount of restraining force. As will be seen in the description of the launch, it was only used once, but it must not therefore be supposed that there was no necessity for providing it.[4]

The arrangement of the checking gear was the same at each of the ways. Attached to the land side of the cradle, by means of bolts, was a strong iron framework which held two large horizontal wheels or sheaves. At the upper end of the ways another sheave was fixed in a strong timber framing; and opposite the middle of the upper end of the ways was placed a large windlass or drum.

This drum was a cylinder, about 20 feet long and 6½ feet in diameter, of solid timbers, strongly bolted together, and secured at each end in a broad cast-iron disc, 12 feet in diameter.

To a point in the framing was attached one end of a 2⅝ inch chain cable; this chain was passed round one of the sheaves attached to the cradle, then round the sheave attached to the upper end of the ways, then round the second sheave attached to the cradle; and its end was coiled round the drum. Thus, as one end of the chain was secured, it was necessary, before the ship could move down the ways, that the drum should revolve, and slacken the end of the chain coiled round it.

Round the discs of the drum were wrought-iron straps; these, when tightened by levers, formed brakes by which the revolution of the drum could be retarded. Gearing was provided with a train of toothed wheels, so that the drum could be turned round by handles, and the chain wound on to it.

The following paragraph is from the commencement of a memorandum by Mr. Brunel on the launching arrangements, written about five weeks before the launch began:—

September 26, 1857.

It is expected that, with the present construction of the ways, the friction and the tendency to descend by gravity will be about balanced; so that when once in motion no very great amount of power (at least, in proportion to the mass to be moved) will be required to keep the vessel in motion, or to check it if disposed to move too quick, or quicker at one end than at the other; still the forces which may be required either to help it on or to check it, though relatively small as compared with the mass to be operated upon, will be very large as compared with forces usually obtained by the ordinary means of rope or chain purchases, and at the first start, or after any accidental or intentional stoppage, a still larger power may be required.

The apparatus which Mr. Brunel prepared for performing the double duty of moving the ship down the ways and hauling her off the cradles was as follows:—At each end of the ship was a powerful chain tackle. One end of a chain cable was secured to a mooring in the river, and it was passed round a large sheave attached to the ship, then round a sheave fixed on a barge about 300 feet from the ship, and the end brought on shore, where it was hauled on by a chain tackle worked by a steam crab. The sheave attached to the ship at the bow was slung by chains about 80 feet from the stem. The sheave at the stern was fixed on the end of the screw shaft. These purchases were intended to be good for 80 and 100 tons respectively, and were to be able to follow up the ship quickly if she moved.

In addition to these purchases, Mr. Brunel desired to have 'the means of bringing a considerable strain to bear in the event of the ship sticking at starting, or at any subsequent time, and particularly at the last;' and he considered that 'nothing under 250 or 300 tons would be of any use for the purpose.' This power he desired to apply to the centre of the ship between the two cradles by means of double crabs and treble purchase blocks on four barges.

One of the double crabs was mounted on each of the four centre barges, and was placed on a platform, elevated so that the blocks of the chain tackle could pass underneath it. This tackle was made fast to a chain attached to the ship; and the mooring chain extending across the river was hauled on by the tackle.

Each of these four crabs and tackle was to be capable of working up to a strain of 80 tons. The strain which Mr. Brunel intended to be able to put on the ship by the river tackle, in the form of a good continuous pull, was in all 500 tons.

Two hydraulic presses were also provided, one at each of the cradles, to overcome adhesion in first moving the ship. Each of these presses should have been able to exert a strain of 300 tons. Therefore, including the force of gravitation, the power which Mr. Brunel hoped to have to start the ship was 2,100 tons, or more than one-sixth of the weight to be moved; and for a continuous steady pull to keep her moving, 1,500 tons, or one-eighth of the weight.

It was at one time thought possible that the launch might be effected in October. But it was found that it would be impossible to be ready before the spring-tides at the beginning of November; and even then, when the time came, there was considerable hurry, and important matters were, as will be seen, insufficiently attended to.

The cradles were put together and wedged up under the ship, and every effort was made at low water to extend the ways as far as possible; so that, by moving her further down the slope, a greater margin might be obtained, to allow for any falling off in the expected level of the tide, or for any miscalculation in the ship's draught.

In the memorandum already referred to on the launching arrangements Mr. Brunel instructed Captain Harrison to superintend the moorings for the river tackle, and to satisfy himself of their sufficiency. A few days afterwards he wrote to Captain Harrison on the subject:—

September 30, 1857.

I fancy (I may be wrong) that you hardly estimate sufficiently highly the forces that we may require to get the ship down if she sticks at all, or to drag the cradle from under her, or to force her off the cradle at the last. She *may* move down pretty easily, and the cradles *may* possibly not stick; but if she does stick at all, it is as likely to require a dead pull of 500 tons as not, and we must not shut our eyes to the real exact amount of strain which may and will come upon purchases and moorings, &c., if this force is required and is exerted; but we must provide for it. The several moorings must really be good for the 80 and 100 tons respectively mentioned in the memorandum—and we must not rest satisfied with the feeling that the moorings are stronger than any generally sold or than common tackle will effect, but must apply purchases that will produce the strain, and if necessary we must strain them to it; and our moorings ought to be beyond a doubt.

We are going to move 11,000 tons, a far greater weight than ever was moved before, and we must not hesitate at providing a clear pull of 500 tons; but bear in mind that 500 tons clear pull is something much beyond what one is accustomed to. The power usually brought to bear with purchases, chain cables, &c., is never measured, but is very small; and we must take care and not be misled by comparison with them. 80 tons is a heavy pull, and nothing under 2 or 2¼ chain will be safe . . .

These are great strains we have to deal with, but they must be had, and therefore we must meet them boldly.

Frequent enquiries were now made relative to the time of launching; and the number of applications which poured in for admission to the yard led Mr. Brunel to write the following letter to the Directors of the Company, and it was published by them in the newspapers. In it he not only removed current misapprehensions as to the nature of the proposed operations, but he also took the opportunity of pointing out that there would be no risk to the ship in the mode of launching adopted, and that, although it might at first be unsuccessful, further power could be applied, and the ship safely launched.

October 23, 1857.

The difficulty of replying to the numerous enquiries made respecting the period at which the ship will be launched seems to render it desirable that some means should be taken of giving the information generally, that it may be uncertain, up to the end of next week, whether the ship will be launched on the 3rd proximo or the 2nd of December, and also of correcting the erroneous impressions which exist as to the nature of the operation, which can only lead to the disappointment of those who are erroneously anticipating a display, on an unusually large scale, of that which is a beautiful spectacle with ships of ordinary dimensions.

As regards the period of the launch I have, for some time past, calculated upon being ready by the first tides of next month, and by the unwearied exertions of those on whose assistance I have depended, with the advantage of unusually fine weather, the principal works required are so far advanced that there seems every prospect of success; but a change in the weather is threatening, the time remaining is short, and comparatively small causes may create such delay as to render it more prudent, if not unavoidable, to postpone the operation until the following available tide, namely, that of December 2. As no mere desire to launch on the day supposed to have been fixed will induce me to hurry an operation of such importance, or to omit the precaution of a careful and deliberate examination of all the parts of the arrangements after all the principal works of preparation shall have been completed, should such postponement prove necessary or be adopted from prudence, everything having been now prepared, the launch would be on December 2.

As regards the nature of the operation, it has frequently been stated, but it seems necessary to repeat it, that the ship will not be 'launched,' in the ordinary sense of the term, but merely lowered or drawn down to low-water mark, to be thence floated off by a slow and laborious operation, requiring two and possibly three tides, and very probably effected partly in the night, and at no one time offering any particularly interesting spectacle, or even the excitement of risk; as I am happy to feel that, even assuming accidents to occur or miscalculations to have been made, rendering the operation unsuccessful—the ship may stop halfway or not move at all, more power or other remedies may have to be applied—but no injury to the ship can result from any failure in the course of proceeding in this mode of launching.

Throughout October an immense amount of work had to be done, and the multiplicity of matters to be attended to pressed heavily on Mr. Brunel and his assistants.

With the check tackle he had reason to be content. The chains, which were the ship's cables, had been very carefully made; and, in addition to the usual tests, pieces had been taken at hazard, and were found to bear a good breaking-strain.

The river tackle was not so satisfactory. In operations that have to be conducted afloat unexpected delays arise and all the work may be suspended by bad weather, and it is moreover frequently dependent on tides. In the present case the work to be done was not easy. Heavy chain cables had to be laid out, and moorings picked up and connected to the tackles; one work having often to wait for the completion of another.

Mr. Brunel had determined that each purchase should be tested by being strained to the utmost stress for which it was intended; but, owing to the delays which had occurred in preparing the river tackle, this was not done.

A few days before the launch Mr. Brunel addressed the following memorandum to all who were to take part in the operation:—

### General Arrangements and intended Mode of Proceeding
October 30, 1857.

It is desirable that all engaged in directing any part of the work should understand the general course of proceeding which it is intended to pursue, so far as may be found practicable; circumstances nay modify these pre-arranged plans, and may compel a total departure from them, but every endeavour will be made to adhere to them.

#### GENERAL COURSE OF PROCEEDING

I propose to commence operations about two hours before high water, or about noon, and to endeavour to get the ship down as quickly as I can into the water, and down to within about 36 feet of the bottom of the ways.

My object in starting at this particular time of tide would be to get the ship into the water, and waterborne to some extent as soon as I could.

I propose to stop short of the end, in order to avoid the necessity of having to knock away all the shores, and clear the cradle at the evening tide, when it would be dark, and to float on the morning tide, when it would be also dark.

I should propose then to stop about 36 feet short of the end.[5]

At low water, although dark, I shall endeavour to knock away the shores of the 20-feet cradles, or as many of them as possible, and clear all from these cradles except the unbolting of the filling-pieces.

If the operations have proceeded easily, and the ways have not sunk much, I shall also knock away all the long shores on the inshore or port side of the ship, so as to leave less to do on the following day.

I shall then prepare at leisure to place the barges to get one pull of 36 feet, or as much more as I can (as I shall not hesitate to pull the cradles 20 feet off the ways) after high water of that night.

Soon after the high water of that night, and when the water has fallen sufficiently to prevent any risk of floating, but while the ship is still waterborne, probably about 4 or half-past 4 A.M., I shall make the last pull; and although it will be in the dark, yet having only one pull to make, and plenty of time to prepare, and no expedition required in the operation, I think it may be easily done.

The ship will then be left till low water, when we shall clear away everything we can from the cradles, and get all ready for floating at high water on the afternoon of Wednesday.

Provided the mechanical arrangements should prove efficient, the success of the operation will depend entirely upon the perfect regularity and absence of all haste or confusion in each stage of the proceeding and in every department, and to attain this nothing is more essential than *perfect silence*. I would earnestly request, therefore, that the most positive orders be given to the men not to speak a word, and that every endeavour should be made to prevent a sound being heard, except the simple orders quietly and deliberately given by those few who will direct.

In a memoranda of 'Particular Instructions,' dated the next day, October 31, there is the following passage:—

> *Starting.*—A strain being brought upon all the purchases, and the holding-back purchase being slack, if the ship does not move, the two presses will then be worked; if she does not then move, or if, when moved, she stops and each time requires the presses, the attempt will be postponed, and more moving power applied for the next time.
>
> If, after being started by the presses, the river purchases are found sufficient to move her, the operations will proceed.

In another part of these 'Instructions' Mr. Brunel again shows that he was not, as has sometimes been supposed, under the impression that the friction would be so small that the only important thing to be thought of was to check the ship from rushing too fast.

On the contrary, he foresaw the possibility of her not moving at all; even with the presses, that is to say, with a force of 1,100 tons over and above the action of gravity. If after moving she stopped, and then required the presses again to move her, this would show that the operation could not be properly carried out, and that the work must be suspended till more motive power was applied. If, again, the river tackle were sufficient to move her, then the work was to proceed, but the friction might even then be so great as to render it desirable to remove all retarding force. He says, in another passage in his 'Instructions':—

> It is very likely that no checking whatever at the drums will be found necessary, but that, on the contrary, it will be found desirable to get rid of any resistance by overhauling the heavy chains through the sheaves.

The best day on which to begin the launch was Tuesday, November 3, as it left two or three days of the high full moon spring-tides for the operations, should they be prolonged.

On Monday, November 2, the chief work remaining to be done was stowing kentledge or iron ballast on the cradles, to prevent the timbers floating when the ship should be moved off them. All the appliances were ready, and, except the river tackle, had been carefully examined. This, as has been already said, had not been tested.

It was now for Mr. Brunel to consider whether, in consequence of the river tackle not having been properly tested, he should postpone the launch till the following month.

It was most important for the Company that the ship should be afloat as soon as possible; and, as any defects which might exist in the river tackle would almost certainly declare themselves in the earlier part of the operation, when nothing worse than delay could be apprehended, Mr. Brunel, after a careful review of all the circumstances, determined to attempt the launch.

On the morning of November 3, the work of putting kentledge on to the cradles was completed by firelight, and the rails were rubbed over with a mixture of oil and black-lead. All the shores and props which supported the weight of the ship had been removed, and she was now resting entirely on the cradles.

Later on in the morning the brakes of the drums were tightened down, and the dogshores were removed from the ways in front of the cradles. Mr. Brunel, who had been engaged from an early hour in examining all the preparations, superintended this operation, and, having satisfied himself that all was clear and ready, returned to the upper part of the yard.

By this time it was crowded with people. The Directors, contrary to Mr. Brunel's expressed wish, and without informing him of their intention, had issued a large number of tickets of admission. A few days before, Mr. Brunel had suggested that four policemen should be obtained, thinking that all they would have to do would be to contend with trespassers. The police force actually present were ignorant of the portions of the yard to be kept clear, and Mr. Brunel had himself to go and assist in ordering visitors away from the neighbourhood of the path prepared for the tackle of the stern hauling gear. The crowd soon became so great that it was almost impossible for the men in charge of the hauling-engine at the stern to see the signals given from the middle of the yard, or for those in the middle of the yard to see what was happening at the stern.

At about half-past 12 o'clock the fastenings of the ship at the bow and stern were let go, and Mr. Brunel ordered a small amount of slack to be given off from each drum. This was done by men turning the handles of the gearing which had been provided for winding the chain on to the drums. The order was then given to haul on the bow and stern tackle, and to pump at the hydraulic presses. It is doubtful what amount of strain was put on by the tackle and the presses, but it was probably not very great.

Presently a shout from the forward cradle announced that it was moving, and almost immediately the stern cradle also started with what appeared to be a considerable speed. The men who had been engaged in turning the handles of the gearing had remained leaning against them. As soon as the ship had moved a few inches, she took up all the slack chain. This made the drum revolve, and the handles of the gearing spun round very rapidly, striking the men, and throwing them into the air. The men who were at the brake-handle next to the gearing ran away. Mr. Brunel, who was standing near the drum when the accident happened, shouted to the men to hold on to the brakes, and ran to the spot. The men who had remained at the other brake-handle hauled it down with the tackle. A great restraining force was thereby brought upon the ship, and her progress ceased; the forward cradle having moved 3 feet, and the after cradle 4 feet 3 inches.

Five men were injured. On the death of one of them it was stated at the inquest, by the foreman of the drum, that, after the slack had been paid out, he had ordered the men to stand clear. Be this as it may, it cannot be denied that the handles should not have been used after the securing chains had been let go; and indeed Mr. Brunel said at the inquest, 'I may blame myself, for I did not anticipate that the handles would have revolved so rapidly.'

After this accident, Mr. Brunel determined to wait till high water before recommencing the operations. In the meantime the gearing was removed from both drums.

A more important change was also made in the arrangements. When the ship moved, the men on the four middle barges became frightened, thinking she was about to overwhelm them; a rush was made, and one man, jumping into a small boat, shoved off, leaving the rest to their fate. A report was at once sent to Mr. Brunel that the men were untrustworthy, and that they would not remain; and that, as the barges would be of no use without the men, the chains had better be dropped and the barges removed. To this Mr. Brunel consented.

It would, however, have been sufficient to take the men off, leaving a tug-boat and a few steady men to keep the barges out of the way of the ship; they would then have been available if required. Mr. Brunel, a short time after he had given the order, ran round the bow with Captain Harrison to countermand it; but it was too late, as it had been already acted upon. As events turned out no harm was done, as the centre barges alone would not have been sufficient to go on with, after the rest of the tackle failed.

The result of these changes was that when the operations were recommenced, the only hauling gear was the bow and stern purchases; the hydraulic presses were also available to start the ship.

At a little after 2 o'clock the signal was given to haul on the bow and stern tackle, the presses being at the same time pumped up. The brakes of the drums were slackened, but kept all ready for tightening.

Not long after the strain had been brought on the tackle, several of the teeth of one of the wheels of the bow steam crab gave way, and the chief anchor at the stern began to drag, so that no efficient strain could be obtained. On this being reported, the operations were discontinued; and, as there was no possibility of getting things ready by the next day, the launching operations were postponed till December 2, the next full moon spring-tides.

As soon as this was known the visitors rushed in on the works, crowding about the cradles and ways; and Mr. Brunel had to postpone those investigations which he wished to make at once.[6] The whole yard was thrown into confusion by a struggling mob, and there was nothing to be done but to see that the ship was properly secured, and to wait till the following morning.[7]

The next day was devoted to an examination of what had gone wrong, and to the consideration of what should be altered before another attempt was made.

At the stern mooring the anchor was bedded into the ground on the further side of the river.

The difficulty with regard to the four centre barges was got over by placing the four crabs with their tackles in the yard, on the landward side of the ship. The four chains attached to the ship, which had before been hauled on directly from the barges, were now passed round sheaves on the barges, and brought back under the ship's bottom to the tackles in the yard.

The chief alteration, however, was in the arrangement of the hydraulic presses. On November 3, there were, as has been said, two presses. Two additional presses were now provided; each of these consisted of two 7-inch cylinders, and was equivalent to a 10-inch press.

With the object of being able to employ the presses continuously during the descent of the ship, they were arranged to point down the ways at an inclination of 1 in 12. The four presses were placed one on either side of the check tackle at the two ways, and were supported by abutments of timber-work. These abutments each consisted of four rows of piles, one behind the other at intervals of about 8 feet. The press abutted against the row of piles nearest to the ship, which were connected by wooden struts to the piles behind them. Long balks of timber of various lengths were prepared to transmit the pressure to the cradles.[8]

The four presses might be considered equivalent, at their full power, to a force of 800 tons; this was so much in excess of the small force that had moved the ship on November 3, that, even making every allowance for the advantage of the fresh lubrication in the first instance, it seemed reasonable to suppose that with this force the ship could be moved down easily.

As the process of moving the ship with the presses would naturally be a slow one, Mr. Brunel determined to proceed with the operations as soon as everything was ready. On November 19 the work was commenced.[9] The bow tackle was hauled upon first, as the forward cradle was more than a foot behind the after one, and the men at the forward presses were set to work. After a short time the timber backing of the presses began to crack and 'cry out'; and, without much stress on them, the abutments were forced back some 3 or 4 inches. The mooring chain of the bow tackle also gave way, although there was not any excessive strain on it. On examining the abutments, Mr. Brunel saw the cause of their failure, and ordered the strain to be taken off. The number of piles was sufficient, but the way in which the strain was communicated to them did not enable them to exert the proper amount of resisting power.

This defect was cured by tying the heads of the rearward piles with bolts to the foremost piles. The ship being secured, each press was tested to a full strain, and the adequacy of its abutments ascertained.

It was different, however, with the river tackle. The chain which had parted was an old river mooring-chain of great size. Much delay in replacing it was caused by dense fogs, which made it almost impossible to work on the river. Moreover, there seemed a fatality about every attempt to get a regular trial of any part of the tackle. When, at last, a trial took place, and a strain was put on, a mooring chain gave way; then this had to be fished up from the bottom of the river, and pieced together, the accident being ascribed to a defective link in the chain. The trials were, therefore, so few that it was only proved by degrees that all the regular moorings were worthless; although they had large chains which ought to have been good for three times the strain put on them.

The stubbornness of the ship on November 19 gave Mr. Brunel great anxiety; not from any fear of being unable to apply sufficient power to move her, but because, on continued consideration of the subject, he apprehended that a serious difficulty might arise, if there should be a prolonged delay at a particular part of the ship's progress.

It has been explained that Mr. Brunel, with a view of obtaining uniformity of bearing over the surface of the ways, had not attempted to support them rigidly on piles, but had rested them on the river-bank. As, however, the foundation of the building slip was comparatively rigid, he feared lest an unequal subsidence might cause injury to the ship, if she were stopped for any length of time before she had completely left the ground on which she was built. He thought that if the ways sank at this point they would assume a slightly convex form, and tend to force upwards the flat bottom of the ship. The main part of the ship's bottom, between the longitudinal bulkheads, could bend in slightly under a heavy upward pressure; but this action could not take place at the transverse bulkheads, as they would not yield without injury. Mr. Brunel shrank from proceeding with the launch without having in reserve such an amplitude of power as would prevent the ship's being stopped at this critical point. This consideration, together with the continued failure of the river tackle under such tests as were applied to it, led him to address the following communication to the Directors:—

November 26, 1857.

We proved two of the presses yesterday afternoon up to the full pressure. A third, the largest, was proved partially; it required some additions, which are nearly completed, and will be in a few hours. The fourth may, I think, also be relied on to the same extent, nevertheless, after a careful examination of the effects of these strains and other circumstances, I have, after a night's consideration, come to the conclusion that our means are too imperfect to justify my moving the ship with them in their present form. The presses would start the ship, but it is evident that if required to be used constantly, that is repeatedly, the piles would become loosened so as to draw and rise; this again might be remedied by loading, but clumsily, and with other contingencies, which I will report, combine to reader it hazardous to depend upon them. My original intention, the right one, was to fit up properly such an hydraulic apparatus as should be fitted to move the ship the whole length of the ways, and to depend upon the whole river tackle only in the event of her moving very easily and for getting her off the ways at the end. From an unwise attempt to economise I determined to dispense with the immediate costly apparatus for pushing, and by sufficient power merely to move the ship at starting or in the event of sticking, and to depend upon the same river tackle to keep her moving down the ways. The power originally calculated upon for the river tackle has gradually, step by step, failed us; the moorings supposed to be sufficient for certain strains have failed us at one-third of those strains, another has parted since our last attempt, and, instead of full 350 tons of power from this source, we cannot now depend upon 200, and this, added to the inefficiency of the pushing power, would risk the sticking of the ship, which might occur exactly at a point which would involve serious difficulty to remedy. I am assuming a combination of adverse circumstances, perhaps not likely to occur, but quite possible; and the conclusion I am compelled to come to is that our apparatus is too defective, and that the original plan of a proper and sufficient hydraulic apparatus, arranged in a complete well-constructed mechanical manner, to push the ship continuously down the ways, ought to have been followed out, and is now the only mode of doing the work safely, that is, without the risk of being involved in a difficulty much greater and more costly.

I have only to add that bad as this report of our condition is, it is at any rate the worst that can be made of it, that nothing whatever has occurred to show that any new difficulty has arisen or anything whatever to create any new difficulty. We could move the ship now if it were wise to do so, but with great doubts whether our pushing apparatus in its present form, imperfect and unmechanical, would continue effective if repeatedly used, and the certainty that our river tackle is far inferior to what is required, and also of doubtful and more than doubtful permanency for repeated strains, it would not be right to commence . . .

Mr. Brunel at the same time determined to obtain, on a large scale, a measure of the deflection that might be expected from the weight of the ship coming on the ways. More than 100 tons of kentledge was piled on a portion of the ways 10 feet square, in such a manner as to give a pressure thereon of about double that which would be produced by the weight of the ship.

It was necessary that this test should not be tried on too small a scale, as a weight resting on an isolated patch would receive support from the surrounding ground, which it could not of course do if that ground was equally loaded. The ways sunk so little under the test as completely to reassure Mr. Brunel, and to show that no serious evil need be contemplated in the passage of the ship from off the place where she was built on to the newly made ways, even though she might be again stopped for some time. He therefore determined to go on at once with the launching operations.

The result of the test was very satisfactory to him, and it enabled him to carry on the work with the same confidence as he had at the first felt— 'that the ship may stop halfway or not move at all . . . But no injury to the ship can result from any failure in the course of proceeding in this mode of launching.'

Shortly before the second attempt to move the ship, on November 19, the experimental cradle had been again put up with a view of obtaining some additional data as to the hauling strain that might be required. The deductions made from them were the same as those obtained in the commencement of the year, and encouraged the hope that the motive power required would not be excessive.[10]

By Saturday, November 28, the four presses had been got ready; and the river tackle, though still far from being beyond reproach, had been got into place, and partly tested.

The brakes were eased, and a small amount of slack was overhauled on the check-tackle chains by the men stationed on the ways for that purpose. As on the previous occasion, the pressure was to be first put on the presses at the foremost cradle.

Arrangements were made for promptly following up the ship if she moved freely. A black board was placed on each cradle for recording the progress of the ship.

Mr. Brunel stood on a low platform in the centre of the yard, as a convenient position from which to watch and command the operations. A little before ten he gave the order to commence pumping, and the men at the hydraulic presses got to work. When the pressure came on the timber framing which formed the abutments, there was considerable noise of creaking and crushing as the several parts subjected to strain came in to their proper bearing. The men soon changed from the large plunger handle to the small one which put on the full pressure; the timbers of the abutments kept on crying out, but it was evident that they were not yielding as they had done before. Presently, while the noise of the timber work was still attracting attention, the man in charge of the measuring apparatus recorded on his black board a movement of one inch; the ship was again in motion.

She moved steadily, but slowly, under the force of the presses, at a rate of about one inch a minute, and as soon as the forward cradle had been moved about a foot in this way, the presses at the after-ways were set to work, and the river tackle was put into operation, first the bow and stern tackle, and then the four middle purchases. All went well with the presses, but the strain had not been put on long, when the stern mooring-chain and one of the two chains at the bow broke; an anchor at the bow had also begun to drag.

Later in the day part of the moorings of the centre barges also gave way. Captain Harrison set to work to repair these defects as fast as they occurred; in no way dismayed that, as he was at work remedying one mishap, the news of another was brought to him. Barges had to be got into place, the broken ends of cables fished up or under-run and pieced together, and this often in the dark; for it must be remembered that the work was going on at the end of November, when the sun rose, invisible for fog, at half-past seven, and set at half-past four.

With the exception of the river tackle, all had gone well; the presses and their abutments had acted efficiently, and the ship had been moved easily down the ways about 14 feet before work was suspended at night.

Though the progress had not been great, there still seemed a reasonable hope that, by pushing on, the ship might be got down to the bottom of the ways in time for floating her off at the next spring-tides, namely, on December 2. Mr. Brunel therefore decided to go on with the operation on the Sunday. Early in the morning the presses and crabs were again set to work. The river tackle soon gave way; and, indeed, there was no reason why it should be superior to that used the day before, as almost all that could be done in the night was to piece together the broken chains, and to replace the anchors. The moorings at the bow and stern began at once to drag, and two of the mooring-chains amidships parted. The hydraulic presses were then the only available power; and, although the full pressure was put on, the ship did not move. This was very disheartening; it was, however, thought that the resistance was due to some exceptional adhesion. Every effort was therefore made to get together the means of giving the ship a first start.

It was not till the afternoon that a large number of screw-jacks and hydraulic jacks which had been procured were got into place; they were then screwed up hard, and the hydraulic presses being set to work, the ship began to move in a manner very similar to that of the day before. There was not, however, much daylight left; and, when night came on, the distance traversed was only about 8 feet. The comparative facility with which the ship moved when once started gave hopes that good progress might be made the next day.

On Monday morning the ship moved without more difficulty than when she had stopped the evening before, and the work went on quite satisfactorily. She continued to move slowly, and by dinner-time had gone about 8½ feet. Three feet an hour was not much, but still if it could be kept up it would suffice to get the ship down by the next spring-tides. Arrangements were therefore commenced for lighting up the ways and pumping machinery, so that the work might be continued through the night. The repairs of the river tackle were pushed on, the ship's anchors, which had just been finished and tested, were laid down for part of the moorings, and some of the chains were replaced with chains lent by the Government and by Messrs. Brown and Lenox.

When work was recommenced after dinner the ship made a short slip of about 7 inches. On pressure being again applied the 10-inch press at the forward cradle burst. This put an end to all work for the day, and it was then determined to replace the broken press and to add two more presses to each cradle, before proceeding with the launch.

The preparations for the new presses were pushed on vigorously, but it was not till the afternoon of Thursday, December 3, that things were again ready for a start.

The pumps were set to work and the tackle hauled upon. The ship made several short slips of a foot or so, and then moved more than 5 feet at one slide. When darkness set in she had moved about 14 feet, in slips of greater or less length.[11]

On Friday, December 4, all was ready early, and during the morning everything went as well as on the day before; but in the afternoon increased difficulty was found in getting the ship to move, and the 14-inch press at the after cradle burst, as did also a 7-inch cylinder of one of the coupled presses.

Notwithstanding the delay due to the bursting of the presses, the ship was moved some 30 feet; but there was no longer any chance of getting her afloat at the spring-tides, and the increased adhesion gave cause for the fear that still more power would have to be applied.

On the next day, Saturday, December 5, the ship made a short slip; but, although the pressure was kept constantly on, no further advance was made until late in the afternoon. Mr. Brunel then tried suddenly letting go the strain on the stern tackle. The sudden relief of the side-way strain on the end of the ship sent a tremor through the hull, which served to destroy the adhesion, and she slid several inches. This operation was several times repeated, and although there were a number of vexatious delays from pushing-pieces giving way and other mishaps, she was moved by the evening a distance of about 7 feet, the resistance due to adhesion being very great.

On Sunday some of the presses farthest from the cradles were moved down the ways nearer to the ship, so as to avoid the necessity of using long pushing-pieces, which required much attention to prevent their bulging sideways. The river tackle now consisted of the bow and stern steam-engine purchases, and two crabs and tackle, one at each end of the ship. The moorings opposite the centre of the ship having proved worthless, it was necessary to lay down new moorings, and it was found more convenient to lay them opposite the ends of the ship.

The next day, Monday the 7th, after the commencement of operations, considerable delay was caused by the failure of some of the feed pipes of the presses. These defects were not cured till after dinner-time, when the operations were resumed, and before dark the ship was moved about 6 feet.

On the following morning several short slips were made, and the ship had been moved about 4½ feet; but, at 10 o'clock, a dense fog came on, and rendered it impossible to proceed. The next day was occupied in re-arranging the tackle and presses.

These were not ready till the morning of Thursday, December 10; when, on the presses being again set to work, and one of the chains being as before suddenly slackened, the ship made one slip of a little over a foot; but, on the strain being again applied, two of the anchors began to drag. As it was now essential to have the river tackle, in order by shaking the ship to destroy the adhesion, and by the drag of the catenaries to increase the length of the slides, Mr. Brunel determined to dispense with the anchors, and to attach the chain cables to piles connected by framework. These abutments or pulling points for the chains were now constructed on the other side of the river, opposite the bow and stern of the ship.

The launching operations last described, namely, from December 3 to December 10, were full of incident. Nor was the scene wanting in that animation which agreeably interests a bystander, the more so if he is not thoroughly conversant with the meaning of all he sees and hears, so that he mistakes a loudly spoken word, loudly spoken merely that it may be plainly heard, for a prompt and urgent command.

The labourers at the pumps relieved the monotony of their work, and shook off the cold, by taking a lively and talkative interest in the progress of the launch, and echoed the orders given them to pump with the 'big plunger,' or 'little plunger' of the pumps, or to 'fleet' the press. This and the singing of the gangs, which were constantly at work moving chains for the repair of the river tackle, or rolling logs of timber on to the ways to serve as pushing-pieces for the presses, gave plenty of life to the operations; and then when the pressure had been got on the presses, and shouts from the bow and stern of the ship passed the word that the river tackle was hauled taut, the order would be given to 'let go' the chain at one end of the ship. Immediately the rattling noise announced that this had been done, and, after a second or two of anxious watching, the ship slid off, the timbers, abutments, and pushing-pieces creaking and groaning as the strain was suddenly relieved. While the ship was in motion, the whole of the ground forming the yard would perceptibly shake, or rather sway, on the discharge of the power stored up in the presses and their abutments. The appearance of the ship moving sideways in these short slips, when seen from the ways, was very imposing.

All these somewhat striking surroundings of the operations were naturally heightened in effect, when the work was being carried on in the early morning or late in the afternoon; and when the timber-framing and the groups of men at work were illuminated by the glare from the open fire which were kept burning near the pumps and presses.

The preparations already described were not completed till December 15. In the meantime Mr. Brunel had been joined by his friend Mr. Robert Stephenson. Mr. Stephenson had not been aware of many vexatious circumstances which had even prevented Mr. Brunel from making full use of his own staff of assistants. Mr. Stephenson expressed to a common friend his regret that Mr. Brunel had not invited him down to the ship, and said that he should have gone down uninvited, but that he thought Mr. Brunel had reasons for not wishing it. On the state of affairs being explained to him, Mr. Stephenson said 'I'll go down to him at once;' he did so, and his arrival at Millwall was very welcome to Mr. Brunel.[12]

Mr. Stephenson agreed with Mr. Brunel as to the expediency of suspending operations until an ample excess of power was applied. Fortified by the support of Mr. Stephenson, Mr. Brunel was prepared to advise the Directors to adopt this course; but, as the preparations for recommencing the work were just completed, it was determined to make a trial on the afternoon of December 15. The presses were all pumped up, and the river tackle hauled on; but, although the force applied was at least 300 or 400 tons greater than that which had last moved her, the ship did not yield, and the attempt was abandoned. After a careful consultation on the depressing result of this day's work, it was determined to make another attempt the next morning, in order to see if any new form of difficulty had arisen; and that after this operations should be suspended, and an ample number of additional presses provided.

The following day, December 16, as soon as Mr. Brunel and Mr. Stephenson had arrived, the pressure was again put on the presses, and the river tackle having been hauled taut, the chains at the bow were let go, and, to the great satisfaction of all present, the ship made a short slide. The record of her movement showed that, although the adhesion was much greater, the retarding force of friction was about the same as before, and that therefore there was no reason to assume the existence of any special obstacle. Another short slip was made; but, in getting up the pressure again, a press was burst, and the work was then stopped.[13]

Mr. Brunel's decision to suspend the launching operations at this point was approved at a meeting of the principal shareholders held the next day. His report to the Directors, and a memorandum of a verbal statement which he made to the meeting, are as follows:—

December 17, 1857.

In my letter of October 23, which was published at the time in the daily papers, I referred to the possible contingency of the power provided to move the vessel down the ways proving insufficient, and the operation then about to be attempted being so far unsuccessful; and, referring to what I considered a countervailing advantage in the absence of risk, I stated, 'the ship may stop half-way, or may not move at all, more power may have to be applied, but no injury to the ship can result from any failure in the course of proceeding in this mode of launching.'

The result has been that after moving the vessel nearly half the distance to low water, it has become necessary to increase very considerably the power which has effected this much, although it had already been much added to during the operation.

This will unavoidably be attended with some expense and delay, but not considerable, as the requisite hydraulic presses can be obtained ready made, and their application is simple, and the result cannot, I apprehend, be doubtful.

I do not mean to imply that I contemplated any such great increase of resistance as probable, such experiments as could be made before moving the ship having given me good reason to hope for a different result; but the possibility of it was contemplated, and I refer to this merely as explaining the statement I now make, that the difficulty is simply one of degree, of more or less power being required, and that nothing whatever has occurred to create any new class of difficulty. The launching ways, about which anxiety had been expressed, and not unnaturally, have stood perfectly and without any settlement or any derangement by being passed over. There is no change of gradient or inclination in the ways capable of producing any effect, as has been supposed; the upper part of the ways having an inclination of 1.025 inches per foot, and the lower part, where the ship now is, one of 1.000 per foot, a difference too small to be appreciable, but which possibly by some mistake of figures may have led to the erroneous impression referred to.

The amount of resistance upon the ways in their present condition and inclination has now been positively ascertained, and an ample excess of power being applied, there can be no reason to doubt the result. I propose to apply that excess by going considerably beyond the amount which the calculation founded upon the results actually obtained would give as the maximum, and to double the power which has last moved the vessel.

*Memorandum of a verbal Report made to the Directors, and a small Meeting of the Principal Proprietors*

December 17, 1857.

That after full consideration of all the circumstances, and assisted by the best advice I could call in to my aid, namely, that of my friend Mr. Robert Stephenson, I considered that the only mode of proceeding, and one which there appeared no reason to doubt would succeed, was to apply considerably more press power; that I proposed to double what we had; that I believed I was able to put my hands upon the requisite presses; that the river tackle so far as it went might now be considered good, but that unfortunately we were obliged to take up the principal part of the chains, which with great kindness and liberality Messrs. Brown and Lenox had lent us, and were now peremptorily called upon to deliver up; but that with their assistance I could replace them . . .

A large number of presses were obtained, the owners for the most part lending them free of charge. Among these presses was the large one, with a 20-inch cylinder, which had been used for lifting the tubes of the Britannia Bridge.

On each of the ways were placed nine presses. The total sectional area of the cylinders at the forward cradle was 1,066 circular inches, and that of the cylinders at the after cradle was 1,358 circular inches; but the Britannia press was not to be worked to its full power, so the total area of the cylinders may be taken as 2,300 circular inches, or 1,800 square inches. The presses might be considered as good for at least 2½ tons on the square inch; this gave a power of 4,500 tons, which, with the 1,000 tons due to gravity, gave 5,500 tons, or equal to nearly half the weight of the ship. The presses were now coupled together in groups, in order to ensure that an equal pressure should be brought on them; and to each of these groups an accurate pressure gauge was attached.

All the presses having been tested, it was determined to recommence the actual operation of launching on Tuesday, January 5.

So much of the water in the pipes had been frozen that it was eleven o'clock before the order was given to the men to pump. When at one group after another the pressure was shown to be one ton on the circular inch, the pumps were stopped. As the backing of the presses continued to yield slightly, a stroke or two of the handles had to be made from time to time, to keep up the required strain. For six minutes there was perfect silence, and then the ship moved, sliding down about 3 inches.

The same process was repeated at the stem cradle once or twice, and then at both cradles. After this the order was given that the pumps should be kept going till she moved. This was accordingly done, and when the pressure amounted to 1¼ ton on the circular inch the ship made a slide of about 4 inches. In this manner she was moved about 5 feet before work was stopped in the evening.[14]

On January 6, there was a singular change in the behaviour of the ship. During the whole of the forenoon she moved gradually, yielding to the pressure at a rate of about an inch in four minutes. In the afternoon, however, she moved in short slides.

During this and the three following days her progress was about 10 feet each day. After this the ship, being to a considerable extent waterborne, was moved with greater ease, and on Tuesday, January 12, 20 feet was accomplished in less than four hours.

By Thursday, the 14th, the ship had traversed a distance of 197 feet at the forward cradle, and 207 feet at the after cradle. It was thought unwise to advance further till the coming spring-tides on the 19th of the month were past, lest an exceptionally high tide might come unexpectedly, and partially float her. As soon as the spring-tides had passed, she was moved on cautiously, a short distance at a time, and the depression of the ways was carefully observed. This was found to be inconsiderable, and the cradles were gradually pushed 25 feet off the ways. As the spring-tides came on, water was run into the ship, to prevent her from floating prematurely.

The upright struts of the cradle on the side next to the river were all removed, and the wedge-pieces had chains fastened to them, with the ends brought on deck; so that, if any of the wedge-pieces got jammed and did not come out when the ship floated, they might be hauled out by the chains.

The river tackle now consisted of two purchases at the bow and two at the stern. To keep the ship, when she floated, from being drifted by the tide or wind, chains were carried from the bow and stern to moorings, by which her movement up and down stream might be regulated. Four tugs were in attendance to tow the ship to her berth, and a floating fire-engine was also ready to pump water into her, should this be necessary owing to any sudden postponement of the launch.

Nothing now remained but to watch carefully for a suitable tide.

It was determined that, if the weather were favourable, the floating should be effected on Saturday, January 30. The tides had been below the average, and on the Friday matters did not look promising; the tides had continued low, and the weather was bad. A careful watch was kept on the tide, observations being taken every half hour, and plotted on a diagram so as to show at a glance the probable height to which it would rise.

The tides showed signs of improvement, and they commenced to pump water out of the ship on Friday night; but, as time went on, the weather did not mend, and the wind was blowing from the south-west against her broadside; therefore in the early morning Mr. Brunel, who was in person attending each turn of affairs, ordered water to be pumped in by the fire-engine. There was hard rain and strong wind; and telegrams which, according to arrangement, were being frequently sent from Liverpool and Plymouth, showed similar weather. This continued throughout the Saturday, and the tide was low; but, when it began to rise in the evening, it gave indications of being a very high one. As soon as the tide reached the Kingston valves, Mr. Brunel had water run into the ship. Although she rested uneasily on the cradles, she remained safely in her position.

In the evening the rain came down in torrents; nevertheless, after midnight the weather mended, and the wind went round to the north-east. As the telegraph gave the same report from Liverpool, Mr. Brunel, encouraged by every sign of fine weather, and having the good promise given by the high tide of the night before, determined early on the morning of Sunday, January 31, to float the ship on that day.

The pumps were immediately started to discharge the water from the ship. The bolts securing the wedge-pieces of the cradles were unfastened at daybreak, and the ship was then ready.

The morning broke with great splendour after the gloom of the previous days, and the tide, as soon as it had turned, began to rise with unusual rapidity. It had been arranged that all the men should be at their posts at the presses and crabs by eleven o'clock; but the tide was not only very high, but exceptionally early; and, although a considerable margin had been allowed, it was not sufficient. Mr. Brunel and his assistants hurried the men to their places, the presses were set to work, and the ship was put in motion down the ways for the last time.

At a little before one o'clock observations taken by levels showed that the ship had ceased to descend, although she was still being pushed forward. Shortly afterwards Captain Harrison, who had gone on board, sent Mr. Brunel word that all the wedge-pieces had floated up on the outer side of the ship; and at twenty minutes past one the stern was seen to be afloat. Mr. Brunel had been loath to haul out the ship by the river tackle, lest the wedge-pieces might get jammed; but, as soon as he was informed that they had floated up, he sent orders for a strain to be put on the bow tackle. This was at once done, and by twenty minutes to two the bow rose from the cradle.

Mr. Brunel then ordered the checking gear to be secured, that the ship might be sooner hauled clear of the cradles, and he went on board. He had scarcely done so when a serious difficulty arose. It has been already said, that in order to keep the ship from moving up or down stream, cables had been carried out to moorings ahead and astern; and both these chains had been hauled up tolerably taut, at least the slack had been taken out of them, so that if required they might be at once available. Now when Mr. Brunel had given the order for the bow out-haul tackle to be hauled upon, Captain Harrison, in order to supplement it, ordered the tug-boats to haul the bow off. This order was by someone conceived to have been given with the object of hauling the ship ahead; and to facilitate this the stern mooring-chain was let go.

By this time, though the tide was still running up the river, its strength was much diminished; and the drag of the chain at the bow of the ship was sufficient to pull her forward against the tide. The paddlewheel on the shore side then came in contact with the upright timbers of the forward cradle. Mr. Brunel ran down from the ship into a boat and examined the place where the wheel was fouled. He then hurried back on board, where, as through some blunder the stern outhaul had also been let go, he had now little but the tug-boats to depend upon. They were of course of but small value for a dead pull as compared with the chain tackle. He had the bow chain veered out and the tugs all set to work, assisted by the tide, to haul the ship up the river.

In about twenty minutes time, the paddlewheel was got clear of the cradle, and this great difficulty was overcome. Fortunately, the tide was an extraordinarily high one, and the time of available high water was long.

The ship had not been moved far towards her berth when another mishap delayed her progress. The barge of the bow purchase came foul of the starboard paddlewheel, and the only way of freeing the ship was to scuttle the barge. When this was done it sank away clear of the wheel, and the ship proceeded to her berth on the Deptford side of the river.[15]

By about seven o'clock she was safely moored; and the cheers of the men, as Mr. Brunel went down her side, announced that the launch of the 'Great Eastern' was at length accomplished.

## NOTE A
### Experiments and Observations on Friction

In January 1857, Mr. Brunel took steps to form an estimate of the amount of hauling or of retarding force that would probably be required in the launch.

Two rails were laid at an inclination of 1 in 12, and upon them an experimental cradle was placed, with three cross-bars similar to those which were to form the under surface of the cradles. The three cross-bars therefore made six intersections on the two rails, and the small cradle was loaded with about 8 tons, so that the weight on each intersection was about equal to that which would come on each of the intersections of the actual cradle. This arrangement was not therefore a model, but a correct representation of a part of the cradles, and which might, with an exception to be noted presently, be taken to exactly represent, by its conduct, the conduct of every similar part of the actual cradles. Experiments were made with one or two kinds of unguents, and, what was a more correct representation of what was likely to occur, with the rails and bars clean but not bright, and without lubrication.

The experiments with lubrication were useful rather as comparing the various lubricants one with another than as representing, by a mere process of multiplication, what would be the behaviour of the ship on her cradles, because, for the reason already pointed out in the case of wooden sliding-surfaces, the lubrication would be more and more rubbed away as more of the cradle passed over it; thus the experimental cradle, when tried with lubrication, represented rather the behaviour of the front part of the cradle than that of the whole. Had the ship herself been moved uninterruptedly down the ways, the state of things would have been something between good lubrication and none at all. As the under sides of the bars were lubricated, any motion of one end of the ship before the other would tend to move the bars sideways over the rails, and so to spread the lubrication, and to pick it up and re-deposit it. Mr. Brunel thought but little of the black-leading of the ways, considering that it would be rubbed off by the leading bars of the cradles; but a very little lubrication on metal surfaces is sufficient; and doubtless, had the ship been moved continuously down the ways, considerable assistance would have been derived from the lubrication which was applied.

The results of the experiment were curious. The generally received notion is, that friction between rubbing surfaces is independent of the velocity; that is to say, that whether a body be moving fast or slow within reasonable limits, the retardation due to friction is the same; that if a body be sliding at a given velocity, whether that velocity be great or small, a drag of a certain number of pounds will keep it moving at that velocity. It was, however, always understood that a greater force was necessary to start a body from rest, to overcome adhesion. The experiments made with the experimental cradle distinctly showed that any rule as to friction being constant at different velocities was untrue. It was evident that, as the speed increased, the power required to overcome the friction became less. No exact records are extant of the experiments made with this experimental cradle before the launch; they were, however, repeated during the launch with great care, and the results very carefully analysed. The experiments showed generally that the tractive force, including the action of gravity, was never more than ⅛, or less than ¹⁄₁₅ of the weight.

Although the experiments showed that the amount of friction in the case of the actual launch would lie between the limits above mentioned, they at the same time indicated that it would not probably approach either of those limits.

Shortly after the commencement of the launch, Mr. Brunel had the experimental cradle and ways re-erected. A very simple arrangement was fitted up, by which the forces at work at each period of the progress of the cradle in each experiment might be deduced. The results of these experiments, which, as may be supposed, were similar to those obtained in the commencement of the year, were most instructive; they showed quantitatively the decided diminution in friction which took place as the velocity increased, and the amount of that diminution. The apparatus was very simple. The experimental cradle, which has already been described, was made to slide down its ways by a chain attached to a suspended weight. The weight employed was generally about 5 cwt. After the cradle had run a certain distance, the weight reached the ground and the cradle proceeded with the momentum it had obtained. The velocity given to the cradle down the ways was measured in the following manner. A long piece of tape was coiled round a reel placed at the top of the inclined rails or experimental ways, so that it could revolve freely and pay out the tape as required. One end of this tape was attached to the cradle, so as to be drawn after it as it ran down the rails. The tape, as it ran off the reel, passed over a guiding board over which swung transversely a pendulum arranged to swing once every quarter of a second. At the lower end of this pendulum was attached a brush which was filled with paint; and as soon as the model cradle moved, the pendulum was set oscillating by a self-acting trigger arrangement. The pendulum in its oscillations made marks on the tape as it ran out at every quarter of a second of time. Thus, by an examination of the tape, could be determined the exact distance which had been passed over by the cradle during each quarter of a second of the time during which it was moving. The rate of progress being thus known, and the actuating force (gravity acting on the cradle and on the suspended weight) being also known, it will be understood that the exact amount of the resisting force, namely, friction, could be calculated exactly, and this for each moment and position of the descent of the experimental cradle.

The following results of these experiments were recorded in terms of the corresponding amounts of tractive force that would be required to produce similar results in the case of the ship and cradles, a weight of 12,000 tons.

| Velocity, feet per second | Force in tons required to move or restrain ship on incline of 1 in 12 | | | | | | Force in tons required to move a weight equal to the ship on similar ways, but on the level | | | | | |
|---|---|---|---|---|---|---|---|---|---|---|---|---|
| | 0 | 0 to 1 | .75 | 1 | 1.5 | 2 to 3 | 0 | 0 to 1 | .75 | 1 | 1.5 | 2 to 3 |
| **Rails and Cradle Bars** | | | | | | Retard- ing | | | | | | |
| No. 1 Ample lubrication | | ... | ... | 60 | 0 | 110 to 200 | | ... | ... | 1,060 | 1,000 | 890 to 800 |
| No. 2 Medium lubrication | | 120 | ... | ... | 0 | 60 | | 1,200 | ... | ... | 1,000 | 940 |
| No. 3 Very little lubrication | 400 | ... | 200 | ... | 0 | ... | 1,400 | ... | 1,200 | ... | 1,000 | ... |
| No. 4 No lubrication | | 560 to 400 | ... | ... | 0 | ... | | 1,560 to 1,400 | ... | ... | 1,000 | ... |

In every case where a velocity approaching to 8 feet per second was attained, whether the ways were lubricated or were quite dry, the model, though there was no tractive force acting on it other than that of gravity down the incline of 1 in 12, rapidly increased its speed till it reached the end of the ways.

These experiments are worthy of note for the contradiction, already referred to, which they gave to the received rules relative to friction. It will be seen by these experiments, and as will hereinafter appear from the results of the movements of the ship herself during the launch, that with different degrees of velocity very great variation in the friction was apparent, amounting to a difference of about thirty per cent in the case of unlubricated surfaces, according as the velocity was nearly *nil* or was 1.5 feet per second, which is a comparatively small velocity. The friction in this case was, on a weight of 12,000 tons, 1,500 tons at a very low velocity, and but 1,000 tons at a velocity of 1.5 feet a second or about one mile an hour, the friction at the very low velocity being fifty per cent greater than that at one mile an hour.

As soon as the ship began to move by slides, the recording apparatus of tape and pendulum was applied to record the nature of the ship's movements. This apparatus was similar to that already described. The tape was attached to the bottom of the ship under her centre of gravity, and recorded the rate of retardation of the ship when left to herself after the motive appliances had, with the exception of gravity, ceased to act; and the amount of friction acting to retard the ship was determined with a very considerable amount of accuracy.

The best experiments made were on December 7, 8, and 10; and the results are very interesting. The dirt and rust of the sliding surfaces had increased the adhesion very much, and a considerable force was necessary to start the ship. There being no good pressure gauges to the presses, it was impossible to decide exactly what was the force required to start the ship, for of course the tape record gave no information on this point; but there is no doubt it was considerable, probably 800 or 1,000 tons in addition to gravity, and was thus far greater than the force to start that had been observed with the experimental cradle.

But the remarkable fact was that, notwithstanding the deterioration of the sliding-surfaces as evinced in the increased difficulty in starting, the friction, when once motion was established, was proved not to be very largely in excess of that which had been exhibited in the experiments.

In the various experiments tried, it was shown that when the ship had a velocity of between 6 and 8 inches per second, the amount of friction was only about 100 tons in excess of the action of gravity down the incline; while as the velocity became less, the friction became greater, till, as the velocity became smaller, the friction increased from 200 to 300 tons in excess of the action of gravity.

The results obtained by the observations made on the motion of the ship having shown that its behaviour when in motion accorded with that of the experimental cradle, there is every reason to believe that if the ship had ever attained a velocity of 1½ to 2 feet per second, which might have happened had the river tackle acted well, the friction would, as in the experimental cradle, have become less than the action of gravity down the slope, and the brakes would have had to be employed to check the motion.

## NOTE B
*Letter from Mr. Brunel to W. Froude, Esq.*

February 2, 1858.

My dear Froude,—It is no news to you to tell you that we have floated, but still you will perhaps feel sympathetically some pleasure in hearing of it from me, as I do in writing to you upon it.

We have in fact gone on well and without mishap since we have resumed operations with plenty of power; we have not gone very quickly because our jumps have been small, or we have gone by a continuous motion—we have had a great deal of this, and all the last 30 or 40 feet I think, or more, has been so, the power being *with* gravity about a quarter of the weight, sometimes less—occasionally, when the water was high, considerably less—buoyancy being of course taken duly into the calculation. Once, when still weighing fully 3,500 tons, and with 1,200 tons of water in her, making 4,700 less some buoyancy of the cradle, she moved so easily that they came running to me from the other cradle, to say that she was moving of herself, and asking what to do.

She certainly had not much more pressure on than we had assumed to be necessary to overcome all the friction of thrust timbers, &c., certainly could not have had above 300 tons of real push to move

$$2{,}000 \text{ tons}, \; 2{,}000 - \frac{2000}{12}(\text{gravity}) = 1{,}833 \text{ tons}.$$

I think that when the load became much lighter on the rails, that the mud and even the sand of the Thames form a lubricator and *rollers* which offered less resistance than the dry rail, or the rail with pressure enough to displace the mud or imbed the sand.

Having at last pushed the cradles beyond the rail, and found her stand well, and moved a few feet more and still stand upright, I waited for a tide, and arranged a good communication with Liverpool and Plymouth to telegraph up wind and weather, morning and night, so as to help in foreseeing a tide.

On Friday morning, at 3 A.M., the tide began to improve, but the wind was still in the wrong quarter for a good tide, and as I think I have told you the tides of this month, and of the whole year of '58, are very poor. Besides this, it blew a gale. I therefore began pumping out the water, but stopped at 1,200 tons; and at 10 A.M., seeing no improvement in the weather, I filled in by the fire-engine about 1,300 or 1,400 more, and gave up the attempt, and very fortunately, as the tide did not rise high enough, and the wind right on her broadside increased.

On Saturday night things grew worse, and the wind at Liverpool and Falmouth finished in the evening still SSW. About midnight the wind lulled, the rain came down in torrents; the wind gradually stole round to the northward, and the tide came rolling up uncomfortably quick. I admitted water as soon as it would run in, and only just prevented her floating; when, if she had, we should have been in a mess, as our wedge-pieces were not yet unbolted from the bottom, nor could we in the night have managed the floating, even if all had been ready, which nothing was. Under these circumstances, the wedge-piece being bolted was perhaps our security, as she all but floated; the stern rose 3 inches though we had 3,000 tons of water in her, the tide rose so high; we began pumping immediately on the turn of tide, and by half-past 6 A.M. the wind was gentle from the NNE, and telegraphs came to announce the same in Liverpool, &c. I therefore determined upon floating.

We pumped away, and at 10 A.M. cleared everything away, and began our preparations. I had fixed 12 o'clock to begin forcing her further down, thinking half-past 12 would have been as early as she would have moved easily, but the tide came rushing up an hour and a half before its time; and, although I hurried the men to their posts, we were rather caught napping. The instant we began pressing, she moved easily, and by a quarter past 1 we had pushed her centre past the ways, and she began lifting with the tide.

I had a centre station with a chalk board, giving me the curve of the tide, and two good levels reading off her stem and stern water marks. Our calculations had proved very correct. In the skew position I had pushed her to, she began lifting at the stern first, as she ought to have done, and her bows soon followed the example. We stopped pushing the cradles, made them fast, and began hauling out the ship—she moved very slowly of course.

When she began to move out, our first difficulty occurred by a little mismanagement, or over confidence, on board; our stern chain was let go, and she forged ahead against the tide by the elastic strain of the bow chain, and *I think* also by the tug-boats pulling too much, which ought not to have been done, and in going ahead her port paddles caught hold of some cradle timber rising, with long 2½ inch bolts holding them to the bottom of cradle. No tug power would have had any effect upon them. But we had two hours before us, with a promise of an unusual tide; I hurried on board, and we succeeded in about twenty minutes in getting a little astern a little out, and getting clear.

We had some mishaps after this, such as fouling one of our barges, getting it jammed under the paddlewheels, and the barge fast by the chain tackle dropt overboard. However, we scuttled and sunk the barge, and got safely across. We had an extraordinary tide, and several assistances from nature to counteract any of our own bungling, and got safely across, and she is now moored in her right place.

Yours faithfully,<br>
I. K. BRUNEL.

William Froude, Esq.

1.  Shortly after the publication of this report, Mr. Brunel received a letter from Mr. G.W. Bull, of Buffalo, U.S.A., encouraging him to adopt the plan of launching sideways, as that was the way in which the large steamboats of the American lakes were launched. In the course of a correspondence which ensued, Mr. Bull gave much information as to the manner in which these launches were effected. He advocated a free launch for the 'Great Eastern.'

2.  For an account of this and other experiments and observations on friction, see note A of this chapter.

3.  The power mentioned here as applied for starting the ship was two hydraulic presses, to overcome adhesion at the first start. The river tackle was relied on to overcome the friction.

4.  See note A, p.283ff.

5.  The reason for stopping was lest the ship should float at high water during the night.

6.  It remained a doubtful point whether or not the bow cradle was stopped by the brakes. Subsequent experience favoured the opinion that it had stopped of itself.

7.  The ship had always been spoken of as the 'Great Eastern,' and this name was specially agreeable to Mr. Brunel. It was not a point on which he set much store, but

his views were pretty well known; and, as the name had not been of his choosing, but had rather grown out of the name of the Company, and the natural association of ideas with his first ship, the 'Great Western,' it might reasonably have been supposed that it would have been adopted. But some fastidious person suggested that the name was objectionable, as consisting of two adjectives. A list of names was prepared and submitted to Mr. Brunel on the day of the launch at the moment when he was busiest. He said off-hand that they might call it 'Tom Thumb' if they liked. The Directors, however, selected the name 'Leviathan,' and so the ship was christened. The new name never stuck to the ship, and she was registered as the 'Great Eastern.'

8.  The action of an hydraulic press is limited to the length of stroke of the ram. Hence, as soon as the ship moved a distance equal to the length of the stroke, it became necessary to relieve the water pressure in the cylinder, to pull back the ram into it, and to insert between the ram of the press and the cradle an additional length of timber. This operation of pulling back the ram and putting in another length of timber was called 'fleeting' the press, a term used in many mechanical operations when the motive arrangements are reinstated in their primary position ready for a further advance in the work.

9.  On this occasion Mr. Brunel had his way, and there were not a dozen spectators in the yard.

10. See note A, p.283ff.

11. At this time the friction when the ship was in motion was not much greater than it had been on previous occasions; but, the adhesion being greatly increased, a large excess of power became stored up in the elasticity of the abutments and the chains. This excess of power, on the ship commencing to move, discharged itself in giving her velocity, and therefore she moved by short slips, instead of by a gradual motion, as had been the case previously, when the resistance due to friction and that due to adhesion had been more nearly equal. For observations on the friction, see note A, p.283ff.

12. Mr. Stephenson went with Mr. Brunel into all the facts relative to the operations and all the experience gained therefrom, and examined carefully the investigations which had been made by means of self-recording apparatus, and which showed that the friction diminished as the velocity increased. So struck was Mr. Stephenson at these results that he urged Mr. Brunel on no account to neglect having the brakes properly attended to; lest the great adhesion should cause a large amount of stored-up power, which might give the ship such a velocity that the force of gravity might exceed that of the friction. The brakes had not been hitherto neglected; but, in conformity with Mr. Stephenson's suggestion, it was thenceforward made a rule that, when a moderate amount of slack had been overhauled from the drums, the brakes should be put down, on a signal being given that the pressure in the presses was rising.

13. Mr. Brunel received a great many suggestions for launching the ship. Puncheons, gas, artillery, and especially levers, were among the appliances recommended. With the desire of behaving civilly to well-meaning persons who were wishing to do him a service, Mr. Brunel adopted a plan similar to that which he had found useful

when the 'Great Britain' was on shore. He had a circular letter printed, thanking his correspondents for their suggestions, but saying that the number of such communications had become so great that it was impossible for him to do more than cause the receipt to be acknowledged, with thanks for the kind intentions of the parties writing.

This plan had, however, the effect of increasing the number of his correspondents, as several of them wrote a second time to express their regret that their letters had had no better effect than to be classed with the numerous communications which Mr. Brunel said he had received on the same subject, which had seemed unworthy of his notice; and they explained that though other people's schemes were, no doubt, worthless, still that their own, if adopted, would launch the ship.

14. The number of the presses being almost doubled, but the resistance of the ship not being much greater than before, the elastic compression of the abutments was less than it had been previously, so that when the ship moved there was much less work stored up to give it velocity; therefore the slides were shorter than they had been before.

15. Among the many congratulations which Mr. Brunel received on the completion of the launch, there was perhaps none which pleased him more than the following letter from Mr. Robert Stephenson, who had been prevented by illness from being present at the concluding operations, the critical character of which he had fully appreciated:—

February 1, 1858.

My DEAR BRUNEL,—I slept last night like a top, after I received your message. I got desperately anxious all day, but my doctor would not permit me to venture so far away as Millwall.

I do, my good friend, most sincerely congratulate you on the arrival of the conclusion of your anxiety.

Yours sincerely,

ROBERT STEPHENSON

A letter from Mr. Brunel to his friend Mr. Froude, describing the floating, is given in note B to this chapter.

# XIII

# The 'Great Eastern': Completion and Subsequent History

Soon after the launch of the 'Great Eastern,' efforts were made to obtain funds for finishing her, and Mr. Brunel proceeded to prepare designs with the view of obtaining tenders for the execution of the decks, skylights, fittings, rigging, &c. He obtained advice from persons thoroughly conversant with this class of work; and a specification was carefully prepared, providing for the completion of the ship in a perfect manner.

Meanwhile it had been considered that, large as the ship was, she might be profitably employed in the American trade, and that it might be expedient to run her on that line for a few voyages before placing her on the Eastern route. Captain Harrison went to America to examine the harbour at Portland, and brought back a favourable report of it.

All efforts to raise the funds for finishing the ship proved unavailing; and it was determined to reconstitute the company.

The new company, which was called the 'Great Ship Company,' was formed towards the end of the year 1858. In the beginning of December, Mr. Brunel was compelled by ill-health to go to Egypt for the winter. On leaving England, he strongly urged the Directors on no account to fail to make a strict contract, distinctly defining the work to be done, and the manner of its execution, as provided for by the specifications he had drawn out. But his advice was not followed.

After Mr. Brunel's return to England in May 1859, he continued to give the greater part of his time to the ship. The difficulties which he had to encounter were certainly neither fewer nor less vexatious than those which had arisen at earlier periods in her history; but they were the last with which he had to contend.

On September 5 he left her in the morning, feeling the commencement of the illness which ten days afterwards terminated fatally.

The ship left her moorings on September 7, and with the assistance of several tugs steamed down the river. She stopped for a night at Purfleet, and again at the Nore, and then left for Weymouth.

On the voyage a serious accident happened, which was made the subject of much misrepresentation.

Round each of the funnels of the paddle engines was what was termed a water-casing, or jacket, consisting of an outer cylinder, about 6 inches from the inner cylinder which formed the funnel. The top of the annular space between the cylinders was at about the level of the deck. From it a stand-pipe was carried up, which, after rising to a certain height, was turned over, and the end brought down into the stokehole. The object of this arrangement was to heat the feed-water before it entered the boiler, and at the same time to keep the saloons cool, through which the funnels passed. The arrangement of the stand-pipe gave this advantage, that when the head of water in the heater and stand-pipe together became equivalent to the pressure in the boiler, the water could be run into the boiler by gravitation. The stand-pipe at the same time, being open to the air at the top, formed a safety-valve to the water-heater.

For the purpose of testing the joints of the jacket with water pressure, while the ship was being finished, a stop-cock had been placed on the stand-pipe, which unfortunately had not been afterwards removed. While the ship was proceeding down Channel, the donkey feed-pumps were not working well, and to ease them it was thought better to cut off the water-heater, and to force the water direct into the boiler. The communication of the water-heater with the boiler was therefore cut off; and, as was afterwards ascertained, the stop-cock at the top of the water-heater had been also closed. The water confined in the heater soon produced steam, and when the ship was off Hastings the casing exploded. The funnel was thrown up on to the deck, and a body of boiling water and steam was driven down into the boiler room, severely injuring several of the firemen, who afterwards died.

That the effects of this accident were confined to one compartment of the ship, was due to the complete protection afforded by the transverse bulkheads

After she arrived at Weymouth the funnel was repaired; but as an outcry was raised against the water-heaters, it was thought desirable, from deference to public opinion, to discontinue their use; although this accident had not in any way proved them to be objectionable, and they are now generally adopted.

While the 'Great Eastern' was at Weymouth Mr. Brunel died.

Many visitors went in the ship when she left Weymouth on a trial trip to Holyhead. At Holyhead she lay in a somewhat exposed situation; and the sudden storm came on in which the 'Royal Charter' was lost. The great advantage of having both paddle and screw was now, for the first time, felt. A portion of the temporary staging erected by the contractor at the breakwater was carried away, and drifted down upon the ship. During the gale her engines were kept going, in order to relieve the strain on the cables. The timbers of the staging got foul of both paddlewheels, and screw; but, as it was always possible to keep one of the engines at work, the ship was saved from drifting.

The season being now too far advanced for a profitable voyage to America, the ship left Holyhead and went to Southampton Water for the winter, where several alterations and additions were made.

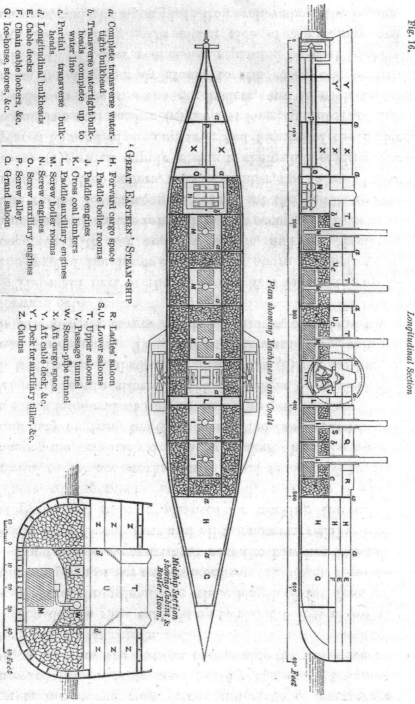

Fig. 16.

Longitudinal Section

Plan showing Machinery and Coals

Midship Section showing Cabins & Boiler Room.

a. Complete transverse water-tight bulkhead
b. Transverse watertight bulk-heads complete up to water line
c. Partial transverse bulk-heads
d. Longitudinal bulkheads
E. Cable decks
F. Chain cable lockers, &c.
G. Ice-house, stores, &c.

'GREAT EASTERN' STEAM-SHIP

H. Forward cargo space
I. Paddle boiler rooms
J. Paddle engines
K. Cross coal bunkers
L. Paddle auxiliary engines
M. Screw boiler rooms
N. Screw engines
O. Screw auxiliary engines
P. Screw alley
Q. Grand saloon

R. Ladies' saloon
S.U. Lower saloons
T. Upper saloons
V. Passage tunnel
W. Steam-pipe tunnel
X. Aft cargo space
Y. Aft cable deck, &c.
Y'. Deck for auxiliary tiller, &c.
Z. Cabins

NOTE.—The masts of the ship, six in number, are not shown on this woodcut.

In Mr. Brunel's report of February 5, 1855, printed above, at p.236, he describes the leading features of the 'Great Eastern' as she was then being constructed, but a more detailed account of them will fitly precede the history of her career as a passenger-ship.

The main arrangements of the ship are shown in the woodcut (fig. 16, p.293).[1]

The ship is 680 feet long, 83 feet wide, and 58 feet deep. Her gross tonnage is 18,915 tons. She is divided into water-tight compartments by ten bulkheads (a and b), all of which, except two (b), extend completely across the ship, and up to the upper deck. These two are complete to 6 feet above the 28-foot water line. In addition there are partial bulkheads (c), which form the ends of coal bunkers, and aid materially in strengthening the flat bottom of the ship. The more remarkable parts of the construction of the ship will be understood by means of the transverse section. The bottom is made double, and between the two skins are webs, running longitudinally. Mr. Brunel considered that the double skin would greatly diminish the chance of such an accident occurring as would cause any of the compartments to be filled with water. The material being arranged in the direction of the length of the ship is all capable of taking part in the strains that are thrown on the bottom, as well as on the top, by forces tending to bend the ship.

Mr. Brunel also made the upper deck cellular, in order to resist the compressive strain that would come on it when the ship was heavily loaded in the middle of her length. Great additional strength to the ship, considered as a girder, is given by two longitudinal bulkheads, 36 feet apart, extending for 350 feet. These bulkheads, with the sides of the ship, form the vertical web plates of the girder. Her structure resembles the tubes of the Britannia bridge; the cellular top flange being connected with the cellular bottom flange by plate-iron webs.

The two skins of the ship, with the web plates between them, forming the cellular bottom of the great girder, may also be considered as a number of smaller girders placed side by side, each resisting the excess of the pressure of the water over the load that may happen to be resting on it inside the ship. The difference of pressure or upward strain is transmitted by the cross bulkheads (a, b, c) from the bottom of the ship to the sides and longitudinal bulkheads.

The double skin extends up to about 6 feet above the water level throughout the whole length of the ship, with the exception of the extreme ends.

The foremost compartment next the bow has two cable decks (E), with capstans and all the necessary riding-bitts, stoppers, and other appliances for working the cables. These arrangements answer well, and the 3-inch chain cables are worked with great facility. The capstans were originally driven by a shaft from the paddle auxiliary engines, but this was found inconvenient, and a small independent engine has been put to work them. The cables are stowed in chain lockers on a deck (F) below the cable decks, and below this (G) are ice-houses and store-rooms. The next compartment of the ship (H) is intended for cargo. It is at present occupied by the forward cable tank.

The main part of the ship, 350 feet in length, up to the level of the lower deck, 34 feet above the bottom, is occupied with her engines, boilers, and coal bunkers. The space above the lower deck was occupied with saloons and cabins for passengers, except at the paddle engine room (J). The boilers, four in number, two in each boiler room (I), which supply steam to the paddle engines, are placed forward of the engines; and forward of the boiler rooms is a coal bunker (K), 20 feet long; and abaft of the paddle engines are the six boilers, two in each boiler room (M), that supply steam to the screw engines (N). These three boiler rooms are separated by coal bunkers (K), 20 feet long. On either side of the boilers and engines, and also upon plate iron arches above the boilers, are bunkers for coal. This will be seen on the transverse section of the ship. Between the paddle engines and boilers is a water-tight compartment, 10 feet long (L), in which are placed a pair of auxiliary engines of sixty horse-power, which pump water out of the ship and also work fire-pumps. There are two other auxiliary engines of sixty horse-power in a compartment (O) aft of the screw engines, intended to keep the screw propeller turning round, either when the ship is at anchor, to relieve the strain on the cables, or when, for any reason, she is only using her paddle engines. They also work bilge-pumps and fire-pumps. Each set of auxiliary engines has two independent high-pressure boilers. Throughout the bottom of the ship there are two bilge-pipes, fitted with valves, with branch-pipes leading to the various compartments of the double skin. These bilge-pipes can be connected with either of the auxiliary engines, and so the water can be pumped out of any part of the ship.

The paddle engines, of 1,000 nominal horse-power, consist of four inclined oscillating cylinders, 14 feet stroke and 6 feet 2 inches diameter, each pair of which work on to a single crank. There are means for disengaging either paddlewheel from the engines.

The screw engines, of 1,600 nominal horse-power, consist of four fixed horizontal cylinders, 4 feet stroke and 7 feet diameter, the two cylinders of each pair working opposite to each other on one crank.

In each boiler room are two donkey engines for supplying the boilers with water, and the main engines are also fitted with feed-pumps. Each of the donkey engines is capable of pumping water out of the ship, and of being used in case of fire.

The screw shaft passes along what is termed a screw alley (P). The weight of the screw rests on the bearing where the screw shaft passes out through the stern-post of the ship. To prevent the water that leaks in through this bearing from penetrating into the screw alley, there is a bulkhead with a stuffing-box round the screw shaft, a short distance forward of the stern-post. To enable access to be gained to the bearing, a tube was provided leading down from the main-deck. This was intended to be fitted with appliances for pumping air in so as to drive the water out, and to admit men to get at the bearing under air pressure.

Another arrangement was also provided for the same purpose. On the after side of the stern-post was placed a ring of india-rubber; and, by pulling in the screw-shaft, the screw was pressed tightly against the india-rubber ring, which prevented the water from entering. By means of this arrangement the stern-bearing was examined and repaired at Southampton.

The screw propeller has four blades, and is 24 feet in diameter and 44 feet pitch. The paddlewheels were 56 feet in diameter, with 30 floats, each 13 feet broad and 3 feet deep.

Over the boiler rooms run two tunnels; one of them (V) serves as a passage to enable the engineers to pass from one compartment to another throughout the part of the ship occupied by the boilers and engines. The openings leading from the tunnel to the engine or boiler rooms are provided with watertight doors, which can be shut in the event of any of the compartments getting full of water. The other tunnel (W) serves as a passage for the steampipe, which leads from the boilers to the engines. Though the boilers are divided into two sets, one for the paddle engines and one for the screw, the steampipes are connected, so that the whole of the ten boilers, or any of them, may be used to supply steam for either of the engines.

In one point a deviation was made from Mr. Brunel's arrangements. It was his intention that there should be no apertures in the water-tight bulkheads except at the tunnel (V), from which the various boiler and engine rooms were to be entered by openings, to be closed by water-tight doors. The tunnel was placed high up, so that in the event of a leak there might be ample time to close the door. The inconvenience of ascending and descending by ladders was, however, considered an evil; and it was found necessary at times to carry coals from one compartment to another. For these reasons, upon the requirement of the Board of Trade, doorways, fitted with sliding doors, which can be closed by handles on the upper deck, have been cut in the bulkheads between the boiler rooms. This arrangement exists in other ships; and in the 'Great Eastern,' even without closing these doors, there are eight watertight compartments.

At the stern of the ship (Y) on the main-deck are arrangements for working cables, similar to those at the bow.

These appliances are required when the ship has to be moored so as not to swing with the tide; and they would allow of the ship's riding by the stern instead of by the bow, which it might often be useful for her to do in narrow waters. On the lower deck (Y′), at the stern of the ship, is a spare tiller and wheel for working the rudder, in case anything goes wrong with the main tiller, which is on the upper deck. The weight of the rudder is carried on the lower deck by a grooved collar, resting on a ring of cannon balls.

The compartment (X) immediately aft of the screw engine room is for cargo.

The saloons and cabins are all in the middle of the length of the ship, where there is least motion. The usual plan of putting the first-class passengers at the stern was not adopted, and they were placed forward. As the smoke generally drifts towards the stern of a ship, the first-class passengers in the 'Great Eastern' are not annoyed by the smoke, or by the dust and smell from the boiler and engine rooms.

The transverse section shows the arrangement of the cabins.

Mr. Brunel intended that the upper saloons (T) should be used as sitting-rooms, and the lower saloons (S and U) as dining-rooms. These were to be lit and ventilated by shafts on either side of the upper saloons, rising up to the skylights on deck. The smell of dinner was thus to be kept away from the sitting-rooms. The cabins (Z) on either side were to be approached from the saloons by passages and steps, as shown on the section. The saloon marked (Q) is the first-class, and (R) the ladies' saloon.

The ship is rigged with six masts. The arrangements of the masts and rigging were especially intended for the Eastern voyages. At the extreme bow and stern are low masts, which carry trysails and staysails. The sails on these masts were chiefly intended for manœuvring the ship. She also has three large masts, the lower masts being of iron. The two foremost of these are square-rigged, and all three of them have trysails and staysails. The aftermost of the three was also made strong enough to be fitted with square yards, in the event of its being desirable to rig it in that manner. Aft of the three principal masts is a large mast; only intended to be rigged with fore-and-aft canvas. The ship has no bowsprit, the stay of the foremost mast being attached to the stem inside the bulwarks.

In the beginning of June 1860, the ship made a trip from Southampton down the Channel, as far as the Start Lighthouse and back, to try her engines; and on June 17 she sailed for New York, under the command of Captain Vine Hall, who had succeeded Captain Harrison. She arrived there, after a prosperous voyage, on June 28, and was received with great enthusiasm. The ship left New York on August 16, and, having called at Halifax, arrived at Milford Haven on August 26.

In the course of these two voyages the stern-bearing of the screw-shaft, which was in white metal, had worn down between two and three inches. With the view of allowing for any depression of the stern-bearing, the lengths of the screw-shaft were not rigidly connected throughout, but the end length, attached to the screw, was coupled to the remainder of the shaft by a universal joint, consisting of two double cranks. Thus the two shafts might be, to a considerable extent, out of line, and yet revolve efficiently.

When the ship returned to Milford Haven, a gridiron was prepared on the beach, and she was grounded on it; and the screw-shaft was drawn in. By this time it had become the general opinion that a shaft cased with brass, and running in lignum-vitæ bearings, was the best. A lathe and machinery for making the shaft revolve were provided, and fixed in the stern of the ship; the shaft was turned, and brass collars shrunk on. The bearing was made with lignum vitæ, and the brass-covered shaft replaced in it. It has since worked well, and has shown no signs of wear.

The ship, commanded by Captain William Thompson, left Milford Haven for New York on May 1, 1861, and returned to Liverpool from New York on June 4, having made an average speed of 13¾ knots on the outward and 14 knots on the homeward voyage.

At this time the Government determined to send her out with troops to Quebec, and she was fitted up for that purpose. She took upwards of 2,500 troops, and about 40 passengers. There were, altogether, about 3,000 persons on board, and 200 artillery horses. Scarcely any of the troops were placed in the regular passenger part of the ship, as they were accommodated in the cargo departments (H, X). Thus a much greater number of men might have been carried in her with perfect comfort. She was commanded on this voyage by Captain James Kennedy, of the Inman service. She left Liverpool on June 27, and arrived at Quebec on July 7. She returned to Liverpool with about 500 passengers in August.

By this time her superiority had become recognised by the regular travellers between England and America. Those who had been in her found that, while they passed other ships rolling and pitching in the sea, the deck of the 'Great Eastern' was so steady that it was difficult to believe that there was a gale blowing;[2] and when, after a continuance of heavy weather, she began to roll, the motion was so slow and easy as to be comparatively unimportant.

When she left Liverpool again, there were a considerable number of passengers, and it seemed as if her success was ensured. She started under the command of Captain James Walker, on September 10, and three days afterwards encountered a severe Atlantic gale.

The ship was behaving well, when one of the boats, which hung on davits outside the ship forward of the paddlewheels, got adrift on the weather side. Fears were entertained that it might foul the paddlewheel, and the captain determined to cut it away. The direction of the ship was altered, in order that the boat might float clear. The ship then resumed her course; but shortly afterwards fell off, with her broadside to the sea.

Relieving tackles having been put to assist the men at the wheel, the tiller was kept hard over, to bring the ship's head to the wind; but with no effect. Towards evening, as the seas beat heavily against her side, first one paddlewheel and then the other was destroyed, being completely torn away from the central bosses. During the night she lay in a helpless condition. The gale had been of some duration, and the waves being large and long, the ship was placed in a very unfavourable position to receive them; and she rolled considerably.

The next morning, when an officer went to examine the auxiliary tiller on the lower deck, he discovered that the rudder-head was twisted short off just above the point where it entered the ship. The rudder was still in its place. The accident had most likely happened on the previous evening, when the ship fell off her course. It had not been noticed by the men at the helm, perhaps because there were so many of them at the wheel and relieving-tackles that they held one against the other; and the broken parts of the rudder-head, grinding together, threw jerks on to the tiller. The fracture of the rudder-head was caused by the badness of the workmanship in the interior of the forging.

Attempts were made to get sail on the ship, but without much success; and with the hope of bringing her head to wind, the screw engines were reversed.

As soon as the ship was driven astern by the screw, the rudder, being uncontrolled, was forced round by the rush of water, and it knocked away the after stern-post.[3] No other harm was done, as the rudder is secured by a pin into the heel of the ship, and by a collar round the rudder-head, attached to the hull above water.

Steps were then taken to get command over the rudder. Chains were wrapped round the stump of the rudder-head inside the ship, and a certain amount of control was thus obtained. A more effective plan was at the same time carried into execution. A man was lowered by a rope from the stern of the ship, who hove a piece of wood, with a line attached to it, through the screw-opening. The wood with the end of the line was caught with a boat-hook; and a rope, and afterwards a hawser, and then a piece of the ship's chain cable were passed through behind the rudder. The two ends of the chain cable were brought together at the stern of the ship, and a large shackle put round both parts of the chain, and shaken down till it held them together. In the edge of the rudder-blade a notch had been made by the rudder striking against the screw, and into this notch the shackle was made to drop. In this way two chains or pennants had been attached to the back of the rudder. One of these was brought to each side of the ship, and they were hauled on by means of the stern capstan.

The ship then turned homeward, and the weather having moderated, she arrived off Cork harbour on the afternoon of September 17. By this time the chains round the rudder had shifted, and were of little service; and before night it began to blow heavily towards the shore. It was dangerous for the ship to remain on a lee shore; and, although the steering-gear was out of order, the captain wisely determined to take advantage of the ship's head pointing in the right direction, and steamed out to sea.

Three days afterwards the ship, assisted by several small steamers; was got safely into Cork harbour, a temporary tiller was attached to the stump of the rudder-head, and she proceeded to Milford Haven, where she was placed on the gridiron, and her after stern-post and paddlewheels replaced. The accident had proved that the original paddlewheels might with advantage have been made stronger, and in the new wheels the bracing was increased.

The ship, under the command of Captain Walter Paton, left Milford Haven for New York on May 7, 1862, and returned to Liverpool on June 11. She left again on July 1, and returned on August 6. Besides the number of passengers the ship accommodated, she carried a considerable amount of cargo; she brought over large quantities of grain and provisions. The custom of carrying this class of freight in steam-ships received a great impulse from the success of the 'Great Eastern' in the traffic. As it was found that the shallowness of the bar at Sandy Hook prevented her taking full advantage of her carrying power, she had on this voyage followed the route along Long Island Sound, so as to arrive close to New York in deep water, and on her return voyage she brought as much as 5,300 tons of cargo in bulk, which with 4,350 tons of coal gave her a mean draught of 28 feet.

She left Liverpool on August 17, and arrived off Montauk Point, at the entrance to Long Island Sound, at about two in the morning of the 27th, to take in the pilot. While stopping, a loud rumbling noise was heard, and presently the ship heeled slightly over to one side. The pilot, when he came on board, said that the ship had passed over a reef of sunken rocks, which was not marked on the charts.

It was at the same time found that many of the spaces between the double skins were full of water. The ship went on to New York, and most of the passengers landed in ignorance of an accident which in any other vessel would have been fatal.

Steps were at once taken to examine the damage, and the divers reported a large fracture in the outer skin 80 feet long and about 10 feet broad. They also discovered afterwards several smaller fractures. It was considered that this damage might be mended while the vessel was afloat, and a very skilful arrangement, contrived by Messrs. Renwick, of New York, was adopted. A large wooden barge was made with a gunwale shaped to fit the ship's side, and two wooden passages leading down into the barge. It was placed so as to cover the large fracture, and was secured by chains passed round the bottom of the ship. The joint between the gunwale of the barge and the ship's side was made water-tight; the water was pumped out, and men and materials passed down through the shafts. By the exertions of those engaged in this difficult operation, the great fracture was repaired, and the ship returned to England in the beginning of 1863.

A gridiron was made at Birkenhead; the ship was placed on it, and the repairs were proceeded with under the direction of Mr. Brereton, who at Mr. Brunel's death succeeded him as engineer to the Great Ship Company. On examination, it was found that fractures had been made in ten separate places in the outer skin.

The ship started again in May 1863, and made three voyages to New York and home. At this time, however, there was a severe competition with other vessels, and the Company could not afford to run the ship unremuneratively for any length of time. In 1864 she passed into the hands of a new company, which consisted almost entirely of those who, from their belief in the capabilities of the ship, had found the money for starting her again after each of her successive misfortunes.

The ship lay idle for some months, and was then chartered by the Telegraph Construction and Maintenance Company.

The 'Great Eastern' had in the course of the four years from 1860 to 1863 made nine voyages across the Atlantic and back. Though this was not the route for which she had been intended, it had given her many opportunities of showing her merits. Adverse fortune had added to these opportunities, and had at the same time demonstrated the necessity of many of the precautions which Mr. Brunel had taken to ensure her safety.

The construction of the hull of the ship has been proved by experience to possess the advantages anticipated by Mr. Brunel.

Its strength as a whole has been proved by the absence of all signs of weakness in the heavy weather she encountered on several occasions, and especially in the gale of 1861, when her rudder was disabled. The strength of the ship has since been fully tested by the enormous loads she has carried while on telegraph cable expeditions.

The importance of the double skin was shown on the occasion of her grinding over the rocks at Montauk Point, when so large a number of leaks were made in all parts of her bottom that no ordinary system of bulk-heads would have saved her from foundering. Moreover, the space between the two skins was sufficient to allow of the outer skin and the webs being crushed to the extent of three feet, while they at the same time acted as a buffer and prevented the inner skin from coming in contact with the rock.

The engines have not had any opportunity of working at high speed in the long voyages for which they were intended, but in the rapid and comparatively short passages made by the ship across the Atlantic they worked with great regularity and success. Although they were commenced seventeen years ago, they are still fine specimens of marine engines, and bear witness to the care taken by their builders in their design and manufacture.

Of the points on which Mr. Brunel laid stress, and which, as he remarked, 'involve no other risk than that of being useless; they cannot do mischief,' many have now come to be considered essential parts of good marine engines. He thought it of great importance that the steam cylinders should be jacketed, especially at the ends; and it was intended that the high-pressure steam for this purpose should be taken from the auxiliary boilers. This plan was, however, not adopted.

Mr. Brunel was also anxious that the steam should be heated immediately before it entered the cylinders, and that fresh water should be supplied for the boilers by using the same water over again. Arrangements for effecting these objects have since been brought into general use.

The advantage of having several cylinders to each engine was shown while the ship was running to America. Part of the valve-gear of one of the paddle-engine cylinders gave way, the engines were stopped for four hours, and the ship was propelled by the screw alone. The cylinder was disconnected and turned back out of the way, and the engines worked efficiently with three cylinders for four passages across the Atlantic.

The average speed of the ship on those voyages in which her performances were fairly tested was about 13½ knots. On two occasions she made the voyage from New York to Liverpool at an average speed of 14 knots; and she maintained her speed in rough weather and head winds to a much greater extent than is the case in smaller vessels.

By having two sets of engines, the ship was saved from serious disaster at Holyhead, and again in the gale of September 1861.

The great handiness of the ship is one of the many beneficial results of the use of both paddles and screw. By working the paddlewheels astern and the screw ahead, she can be kept from moving forward or backward, and at the same time the stream of water from the screw, acting on the rudder, makes her answer her helm and turn round on her centre. By modifying the speed of the two engines, she may be made to creep slowly forward; and as the rudder is in the full rush of the water driven back by the screw, the ship has practically as much steering power as she has when moving rapidly. The importance of this power of controlling her when passing through narrow channels, in entering and leaving port, can scarcely be over-estimated.

The career of the 'Great Eastern' since the formation of the present Great Eastern Steam-Ship Company, has been prosperous.

In the commencement of the year 1864, when Mr. Cyrus Field had succeeded in reviving the project of laying the Atlantic cable, the Telegraph Construction and Maintenance Company took the contract for laying it, and hired the 'Great Eastern.' She was brought round to Sheerness, where the cable tanks were fitted into her. One tank was placed in the forward cargo compartment, and another in the after cargo compartment; and the largest tank was placed over the middle screw boiler room, the funnel being removed. Therefore, during the cable-laying expeditions, the ship only used eight of her boilers.

The history of the laying of the Atlantic cable is well known. The 'Great Eastern' started from Valentia on June 23, 1865, under the command of Captain (now Sir James) Anderson, and the cable was laid more than half way across the Atlantic; but, on hauling in to recover a fault, it was broken, and dropped to the bottom of the sea.

The grappling tackle was not sufficiently strong. The cable was three times partially raised, and each time lost; and the expedition returned to England defeated, but with the knowledge that ultimate success was certain.

The engineers and scientific men on board the 'Great Eastern' drew up a memorandum as to the results of this expedition, and, among other things, stated—'That the steam-ship "Great Eastern," from her size and constant steadiness, and from the control over her afforded by the joint use of paddles and screw, renders it safe to lay an Atlantic cable in any weather.'[4]

Sufficient additional cable was made to lay a second one and to finish the old cable when it should be recovered. The ship started again on July 13, 1866, and laid the cable across the Atlantic without the slightest mishap. She then returned, and after three weeks of hard work, the end of the cable which had been lost the year before was picked up, and completed to Newfoundland.

On her return to England there did not seem to be any immediate employment for her in cable laying, and the tanks were taken out. In the following year a company was formed in France to charter the ship, and to work her between New York and Brest during the French Exhibition. She made one voyage, from Liverpool to New York and back to Brest and Liverpool; but the undertaking was a commercial failure.

She remained at Liverpool till October 1868, when it was proposed to lay a cable from France to America; and she came round to her former berth at Sheerness. Tanks were made in the fore and aft cargo spaces, and a very large tank, 75 feet in diameter, was placed amid-ships. The bulkheads were cut partially away to make room for it, the ship being strengthened above and below. She started on this expedition from Brest under the command of Captain Robert Halpin, and encountered very heavy weather. The cable was laid successfully.

The ship then returned to England, and was fitted out for the greatest adventure she has yet undertaken. She was to proceed with a full cargo of cable round the Cape of Good Hope to Bombay, to lay it thence to Aden, and from Aden a portion of the way up the Red Sea. With her cable and coals, on leaving England, she drew 34 feet 6 inches, with the enormous displacement of 32,724 tons. She laid the cable with perfect success, and returned to England.

Throughout all these cable-laying expeditions, and especially in the work of picking up the Atlantic cable of 1865, the good qualities of the ship have been fully exhibited.

The later voyages of the 'Great Eastern' were undertaken for the accomplishment of a work in which Mr. Brunel had felt a great interest.

In 1856, Mr. Cyrus Field came over to this country in order to consult English engineers and scientific men upon his project of an Atlantic cable. It is stated in a work written by Dr. Field,[5] that—

> From the beginning Mr. Brunel showed the warmest interest in the undertaking, and made many suggestions in regard to the form of the cable and the manner in which it should be laid. He was then building the 'Great Eastern,' and one day he took Mr. Field down to Blackwall to see it, and said, 'There is the ship to lay the Atlantic cable.'

It appears, however, from drawings in Mr. Brunel's sketch-books that at this time, and again in 1858, he thought that it would be better to have a vessel specially built for the work.

He was throughout sanguine as to the ultimate success of the undertaking, as is shown by the following extract from a letter written in December 1856:—

> I would suggest a more moderate expression of doubts of the successful results of the American cable. The impossibility of running steamers profitably over the surface of the same sea was, though it is now denied, asserted and proved from established facts just as clearly as the impossibility asserted now to exist in respect of the electric telegraph. It is a pity in these days to lay down any such dogma. Every day's experience proves that nine-tenths of them are refuted; that the circumstances do not prove to be such as are assumed, or the difficulties are overcome; and however correct the arguments may have been, the result is not as predicted.

The 'Great Eastern' has not yet been engaged on the work for which she was originally designed by Mr. Brunel; but her employment in the promotion of great scientific enterprises has been an occupation worthy of her connection with his name.

## NOTE
### Dimensions of the 'Great Eastern' Steam-ship

| | | |
|---|---|---|
| Extreme length | 693 | feet |
| Length between perpendiculars | 680 | " |
| Breadth | 83 | " |
| Depth | 58 | " |
| Greatest draught of water | 30 | " |
| Registered tonnage | 13,343 | tons |
| Gross tonnage | 18,915 | " |
| Displacement at 30 feet draught | 27,419 | " |

### DIMENSIONS OF PADDLE-ENGINES
(1,000 nominal horse-power.)

| | |
|---|---|
| Number of cylinders | 4 |
| Diameter of cylinders | 6 feet 2 inches |
| Length of stroke | 14 feet |
| Number of boilers | 4 |

### DIMENSIONS OF SCREW-ENGINES
(1,600 nominal horse-power)

| | |
|---|---|
| Number of cylinders | 4 |
| Diameter of cylinders | 7 feet |
| Length of stroke | 4 " |
| Number of boilers | 6 |

1.  The principal dimensions of the ship and engines are given in the note to this chapter, above
2.  During the last-mentioned voyage, returning from Quebec, a north-easterly wind blew with a velocity of from 30 to 40 miles an hour. For the first 24 hours the ship paid no attention to the sea; the next day, the wind remaining the same, but the sea having got more swell on, the ship began to roll slowly and sedately, the rolls gradually increasing up to about 9 degrees on either side of the perpendicular, and dying out again and then recommencing.

3. Between the screw and the rudder was what was termed the after stern-post, which had no duty to perform except to steady the heel of the ship, into which the rudder was stepped.

4. Before the ship was finished, Mr. Brunel had her stability and rolling carefully investigated. Calculations were made to ascertain the position of her centre of gravity and the other necessary elements for determining her stability and period of rolling.

He also had a model made, with arrangements for altering the levels of weights placed inside. By this means the results of the calculations were verified. It was determined that the ship would make a single roll from one side to the other in about six seconds. While she was in the Thames a steamer struck the hulk alongside and gave the great ship a slight impulse. Mr. Brunel, who was on board, took advantage of the opportunity to observe the period of the roll she made, and was pleased at finding it agree with the calculated period. It is to the investigations initiated at that time by Mr. Brunel that are due the great steps since made in the knowledge of the laws which govern the rolling of ships. Had Mr. Brunel lived he would no doubt have taken the same pains to record the rolling of the 'Great Eastern' as he had in the case of the 'Great Western' when, in 1839, he sent an assistant to America and back, who took observations of the rolling and pitching of that vessel in several voyages. These observations were made by a simple angle-measuring instrument, adjusted by the visible level line of the horizon, and not by the fallacious method of noting down the swings of a pendulum.

The 'Great Eastern' remains almost perfectly steady in ordinary rough seas. When the seas become very long, so that their period is nearly the same as that of the ship, she rolls; though the number of degrees on either side of the perpendicular is not large. By stowing the weights of cargo high in the ship, the tendency to roll has been much diminished, and when engaged in cable laying, with the enormous weights in the cable tanks all placed above the lower deck, she is remarkably steady.

5. *History of the Atlantic Telegraph:* New York, 1866.

# XIV

# DOCK AND PIER WORKS

MR. BRUNEL's dock and pier works are interesting, not only in their general features, but also in the details of their construction; and the plans he made for large docks at Monkwearmouth in 1831, which he carried out on a smaller scale shortly afterwards, were among the earliest of his independent designs.

With the exception of the gates of this dock, which are of timber, all the dock gates Mr. Brunel constructed are of wrought iron. This material had been employed in ship-building before Mr. Brunel adopted it for the gates of the new lock at Bristol, and it was also beginning to be extensively used for bridge girders.

At the same time that he introduced the use of wrought iron into dock gates, he constructed them with a large amount of buoyancy, in order that they might be moved easily, while being opened and shut.

The dock and pier works which he constructed are at Monkwearmouth, Bristol, Plymouth, Briton Ferry, Brentford, and at Neyland, Milford Haven. They will be described in this order, which is nearly that of the dates of their construction.

### Monkwearmouth Docks

The town of Monkwearmouth is situated at the mouth of the river Wear, on the north side, and opposite to the towns of Sunderland and Bishopwearmouth, which extend for about a mile along the south side. In order to improve the entrance of the river, and to diminish the sand-banks which lay near its mouth, piers were proposed as early as the middle of the last century, and were partly built on both sides of the river before the year 1800. From that date until 1831, although the question of making docks had been considered, and designs proposed by different engineers, no steps had been taken for their construction, and the only works executed for the improvement of the port were the extension and alteration of the piers already existing. In 1831 designs for docks to accommodate the increasing traffic were made simultaneously by Mr. Brunel and Mr. Giles. Mr. Brunel's docks were to have been on the north side of the river, and to have had an area of 25 acres, with quays, warehouses, &c. Mr. Giles's were to have been on the south side.

Neither of these schemes was approved of by Parliament; but shortly afterwards a private company was formed for the construction of a dock on a plan designed by Mr. Brunel, though on a much smaller scale than his scheme of 1831, the dock being only about 6 acres in area, with a tidal basin of about an acre and a half. The company encountered considerable opposition from the authorities of the town of Sunderland, but succeeded in obtaining a royal charter for the construction of the dock. They subsequently obtained an Act of Parliament empowering them to make the entrance from the dock to the river. The dock was constructed, and eventually became the property of the North Eastern Railway Company, to whom it now belongs; they have erected coal drops along the quay, and have made it a shipping place for collieries connected with their railway.

The work was begun in 1834, and the dock and tidal basin occupy part of the site chosen by Mr. Brunel for his larger scheme of 1831.

The quay wall was built with a curved batter, the chord line joining the top and bottom having an inclination of 1 in 5. The masonry was carried up in courses, and made solid by filling every part thoroughly with mortar. A course at the face and at the back of the wall was built up; an abundance of mortar was then spread in the heart of the wall, and the stones built in the mortar. Thus no crevices could be left in any part of the work, and the back of the wall was soundly built throughout.

The entrance to the dock is 45 feet wide, with side walls of the same profile as the quay wall. Except at the gate floors, there is a segmental invert of dressed stone of such curvature that it is 6 feet 6 inches lower in the middle than at the side walls. The gate floors are formed with inverts, curved to correspond with the under sides of the gates.

The masonry of the entrance was executed within a four-sided coffer-dam, the sides of which were slightly convex outwards. This coffer-dam was constructed in the usual way, there being two rows of close piling with puddle between them; and it was strengthened by internal horizontal shores which connected the opposite sides, and by diagonal bracing. The piles were driven until they met with so much resistance as to render it unsafe to drive them farther. When the ground inside the coffer-dam was excavated, it was found that the piles had been driven into sand and gravel, and that, to enable the masonry to be built on a good foundation, it would be necessary to excavate about 7 or 8 feet below the piles. They were therefore driven down gradually, as the ground was removed from the inside, until the requisite depth was obtained. The whole coffer-dam was thus an immense caisson, the sides of which were lowered by gradual driving, instead of being simultaneously forced down by weights.

The masonry of the walls of the tidal basin is similar to that of the walls of the dock; some parts of the foundation were laid by means of a diving-bell.

In the entrance between the dock and the tidal basin there is a pair of gates pointing inwards, which serves to retain the water in the dock during the fall of the tide, and there is also a pair of storm gates pointing outwards, which protects the inner pair from the force of the waves.

The construction of both pairs of gates is similar.

The two leaves of each pair meet at an angle of 125°. Each leaf is about 30 feet long; the bottom beams are curved to the form of the segment of a circle; the height at the meeting-post is 27 feet, and at the heel-post 22 feet (see woodcut, fig. 17). This arrangement is, to a greater or less extent, followed in the dock gates Mr. Brunel afterwards constructed. By raising the pivot, the gate floor can be made of ample strength, and the cills and heel-posts are free from mud and deposit. The gates are constructed of horizontal beams of yellow pine timber, 21 inches thick, placed close together for a height of 12 feet above the bottom. Above this there are beams of timber and of cast iron at intervals. The whole is planked over on the inner side with 4-inch planking. The heel-post and meeting-post are socketed into cast-iron uprights, which also receive the ends of the horizontal beams. To preserve the gate from any change of form, a diagonal iron tie-bar extends from the top of the heel-post to the timber beams forming the lower part of the gate.

Nearly under each meeting-post is placed a bevelled cast-iron wheel, 18 inches in diameter, which supports part of the weight of the gate.

There are four sluices in each leaf, placed in pairs, with a small interval between them. Each pair of sluices counterbalances the weight of the other pair by being attached to opposite ends of a lever at the top of the gate. A screw works in the segment of a large worm-wheel formed on the end of the lever, and, being turned round, opens and shuts the sluices. After the timber work of the gate had been fitted together, it was taken to pieces, and subjected to the preserving process called Kyanising, which consists in immersing the wood in a solution of corrosive sublimate. This process has been so successful, that when the gates were

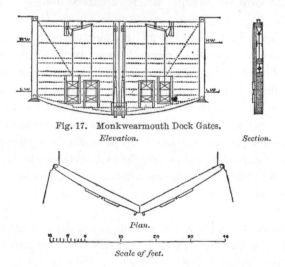

Fig. 17.   Monkwearmouth Dock Gates.
*Elevation.*                                    *Section.*

*Plan.*

*Scale of feet.*

recently taken out for examination the timber was found to be nearly perfect, only slight surface repairs being required in one or two places.

The great bulk of light wood at the bottom of these gates gives them a certain amount of flotation at all times of tide. After the gates had been in use many years it was found that one of the wheels had been detached for some time, but the buoyancy of the gate had prevented any mischief resulting. The buoyancy of the lower part of the gate is somewhat analogous to that of the air-chamber which Mr. Brunel introduced afterwards in his wrought-iron dock gates.

### Bristol Docks

About the year 1804 that portion of the river Avon which flows in a serpentine course through the city of Bristol was enclosed, and the water in it retained at a constant level, a new cut or shorter channel being made for the river. The portion separated, called the Floating Harbour, or Float, is about two miles long and 100 yards broad. At its lower end it is connected with the river by the Cumberland Basin, a half-tide basin, with two locks, and at its upper end by a feeder, which brings the water of the Avon into it, the river in dry weather being stopped from passing into the new cut by the Neetham Dam. About half-way up, the Float is entered on the north side by the river Frome; and, a little above this junction, it is crossed by the Prince's Street drawbridge, which divides it into two parts. About 170 feet above the bridge the Float is connected with the new cut by another basin with a lock, called Bathurst Basin.

Mud and other deposits had accumulated to such an extent in the Floating Harbour, that at the end of the year 1832 the directors of the Dock Company employed Mr. Brunel to suggest remedial measures.

In order to effect his object at the least possible cost, he proposed certain works, together with an improved system of managing the water of the river, so as to allow more of it to pass through the Floating Harbour, by means of which great benefit might fairly be anticipated.[1] He remarked that,—

> By systematically following this course, the object of which is simply to keep in continual action all the means, however small, which can at the moment be brought to bear, and thus day by day to remove or neutralise, or merely diminish (as the case may be), the continual deposit which is going on—in fact, by applying a constantly acting remedy to oppose a constantly acting evil—I have little doubt that the formation of shoals similar to the existing ones may be entirely prevented, or at all events that they will be of such a nature as to be easily removed by two or three yearly scourings, and without that time and labour which are now expended with so little effect.

It should be observed that the yearly scourings, which became so objectionable to the trade, were not introduced by Mr. Brunel, but were part of the original arrangements of the docks.

After the reception of Mr. Brunel's report, the Dock Company executed the works which he required, namely, the Sluice, Trunk, and Drag-Boat; but his other recommendations, as to scouring and increased supply of water, were only acted upon to a limited extent.

In 1842 the Directors again asked Mr. Brunel to report, in conjunction with Captain Claxton, upon 'what further measures are requisite for keeping the Floating Harbour more clear of mud than it has been for a few years.'

Mr. Brunel thereupon made a report to the Directors. After having referred to his previous reports of 1833 and 1834, he remarked that 'the efficiency of the whole system then recommended and adopted, and subsequently partially carried out, depended entirely and was founded on the supposition of the then existing mud-banks and shoals being first removed, and the Float deepened at once to the full extent required,' according to the plans which he had pointed out; and that 'the increasing the supply of water through the Float was one of those means on which he had most insisted' as necessary for keeping it clean and preventing its becoming a settling reservoir. He then continued,—

The sluice at Prince's Street Bridge, the trunk—or syphon, as it was originally, and perhaps more correctly, called—at the underfalls, and the drag-boat have alone been brought into operation. These were originally intended as mere aids, which, in conjunction with *the increased supply of water*, were expected, after the complete deepening of the Float, to be sufficient, with *two or three yearly scourings*, to keep it to the required depth: these were (perhaps unfortunately) found so effective as to induce a hope that they might be depended upon solely for the removal of the evil. The permanent interests of the port were, I cannot but think, sacrificed to temporary convenience: the scourings which were required as a preliminary step to restore the Float to its original state, or to that which was said to have been its former state, and which is now required, were indefinitely postponed.

A material improvement being notwithstanding soon perceptible from the first effects of the drag-boat and the removal of mud through the trunk, the periodical scourings which formed part of the system approved of for adoption were in a great measure given up to the objections of the traders. The precautions actually necessary against admitting into the Float the tide water of the Avon, heavily charged with mud, were gradually sacrificed for the same reasons. From all these and many other, but very similar, circumstances no further progress has been made since the first improvement, which was felt to be, and which unquestionably was at the time, very considerable. For until this period it had been the general practice to lighten all deeply laden vessels at the entrance of the Float; and, notwithstanding this precaution, it must be in the recollection of everybody that it was a common sight to see several large vessels aground at various shoals in the Float, and unable, without further discharging the cargoes, and without great consumption of time, labour, and ropes, to get up to the quays. For several years past the grounding of the deepest vessels has been the exception, not the rule, but during all this period it has been one continued and almost vain attempt to struggle against the old difficulties with insufficient means.

.    .    .    .    .

For the removal of such deposits as will still be formed, I propose two means; and first as regards those deposits which are continually going on. These are formed almost entirely of mud, which, from its want of consistency when first deposited, the great quantity and the large surface over which it extends, as well as the great depth of water, cannot, as I have frequently explained, be easily or economically removed by the ordinary process of dredging: for this reason I originally proposed the drag-boat in conjunction with the trunk at the underfalls, and which has, so far as it has been applied, completely answered my expectations.

I should now propose a similar arrangement for the upper float.

.    .    .    .    .

I should propose to make the additional drag-boat thus required of rather greater power than that now in use, and to construct it so that a chain of dredging-buckets could be hereafter attached to the shaft; and, secondly, for the purposes of deepening the hard bottom of the float, and of removing those banks of hard materials which have either always existed, or have been allowed to accumulate, dredging must be resorted to. But I should be disposed to attempt hand dredging or spooning in the first instance; for, although the depth of water is great, I believe the work could be executed as cheaply, or nearly so, as by the steam dredging, as the original outlay of capital would of course be much less, while the facility of working at several points at once, or of moving from one berth to another as the convenience of the trade would best allow, would be much greater; the operation also would be much more under command, which, taking into consideration the possibility of undermining the foundations of the quay walls or buildings, is not an unimportant advantage.

When the required depths are once obtained, the natural deposits even of the harder description may probably be easily removed by the drag-boat or by the occasional use of hand dredging.

The recommendations contained in this report were not adopted.

In 1839 Mr. Brunel, in a report to a Committee of the Council of the City of Bristol, suggested several improvements connected with the Port, almost all of which have since been undertaken. He proposed to straighten and widen the course of the river, and to make new locks, both from the river to Cumberland Basin and from that basin to the Floating Harbour, or, as an alternative, to construct docks at Sea Mills, a creek on the Avon, about two miles below Bristol.

He also proposed to construct a large pier at Portishead. The rise of tide is there sometimes 45 feet, and the velocity of rise or fall as much as 10 feet per hour. There is also a great deposit of mud by the Severn. Mr. Brunel considered that these circumstances rendered a fixed structure undesirable, and he therefore recommended a floating pier. He said:—

> I propose two or three vessels of 300 or 200 feet of length, built of iron, as the material cheapest and best adapted to the purpose, of 16-feet or 20-feet draft of water, and about 30 feet beam, moored close stem and stern, so as to form one continuous floating body. Any steamboat or other vessel alongside will of course be on the same level as the pier; the passengers, on disembarking, will at once be on a level platform or deck, under shelter, where the luggage or goods can also be placed; and the communication with the shore will be effected without steps . . . Such a pier would afford stowage for almost any quantity of coals, fresh water, and general goods, which could be stored here for embarkation.

In 1847, after an Act had been obtained for making a railway from Bristol to Portishead, with a pier at the latter place, Mr. Brunel designed the pier on the plan described in his report of 1839; but the project was not carried out.

As has been already mentioned, the communication between the lower part of the Floating Harbour and the river Avon is through the Cumberland Basin; between this basin and the river were two locks, made at the time that the Floating Harbour was constructed.

Owing to the increased size of merchant vessels, it had long been in contemplation to enlarge the entrance. At the time when the 'Great Britain' was built, the northern lock was so narrow that a portion of the upper masonry had to be removed in order to give room for the ship to pass from the basin to the river on a spring tide. It was then felt that the enlargement of one of the locks could no longer be delayed, and Mr. Brunel was asked to adapt the narrower or southern entrance lock to the passage of the largest vessels.

Between the two locks was a pier, from which vessels were guided, and the gates opened or shut. The elongation of the lock was limited by the length of this pier, as it could not be extended towards the river without diminishing the area of entrance, nor could it be extended upwards without lessening the area of Cumberland Basin. Mr. Brunel, although hampered by this restriction, succeeded in obtaining a lock of considerable length. He constructed the gates of a single leaf, and placed the upper gate outside the lock so as to shut against the upper end of the middle pier, and to swing back when opened into a recess in the side wall of the Cumberland Basin. He thus avoided the necessity of finding room on the pier for the machinery to open one of the leaves of the upper gate. Had the gate been in two leaves, the lock would have been shortened from 30 to 40 feet. At the lower end he placed the gate as near the river as possible; and, lest the end of the middle pier should not be strong enough to withstand the pressure, he secured the quoin stones, against which the gate closed, by horizontal wrought-iron bars at different levels, built into the side wall of the lock.

The lock is 262 feet in length between the gates, and 54 feet wide at the narrowest part.

The masonry is of plain character, all the part below the ashlar coping being of ordinary fitted rubble of great thickness, solidly built with hydraulic lime mortar. The ground behind the wall consisted of a wet silty clay, causing a great pressure against the masonry. The under part of the body of the lock is formed to a semi-oval cross section.

The works were commenced by the construction of coffer-dams at each end. In 1846, when the masonry was approaching completion, a very high tide took place, and a portion of the upper dam gave way. As some work still remained to be done at the sill and apron of the lower gate, Mr. Brunel decided to make a brick dam in the middle of the lock, where the masonry had been completed. This brick dam was a horizontal arch built on the bottom of the lock, up to the level of the water in the Floating Harbour. The abutments were formed by the masonry of the lock walls, which was notched to receive the bricks of the arch rings. The dam was 28 feet high, only 8 feet thick at the bottom, and 3 feet thick at the top. It was set in Roman cement, and was completely water-tight. It was easily and rapidly made, and the cost was small, as compared with what would have been the cost of repairing the upper dam.

In this lock the chief point of interest consists in its being the first in which wrought-iron gates were introduced, these gates being at the same time made buoyant.

Floating caissons had been previously used at the entrances to graving docks, and in similar situations; indeed at Bristol, a caisson had long been employed at Prince's Street Bridge, to separate one part of the Floating Harbour from the other. The buoyant gates of the Bristol Docks differ essentially from these vessels, inasmuch as, instead of requiring to be floated into their places, they turn on a hinge, and do not rise or fall vertically.

The gates are provided with wheels, but only a small part of the weight rests on them, as the gates are rendered buoyant by large air-chambers, formed in the lower part of them.

The upper and lower gates are alike in construction and dimensions, so that it is only necessary to describe one of them. (See woodcut, fig. 18.)

The extreme length of the gate is 58 feet, and the extreme height 29 feet. In the middle it is about 10 feet wide, the width diminishing to 3 feet at the top. In plan it is curved to resist the pressure of the water. The gate when closed is not at right angles to the direction of the length of the lock, but is at an angle of about 12°. The length is thus scarcely increased, while the travel in being opened and shut is much reduced. The top is at the level of the water in the Floating Harbour; so that, when the tide falls, the water in the Cumberland Basin may be retained at the same height as in the Float.

The air-chamber is formed by two water-tight decks of wrought-iron plating, one at the level of half-tide, the other a short distance above the bottom of the gate. Above the deck forming the top of the air-chamber, the water, as the tide

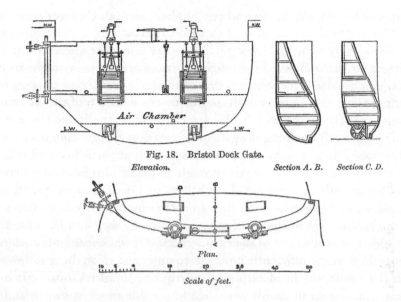

Fig. 18. Bristol Dock Gate.

Elevation.      Section A. B.    Section C. D.

Plan.

Scale of feet.

rises, flows freely into the interior of the gate, through openings in the face next the dock, so that when the water is level with the top of the gate, the part above the air-chamber is full of water, which flows out again if the level of the water falls.

In ordinary working the gate only needs to be opened or shut when the water is above the level of half-tide, and therefore at these times the whole of the air-chamber is immersed. To whatever height the water rises above this level, the buoyancy remains almost the same, the only change being caused by the displacement of the iron of the upper part. This displacement, when the water is level with the top of the gate, amounts to about six tons.

The size of the air-chamber is so arranged that when the water is level with the top of the gate it is just afloat; and at half-tide, when the water is at the level of the top of the air-chamber, there is a weight of about six tons on the wheels.[2]

The gate is provided with a sluice, by which water may be admitted into the air-chamber, or allowed to escape when the water outside is at a lower level; there is also a pump, by which leakage water may be extracted. The volume of the air-chamber may thus be altered at will, and the buoyancy may be modified so as to counteract the effect of the weight of any mud which may be deposited upon the decks of the gate.

There are in each gate two very large double sluices, which are used for working the lock, and for lowering the water in the Cumberland Basin to meet the tide. They are also used for scouring away the mud. The shutting pieces of the gates, which bear against the granite masonry and form a water-tight joint, were made of Honduras mahogany, a very durable wood for the purpose. The timber is bedded in creosoted felt and bolted to the gates; and the pieces are still sound, after the lapse of more than twenty years.

Underneath the gate are two wheels, slightly conical, 3½ feet diameter and 1 foot wide, which travel on level cast-iron rails let into the masonry of the gate floor. There is no heel-post such as is usual in dock gates, but the gate is hinged to the masonry by wrought-iron collars and a wrought-iron pin about 8 inches diameter, which is passed through them. Any portion of the weight of the gate which is not supported by the flotation of the air-chamber is borne entirely by the wheels, the arrangement at the hinge being merely for the purpose of retaining the gate in its position and guiding it in opening or shutting. The gate is moved by chains, which are attached to the barrels of powerful crabs, and conveyed through passages formed in the masonry; at the lower ends of these are chain rollers or broad sheaves, round which the chains pass. The sheaves are fluted on the circumference, to ensure their turning readily. When the gates are nearly afloat, they can be moved with great ease; but at the time of their construction it was considered essential to provide machinery sufficiently powerful to open and close them at low water, when the whole weight of each gate, nearly one hundred tons, rests on the wheels. The crabs and chains were therefore made much stronger, and were more difficult to move, than if they had been merely designed to work the gate under ordinary conditions.

The gates were constructed in Bristol at the Great Western Steamship Works in 1847. After the lower gate had been tested and proved to be water-tight, it was launched, and floated with the front surface nearly level. The positions it would assume under different conditions had been calculated beforehand. Before it was fixed, the gate was made to float nearly upright by the admission of water; it was then towed to its place, and brought into correct position. The hinge-pin was dropped through the collars, and by admitting water into it, the gate was sunk, so that it rested upon its wheels. This operation had to be performed at high water, which only lasted for a short time.

In fixing the upper gate, the water was kept up to the proper level by the lower gate, and therefore there was no need to do the work quickly.

### Plymouth Great Western Docks

As it was considered probable that, on the completion of railway communication, mail packets and other large ocean steamers might make Plymouth their port of departure, a company was formed in 1846 for the construction of a dock in Mill-bay, a large inlet in Plymouth Sound near the entrance to the Hamoaze.

The bay was already protected from the prevailing winds by a pier at its eastern side constructed by Mr. Rendel.

It was decided to form a wet dock and graving dock at the inner end of the bay, and to make quay walls along the side of the outer part, to join the existing pier. A floating pier for large vessels afterwards became part of the scheme. In 1847 preliminary trials were made as to the best means of excavating in deep water the limestone rock of which part of the bottom of the bay consisted; some quay walls were also constructed and made use of by the shipping.

At the end of 1851 a contract was entered into for the execution of the whole of the works remaining to be completed; the most important of these were the entrance, the graving dock, and the completion of the floating basin.

The progress of the undertaking was much facilitated by Mr. Brunel giving his sanction to the proposal of the contractors to form a stone and earth embankment across the mouth of the bay, instead of employing the usual timber coffer-dams. This embankment, which completely answered its purpose, was finished by the middle of 1853.

The works now proceeded steadily until they were completed, and at the end of the year 1856 the dock was opened.

The middle of the embankment was cut through, and an entrance channel dredged to a level of 8 feet below low water. The remainder of the dam, being partly protected by masonry walling, was used as a quay, and served for the protection of the outer basin.

Mr. Brunel had prepared the foundation, and had intended to build a pier to shelter the entrance gate from the sea, and to assist in pointing long vessels. The want of this shelter was felt in the gale of October 1857, when the gates were thrown down. In order to avoid any future casualty, storm gates, or framed struts reaching down to low water level, were placed immediately behind the entrance gates, so as to support them against heavy seas.

The dock, which is of oblong shape, has an area of 13 acres, and a length of quay wall of 3,490 feet. The greater part has a depth of 22 feet below high water at spring tides, or 16 feet at neap tides; the remainder is 4 feet deeper, or the same depth as the channel in the outer basin.

The walls are 8½ feet thick at bottom, and 3½ feet at top and are built to a curved batter. They reach generally to a depth of 5 feet below low water, and rest on concrete 12½ feet thick, carried down until a rock foundation is reached, which in some places is as much as 40 feet below the bottom of the dock.

To accommodate the larger paddlewheel steamers, the entrances to the basin and graving dock were made 80 feet in width.

The dock is entered through a short passage, the sill being 13 feet below low water. The masonry of this entrance has curved battering sides, with a segmental invert.

The entrance is closed by a pair of wrought-iron buoyant gates, which meet at an angle of 127°. Each gate is 48 feet long, and weighs seventy-five tons, the breadth being 8 feet in the middle, curving to 2 feet at the ends. There are six horizontal decks and four vertical bulk heads. The depth at the heel-post is 22 feet, and at the meeting-post 35 feet (see woodcut, fig. 19).

In each gate is an air-chamber, the top of which is at half-tide level; and its volume is such that when the gate is wholly immersed there is a small downward pressure.[3]

Under each gate, near the meeting-post, is placed a wheel, which supports part of the weight. This wheel is so arranged that it can be easily removed for repair. The heel-posts are of cast iron, planed and ground in place against

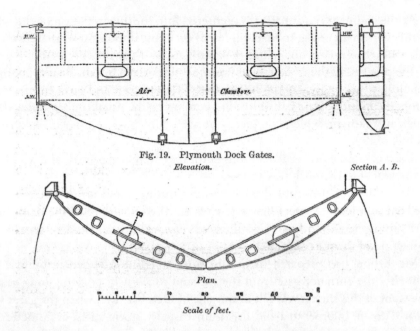

Fig. 19.  Plymouth Dock Gates.

*Elevation.*                                              *Section A. B.*

*Plan.*

*Scale of feet.*

the polished surface of the granite hollow quoins so as to form a water-tight joint.

There are large scouring culverts behind the side walls of the entrance; but for the purpose of regulating the level of the water in the dock, and of discharging a large volume of water readily, without having to overcome the friction of ordinary sluice valve faces, each gate is furnished with a cylindrical valve, of the following description:—

From an opening in the side of the gate next the dock there is a large curved pipe or sluice-way, which terminates inside the body of the gate, with a circular horizontal orifice about 5 feet in diameter. The opening is covered by a short length of vertical pipe of the same diameter, reaching above high water, the bottom edge making a water-tight joint. This pipe can be raised or lowered by a screw at the top of the gate. When raised a short distance it allows the water from the dock to flow out between the bottom of the movable pipe and the orifice of the sluice-way into an isolated compartment of the gate, and to escape by an opening provided in the outer face. The movable pipe is guided by rollers, and from the construction the pressures on it are balanced.

The entrance to the graving dock, 80 feet wide, is closed by a pair of gates of the same dimensions and construction as those of the entrance. This dock is 380 feet long, 92 feet wide, and has a depth of 28 feet over the sill.

The floating pier was erected in Mill-bay in 1852, to accommodate the steam-shipping trade. It consists of a pontoon with a bridge leading to it.

The pontoon is a large wrought-iron vessel of nearly rectangular cross-section, 300 feet long, with 40 feet breadth of beam, and a depth of 17½ feet. It is loaded so as to draw 10 feet of water, and is capable of storing 4,000 tons of coals.

To connect the pontoon with the shore there is an iron bridge, working upon hinges, in two spans of 125 feet each. The connection between the two spans was supported on a timber pier, and was made so as to be movable vertically, and capable of being adjusted by a crab and counterbalance weights. By means of this arrangement a uniform gradient over the whole length of the bridge could be obtained at any time of tide, for the convenience of heavily laden carts passing to and from the pontoon.

### Briton Ferry Docks

In 1846 a company was formed to establish docks in Baglan Bay at Briton Ferry, near the mouth of the river Neath; but nothing was done until 1851, when the necessary powers were obtained under an Act of Parliament. From the vicinity of the proposed docks to the South Wales and Vale of Neath Railways, a large amount of trade was anticipated; and in 1853 an Act was obtained for the South Wales Mineral Railway, which was intended to terminate at the docks, and was expected to bring a large traffic in coals and other minerals.

The dock works were begun soon afterwards, but the earthwork was not completed, nor the construction of the gate commenced, until the year 1858. The docks were completed, after Mr. Brunel's death, by Mr. Brereton, and were formally opened on August 22, 1861.

They consist of an outer tidal basin of about 7½ acres, and an inner floating basin of about 11 acres, with a depth of water of 27 feet at spring tides, and 16 feet at neap tides. The two basins are connected by a passage or entrance of 50 feet in width, with curved battering walls and an invert, closed by a gate of a single leaf. An important advantage of single gates is that the sill and quoins may be in one plane, and that the troublesome and costly fitting of the hollow quoins is avoided. The sill was laid 6 feet below what was then the low-water level, as it was thought that future improvements might reduce the bar at the entrance of the river Neath to that level. This has already been nearly accomplished.[4] The total length of the docks is a little over half a mile, and the average breadth is about 400 feet. They are connected with the South Wales Railway by branch railways.

With the exception of the walls dividing the two basins near the entrance gate, the sides of the dock are not constructed with masonry quay walls, in the ordinary manner, but are formed in a very inexpensive manner of slopes pitched with furnace slag, obtained from the copper smelting works on the Neath river, with jetties at intervals for the shipping. Besides being suitable for the soft clay in which the dock is made, this plan is specially adapted for mineral traffic, as the work of loading or discharging a cargo of minerals can only be properly carried on at the point where fixed machinery is provided for the purpose. This machinery may be placed as conveniently on projecting jetties as on a dock wall. The traffic, which consists mainly of coals and metals, is accommodated at the jetties, which are furnished with cranes for loading and unloading the vessels employed in the metal and iron ore trade, and with tipping-frames, which discharge the coals into the ships. These cranes and tipping-frames are worked by hydraulic machinery.

In order to facilitate the entrance of vessels from the river Neath to the tidal basin, and to protect it from the sea, two pier-heads were built, one at each side of the basin at the point where it joins the river. These piers are of timber piling and framework, filled in with copper slag; the entrance between them is 300 feet wide. They were constructed after Mr. Brunel's death.

The gate is a wrought-iron buoyant gate, with five vertical partitions or bulkheads and six decks. The length is 56 feet; the depth in the middle is 31 feet 6 inches, and at the sides 26 feet 6 inches. The breadth in the middle is 9 feet, and is curved to 2½ feet at the ends (see woodcut, fig. 20).

The air-chamber, which is similar to that of the Plymouth gates, is placed so that the top is at the level of high-water neap tides, about half-way up the gate.

There are two sluices at low-water level, each having an area of 8 square feet.

The entrance invert being subject to the influx of sand from the outer basin, and to the deposit of coal rubbish dropped into the dock, Mr. Brunel decided in this gate not to use wheels, but to make the hinge and heel-post strong enough to carry the whole weight of the gate, even if it were unsupported by the buoyancy of the air-chamber.

The heel-post is a massive piece of cast iron; the bottom part is bored out, and into it is fitted a cast-iron cylindrical pin, 1 foot 6 inches diameter and 7 feet 6 inches long. On this are ground discs of steel, lubricated with oil, whereon the gate rests and turns. Thus the surfaces exposed to friction are above the sand or grit at the bottom of the dock. The lower end of the pin fits into a cast-iron socket fastened to the masonry, and is prevented from turning round by being made hexagonal. The sides of the hexagon have sufficient play to enable the gate to adjust itself, when shut, to the masonry sill, so as to be water-tight. The top of the heel-post works in a brass bush, 18 inches diameter and 15 inches broad,

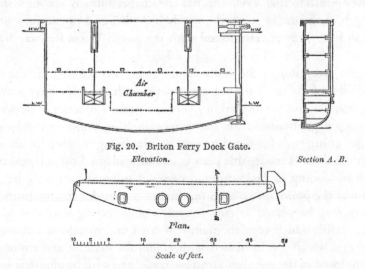

Fig. 20. Briton Ferry Dock Gate.
Elevation.                                    Section A. B.

Plan.

Scale of feet.

enclosed in a massive wrought-iron collar, which is strongly fastened by anchor chains to the masonry. After the gate was completed the strength of the hinge was tested by moving the gate before the water was admitted into the dock. The only resistance to the motion of the gate is the slight friction at the hinge.

No coffer-dam was used in the construction of this work, but advantage was taken of a large bank of slag and earth enclosing a portion of the site of the dock. This was extended and raised, and a sea dam formed. The dam was cut through when the works were completed, and a channel dredged to the depth of 6 feet below low water.

### Brentford Dock

In 1855 an Act was obtained for making a dock on the Thames at Brentford, and a railway to join the Great Western Railway at Southall. The dock has an area of about 3½ acres.

The works were begun in July 1856, and were completed, and the dock opened, three years later.

The walls are founded in the London clay, which here underlies a bed of gravel of some thickness; from this there was a considerable influx of water.

The chief peculiarity of the dock is the form of construction adopted for the sides. Piers of brickwork, 10 feet long and 2 feet 3 inches thick, are placed at right angles to the sides of the dock at intervals of 26 feet. The backs of those piers are connected by horizontal arches, carried up with a curved batter. The piers are about 20 feet high, and arches are turned upon them, which support the front part of the quay, and meet the horizontal arches at the backs of the piers. Thus the sides of the dock consist of a series of vaults, arched over at the top, and also at the back towards the pressure of the earth.

The thickness of the horizontal arches which form the bulk of the wall is only 3 feet, but these are so strengthened by the piers in front, that a wall strong enough to resist the pressure of the earth behind it was obtained by means of a very small quantity of brickwork.

Along one side of the dock the piers are 31 feet long, in order that coal barges may lie with part of their length in the vaults between the piers while their cargo is being put on board. By this arrangement the barges have their longest dimension at right angles to the side of the dock, and a much greater number can be accommodated than if each occupied a space alongside the quay wall. The contents of the coal trucks are tipped into the barges through sloping shoots.

The entrance has a clear width of 30 feet, and is closed by a single wrought-iron buoyant gate, which, like the Bristol gates, is, when shut, not quite at right angles to the entrance.

The gate is 33 feet long, 19 feet high, 2 feet 6 inches wide at the middle, curved to 1 foot 6 inches at the sides, and weighs sixteen and a half tons. It is divided into compartments by four decks and two vertical bulkheads. The air-chamber occupies the whole space below one of the decks, 7 feet 6 inches above the bottom (see woodcut, fig. 21); and there are two sluices, each having an opening of 4 feet

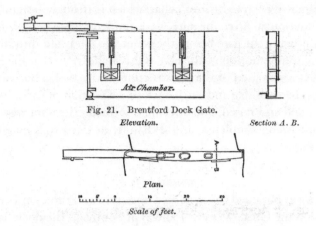

Fig. 21.   Brentford Dock Gate.

*Elevation.*                    *Section A. B.*

*Plan.*

*Scale of feet.*

by 2 feet. This gate, like that at Briton Ferry, has no wheel under it, the weight being carried upon the pivot.

In order to avoid side strains upon the pivot and top collar, a counterbalance arm is fastened on the top of the gate. This is formed of two cast-iron girders, bolted together and enclosing weights between them. The ends of these girders project beyond the heel-post over the quay, as in canal lock gates, and carry the machinery by which the gate is turned, as there are no crabs, chains, or chain rollers. Instead of these, a cast-iron circular rack is fixed on the top of the masonry, in which a pinion works, turned by gearing fixed to the end of the counterbalance.

This gate turns with remarkable freedom, and the current of water running into the dock on a spring tide opens it completely.

### Pier at Neyland, Milford Haven

The South Wales Railway was originally intended to terminate at Fishguard, on the north coast of Pembrokeshire, with the view of securing a large quantity of Irish traffic, the distance across the channel to the Irish coast at Wexford being only 60 miles, less than the distance from Holyhead to Kingstown.

It was, however, ultimately decided to form a terminus, which would accommodate the ordinary Irish traffic, and would not require such an extensive outlay on harbour works as would have been necessary on the northern coast.

With this object, the inlet or natural harbour of Milford Haven was examined, and the South Wales Railway was carried to Neyland Point, opposite Pembroke, where the position is sheltered, and there is deep water at all times of tide for the largest vessels.

The pier at Neyland, or New Milford, which was made in 1857, consists of a timber viaduct, with a pontoon at the end, 150 feet long and 42 feet beam, loaded so as to draw about 7 feet. There is a depth of 16 feet alongside it at low water, and it is connected with the shore by a landing bridge. The pontoon is made of wrought iron, and has three transverse and two longitudinal bulkheads. It is moored by chain cables, which pass through two large hawse pipes, extending from the bottom nearly up to the deck, with cast-iron mouthpieces at their lower ends. The cables passing through these are anchored firmly to the ground at a considerable distance from the pontoon.

The pontoon was intended to be the centre of several others, which were to be moored in deeper water.

The rise of tide being sometimes as much as 25 feet, it was necessary that the landing bridge should be of considerable length, in order that there should be a moderate inclination at all times of tide. It is accordingly made in one span of 205 feet. It consists of two plate-iron side girders, of the uniform depth throughout of 14 feet, and width of 2 feet 6 inches. These are placed 12 feet 6 inches apart, the roadway being between them. The ends of the girders which rest on the pontoon are provided with cast-iron wheels, 1 foot 6 inches in diameter.

The pier has since been extended by additional pontoons, which were those used in the floating of the Saltash Bridge.

1.  It may be desirable to give a short explanation of these works, which consisted principally of a sluice, a trunk, and a drag-boat.

    The Sluice was made in the abutment of Prince's Street Bridge, and was intended to create a scour after the water had been let off from either side of the float, the opening at the bridge being closed by a caisson which had long been in use when (as was formerly the case) the upper part of the float was scoured through Bathurst Basin.

    The Trunk, near the entrance from Cumberland Basin, is a wooden culvert between the floating harbour and the river. As much of the mud as could be dragged there was deposited at the entrance of the trunk and, when the tide was low in the new channel of the Avon, sluices were opened, and the water rushed through from the floating harbour, carrying the mud with it.

    The Drag-Boat was fitted with a steam-engine which worked a large windlass with three compartments, round two of which chains were passed and fixed to posts on the quays, and the boat was dragged backwards and forwards. The third compartment of the windlass worked a chain which elevated or depressed a scraper, attached to a long pole at the stern, and secured from swerving by a chain-bridle which passed under the boat. The scraper stirred up the mud, and deposited the more solid parts at the entrance of the trunk.

    Mr. Brunel also desired that the float boards of the Neetham Dam should be put into proper working order; and that they should be altered so that, in times of land floods, the whole or a considerable portion of the excess of water should be retained, and passed through the feeder; and that even the water of spring tides should be allowed to pass the dam, and then be stopped back for the same purpose.

2.  The advantage of forming an air-chamber in the lower part of the gate, and allowing the water to enter above it, is that when the size of the air-chamber is properly

      adjusted to the weight of the gate, there need not come on the wheels, while moving, more than a trifling amount of weight.

3.   The arrangements of a buoyant gate have been explained above, p.314.

4.   In the year 1856 and afterwards, Mr. Brunel was engaged in improving the navigable channel of the river Neath, at its embouchure into the Bristol Channel. A bank of furnace slag, for directing the course of the river, had been made previously by Mr. Palmer, and continued as far seawards as it was then thought could be done with safety. Mr. Brunel carried a training bank still farther, and succeeded in cutting off a bend of the river; Mr. Brereton has since extended the navigable channel in a straight line to low-water mark, a distance of two miles; and the bar has been lowered to within 1 foot of the level of the dock sill.

# XV

# MISCELLANEOUS WORKS

THERE are several matters of importance in which Mr. Brunel was engaged, which could not under any system of classification be introduced into the preceding chapters; these are therefore collected under one head of 'miscellaneous works.'

The first of these is his connection with the Great Exhibition of 1851. He was from the beginning one of the most zealous supporters of this undertaking, and was appointed a member of the Committee of the Section of Machinery, whose duty it was to classify the objects to be exhibited in that department.

Upon the question of awarding prizes to exhibitors, Mr. Brunel held a very decided opinion adverse to the plan ultimately adopted. In a letter to the Chairman of the Committee, dated March 11, 1850, he writes:—

I am sorry to say that I am obliged to leave town to-night.

We are summoned to-morrow, it appears, on the subject of prizes. Not being a member of the Commission, I have perhaps no right to express an opinion upon a principle which seems to have been adopted—that of giving prizes—but as applied to machinery I suppose I may. I strongly disapprove of any prizes being offered in our section.

1. I believe it is quite unnecessary.

2. I believe it will be impossible to define beforehand the subjects for which any limited number and amount of prizes are to be promised, the subjects are so indefinitely numerous; and like the building, however large it may be made, will not be large enough to hold all that is sent, so as regards the prizes, however numerous the subjects, they may very likely not embrace the very things which turn out to be most deserving.

3. I believe it will be impossible to distribute any limited number of prizes with justice, and quite impossible to satisfy the public.

Two machines for the same purpose may be remarkable—one for its ingenuity and beauty of workmanship, but of doubtful practical economy in application; the other clumsy, and not well made, but apparently likely to have the germs of much good—there are thousands, or rather an infinity, of shades of degrees and qualities of merit.

And lastly, I believe the prizes will be mischievous, as conferring undue advantages in many cases upon a thing well displayed, and well got up, and will be sought for and obtained for puffing purposes. The opportunity of exhibition I believe will be quite sufficient to induce all the competition we can desire.

I think money prizes quite a mistake, and medals or distinctions pretty nearly as bad. I hope you hold the same views, but I send you mine.

Mr. Brunel's views found no favour at the time; but subsequent experience has convinced those best able to form a sound judgment in the matter, that 'no prizes of any kind should be awarded' in International Exhibitions.[1]

Mr. Brunel was also a Member of the Building Committee; and he accepted the office of Chairman and Reporter of the Jury for Class VII, on Civil Engineering, Architecture, and Building Contrivances.

He took a very active part in the proceedings of the Building Committee. Designs were invited, and two hundred and forty-five were sent in. None of these were considered satisfactory by the Committee, and they submitted to the Royal Commission a design of their own, the principal feature of which was a dome 200 feet in diameter.

Mr. Brunel was responsible as a member of the Committee for the plans prepared by them, and as regards the dome may be said to have designed it himself, but he expressed strong objections to the substantial and expensive buildings which it was proposed to erect in brickwork. His idea was that the building should be in what he called the 'railway shed style;' and he wished to produce effect rather by the construction of the roofs, &c., than by any architectural elevation.

When, therefore, the plans of the Building Committee failed to meet with public approval, and the late Sir Joseph Paxton submitted his well known design, Mr. Brunel gave it his cordial support, and defended it against its detractors. He thus spoke of it in the report of the Jury of Class VII.

> As regards Mr. Paxton's claim, amid the competition of the whole of Europe, he proposed that mode and form of construction of building which appeared on first sight, and has since proved to be, the best adapted in every respect for the purpose for which it was intended. The design possessed this merit of fitness for its object in a singular manner. There was no startling novelty in any one point which could lead astray the judgment of those who had to determine upon the choice of plan, or which could in the first instance obtain, still less permanently secure, the good opinion of the public. As regards the form of outline, which is most simple, several designs nearly resembling it had been submitted in the general competition. As to the material, several proposals had been previously made to cover the whole area to be enclosed with glass,

and iron would of necessity be employed for the framing; but in the combination of form and materials, in the particular mode of applying those materials, and in the adaptation of the forms to be selected to their convenient use, as well as in the various details by which the whole was rendered perfect, the design was entirely distinct in character from all that had been proposed, and appeared at once to have the one single merit of being exactly that which was required for the purposes in view. The design as realised has completely fulfilled every condition of utility.

The award of Council Medals (the highest prize given) was recommended to Sir Joseph Paxton, and to the contractors, Messrs. Fox, Henderson and Co.

In a later part of the report, in announcing the recommendation of a Council Medal to His Royal Highness Prince Albert, for the model dwelling houses which were erected near the Exhibition building, and exhibited by the Prince, Mr. Brunel spoke in emphatic language of the magnitude and importance of the results which would follow from the introduction of improved dwellings for the working classes.[2]

When the Crystal Palace Company was formed in order to purchase the Exhibition building and erect it, with additions, at Sydenham, Mr. Brunel took a great interest in the project, and frequently went down to examine the progress of the building and gardens, and the beautiful architectural courts which were to be the chief attraction in the interior of the Palace. The water towers, which are so conspicuous a feature in the building, were designed by him.

The towers are 284 feet high, and carry near the top tanks 47 feet in diameter and 38 feet high, holding 1,200 tons of water.

The foundations required great care in their construction. The tanks had to be placed at a height of more than 200 feet, and the towers, which, with their load, weighed fully 3,000 tons each, had to rest on the sloping side of a clay hill. There was also the possibility that by the bursting of a pipe a large quantity of water might be suddenly discharged, and so cause a slip in the surrounding ground. Mr. Brunel carried the foundations down to a considerable depth, forming a large base of Portland cement concrete, and placing on it a cone of brickwork in cement, rising up to the ground level. The towers are twelve-sided, with two hollow cast-iron columns at each angle. The height of the building below the tanks is divided into ten stories, and at each floor there is a strong wrought-iron diaphragm, or shelf, 5 feet wide. The columns are also connected by strong diagonal bracing in the sides of the tower.

The tanks are made of wrought iron, and the water pipes are placed in the interior of the tower. Mr. Brunel did not think it would be prudent to form any of the columns of the towers into pipes, lest the expansion due to the temperature of the water should cause unequal support to be given to the tanks.

In July 1855, the pipes were proved, and the towers were completed shortly afterwards.

The remainder of this chapter will relate to matters which have but little in common with the subjects of the earlier part of it; but the change is hardly less marked than that which took place in the nature of the questions which occupied public attention within a few years of the close of the Great Exhibition.

## Polygonal Rifle

In October 1852, Mr. Brunel consulted Mr. Westley Richards, of Birmingham, as to the manufacture of a rifle 'for the purpose of determining whether there was anything in a crotchet he had upon the subject.' The rifle was made by Mr. Westley Richards according to Mr. Brunel's directions, and finished in May 1854. Many experiments were tried with it, at Birmingham and at Manchester, in the spring of 1855, and afterwards at Woolwich; and its performances obtained great notoriety.

The history of this rifle, and the objects Mr. Brunel had in view in its design, will be understood from the following letters to Mr. Westley Richards:—

### I

October 25, 1852.

I have long wanted to try an experiment with a rifle, for the purpose of determining whether there is anything in a crotchet I have upon the subject, but I have been deterred from attempting it from the feeling that in these abominable patenting days (I hate patents) the chances were, that if, in the progress of my experiments, any new result, good or bad, were observed, or a workman should think he *saw* something, a patent would be taken over my head, and, to say the least, I should be stopped in pursuing my own investigation, as has happened to me more than once.

I have also been deterred by my not knowing whether the existing machinery for rifling barrels would enable me to obtain an increasing or varying twist from one end of the barrel to the other, as this would be necessary to make the experiment, and I should not care to incur the expense of a machine on purpose. My introduction to you, through our mutual friend Whateley, induces me to make the enquiry whether your apparatus or mode of rifling enables you to give such a twist, and if so, whether you could and would make me a barrel. If so, I will trouble you with an explanation of my scheme, as I should have no secrets with you.

### II

February 7, 1853.

I take this opportunity of mentioning again the subject I once wrote or spoke to you about. I want a rifle barrel made octagon shaped inside, the octagon having a twist rather more than usual, and an increasing twist, say twice as much at the mouth of the piece as at the breech. Can you make me such a barrel for an experiment? I will explain to you the object when we meet, as it can only be done *viva voce*.

## III

[The following letter was written to Mr. Westley Richards in answer to a request that Mr. Brunel would permit him to obtain a license from Mr. (now Sir Joseph) Whitworth to make rifles of a polygonal shape. Mr. Whitworth had obtained a patent for improvements in cannons, guns, and fire-arms, in February 1855, and in his complete specification, dated May 30, 1855, he had for the first time claimed—

Firstly, the several combinations of parts forming, when put together, the barrels of ordnance, or fire-arms, having a polygonal spiral-shape. Secondly, the use of the spirally-shaped segments. Thirdly, the adoption of the polygonal spiral for rifled ordnance and fire-arms. Fourthly, the combination of parts forming the breech-loading apparatus.]

November 26, 1858.

I am obliged to you for your communication on the subject of the octagon gun, and in acknowledging your courtesy and gentlemanly feeling I would add that it is only what I always felt I could rely upon from you.

I beg you will not hesitate to take out a license from Mr. Whitworth for octagon guns (if, as a matter of business, you think it convenient to do so) on account of any prior claim which you may know I could set up, and if you get a license for a nominal consideration, as I understand you can, of course as a man of business you should do so. I have no intention of interfering with Mr. Whitworth's patent, even to indulge the feeling I have against all patents and protective laws, which I consider have become the curse of the day, and the sources of the greatest injury to inventors and manufacturers, and still more to the public; and I should also be very sorry even to annoy my friend Mr. Whitworth merely for the sake of showing that I had previously made the gun (at least you made it for me), and I believe others have preceded me, which he has patented; and I assure you that I shall not consider your taking out a license as in any way a denial of this fact, of which you are cognisant.

I have never seen Whitworth's patent; what is it exactly that he does patent? It cannot be merely the polygon, because, even if nobody had preceded me, that would have been already a copy of mine, which not only was made before he began his investigation, but was lent by me, at your request, to him, I think before his patent. My rifle is, I am told, doing quite wonders at Woolwich, and I begin to think there must be something in the principle which I intended to introduce into it, and which is totally different from what I understood to be Whitworth's. I sought to use a comparatively loose ball, but which I thought would centre itself; both in position and direction, to the axis of the barrel by the peculiar action of a polygon within a polygon acted upon by an increasing pitch, and it really seems from the results as if my theory was correct.[3]

*Gunnery Experiments*

In 1854 Mr. Brunel took up warmly the question of improvement in large guns, which was then attracting the attention of several scientific men.

The friendship which had for some years previously existed between him and Mr. (now Sir William) Armstrong, of Newcastle-on-Tyne, gave Mr. Brunel opportunities for discussing these matters, with a view to their being carried into practical effect.

The following extracts from a letter written by Mr. Brunel, in April 1855, to Mr. James Nasmyth, explain generally his opinions at that time upon the construction of large guns:—

From what I have observed of the operation of fractures under *sudden*, quick-acting forces—such as bursting of guns, and fractures under blows, as in our railway smashes—I have arrived deliberately by observation at the conclusion, which every mechanical-minded man arrives at more or less by intuition, that homogeneity and equality of tension and of elasticity in the parts are necessary for strength to resist a violent strain applied suddenly in its full force, which I will call a blow. My impression, from the result of observation, is, that this operates much more than is generally assumed.

If you suppose a bar, say an axle of uniform section and uniform quality in every respect, it will bear bending into extraordinary forms even by a blow; and if you assume that portions of it become more tenacious and stronger, but remain equally elastic, the ultimate strength of this bar will not, I think, be materially increased or diminished; but if you suppose the elasticity of these portions either increased or diminished, I believe the ultimate strength of the bar under a blow is diminished. In like manner, I imagine that in the section of a gun barrel, if portions are more or less elastic than others, or at all different in their character, not only many points of fracture may be determined on, but that the whole may be rendered much less able to resist the violent explosion. The strain produced by the explosion and the plane of fracture is almost certain to be in a plane passing through the longitudinal axis, and therefore I had assumed that one would avoid as much as possible having any variation of quality which fagotting must produce to some extent in planes in this direction. To attain this end, I had endeavoured to scheme some way of welding up 'cheeses' or discs, which might be hammered up splendidly homogeneous of the full diameter and of a considerable thickness, and I wish that you would scheme the best way of welding them together. I should suppose that the centre surface might be welded, and wedges welded in all round, or some other mode devised, bearing in mind that the strain in any plane transverse to the axis is small, only that arising from the recoil of the breech and the friction of the shot.

I have also an impression that something harder than ordinary wrought iron is wanted for the inner surface to resist the explosion. This you might give probably in fagotting up. I am trying the effect—as much for the amusement of the thing as with any great expectation—of a cylinder of hardish material wrapped round with iron wire, laid on with a certain amount of tension proportioned to the diameter. Such a barrel ought to be strong—whether practically successful is another thing.

The scheme of making a gun with the barrel wrapped round with wire, which is referred to in this letter, was one which Mr. Brunel and Mr. Armstrong were very desirous of making the subject of actual experiment. Whether or not it would under their hands have become practically successful, could not be ascertained, as they were obliged to abandon the project, in consequence of the wire covering being patented, in May 1855, by Mr. Longridge.

The following letter to Mr. Armstrong relates to the same subject, and is interesting not only as showing Mr. Brunel's correct appreciation of a principle which is the essence of the coil system of constructing guns, but as further illustrating his objections to the patent laws:—

June 8, 1855.

Have you ever done anything towards my experiment of the wire gun? I have been anxious for some time past to learn about it, but have waited to see you; to-day I learn that Longridge is taking out a patent for it. I daresay it is his own idea, and I only regret it, as I suppose it will now prevent my pursuing it; and I think it likely that with your assistance we should have succeeded in making at least as good a gun as he will. The principle I am disposed to think good; the success would depend upon the practical application, and but for these patents, the more competitors the better for the public. As it is, competition is destroyed. Let me know if you had done anything. Pray let me know also what you are doing about your own, in which I feel equally interested.

Mr. Brunel had also considered the advantages of making the bore of the gun polygonal, with a projectile shaped to fit it. He had a portion of cannon tube and a projectile made by Mr. Armstrong in the beginning of 1855, but he did not himself pursue the question further.

Indeed, after the middle of the year 1856, when the works of the 'Great Eastern' steam-ship began to occupy a large portion of his time, Mr. Brunel was unable any longer to take part in gunnery investigations; but he watched with unabated interest the proceedings of those friends who have continued their experiments, with the great practical success of which he lived to see only the beginning.

### Floating Gun-Carriage

The plan of a gun-boat, or, as it would be more correctly called, a floating gun-carriage, which Mr. Brunel designed for an attack on Cronstadt and other Baltic forts during the Russian war, is clearly described in the following memorandum, which he drew up for the information of the Admiralty:—

December 20, 1855.

The principle is simply the fixing a very heavy gun in a floating shot-proof chamber or casemate, exposing the smallest possible surface; that surface to be of such a form as to be struck by shot only at a very oblique angle; and the gun being a fixture, with the means only of elevating and depressing to an extent of

10 or 12 degrees, but with no lateral motion, the port or embrasure need be only of the size of the muzzle of the gun, so that the gun, the men working the gun, and everything on board will be perfectly protected.

The gun will be directed by elevating the breech, and by slewing the vessel slightly and slowly backwards and forwards across the line of aim, by means afterwards explained.

The men loading the gun will simply load as quickly as they can, and when the gun is loaded push out a trigger.

The governor or person directing the gun will stand behind the hood or chamber, looking direct at the object through a telescope of low power, fixed horizontally in the axis of the vessel, and made to move vertically parallel with the axis of the gun, and mounted with reflectors; so that both telescope and man are completely under cover, and he, keeping the vessel truly in range and the elevation correct, will only touch the trigger whenever his line of sight crosses the object.

The vessel will carry a small engine, of power sufficient to drive it for a short time at a good speed, any eight or nine knots, and at other times to keep up a small forward motion to counteract the recoil, and to keep the vessel's head moving a few degrees right and left across the line of range.

A sufficient portion of the vessel to contain and to float the gun, ammunition, and engine, will be shot-proof.

A fore-body and after-body, the top of which will be *à fleur-d'eau,* or a few inches under water, will be added, to give such a form of entrance and run as will admit of the vessel attaining the speed mentioned; but these parts will be mere shells, and may be full of water, and if damaged by shot will not affect the buoyancy of the float, besides which, not being above the surface of the water, they cannot be much exposed to injury.

The mode of propelling may be by a screw, but I prefer the 'jet,' which, whether an economical mode of propelling or not, is a sufficiently good one for this purpose, and exposes nothing whatever to be injured by shot.[4]

Whether propelled by jet or not, I should have two small lateral jets for directing the vessel, such jets being governed by two cocks handled by the gunner.

Such a mode of directing the aim by a man under cover looking through a telescope, with one hand directing the gun and the other on the trigger, will admit of an almost unlimited degree of accuracy.

The gun being in a perfectly shot-proof casemate, machinery may be adapted to expedite the loading of the gun; and it is not difficult to make a mechanical arrangement by which the shot and cartridge shall be lifted up to the gun, inserted, and rammed home, at a rate far exceeding anything that can now be done by hand; and as the weight and clumsiness of the gun, the carriage, and machinery are of no object, I think I can make a breech-loading gun capable of carrying 12-inch solid shot with a full charge, which may be loaded and discharged at the rate of two or three per minute; but the principle of mounting a gun in such a float is equally applicable to a common gun, which might still be loaded mechanically.

A few loopholes may be provided through which a fire could be kept up from a couple of heavy swivel rifles, carrying, say 6-oz. shot, which would pierce any mantelets or other cover likely next year to be provided against ordinary rifles.

A battery, say of twelve such guns, should probably have also two, or perhaps three, shot-proof vessels of about the same size without guns, but pierced with a longitudinal fin or ridge, like a wall, standing, say 10 feet above the water, and 50 or 60 feet long, strong enough to stand the direct blow of heavy shot at long range, or the oblique blow of the same shot at short range, and which could be placed as screens or traverses to cover the flanks of the battery against distant shot. Against vertical fire I cannot suggest any defence: the point of attack must be selected to avoid it.

The covering vessels may be provided also with loop-holes for heavy swivels.

There should also be two or three small and comparatively light, but shot-proof vessels, to run in and bring out a disabled gun-boat.

These last-named auxiliary unarmed boats form an essential part of the system. In all probability the enemy have by this time thrown stones and other obstacles, and placed infernal machines round the detached fort, to impede a close approach. They cannot, however, have covered a very large surface, so that, with some previous sounding, an approach may be found and a position taken up.

The auxiliary boats should therefore have strong bottoms under the engine-room, and the rest of the body be so subdivided into compartments that they would be proof against serious damage from rocks and infernal machines, and be able to run in under fire and ascertain if obstructions exist, and find the channel if they do.

A battery of such guns could be placed at various points out of range, say at 3,000 yards, at which distance they would hardly attract attention by daylight, and would not be visible in the twilight of night, and could then be concentrated in a few minutes at the point selected for attack within safe breeching distance, say 250 yards; and, if twenty-four shot per minute, of 200 lbs. to 250 lbs. each, thrown with a full charge at 250 yards, can be directed against a small surface of any stone wall yet built (which is pierced with embrasures), the effect ought to be great and rapid. I believe, moreover, that the means of directing the aim will be so effective that if the embrasures can be seen a shell or shot may always be sent in with certainty at 250 yards, and the enemy's guns dismounted.

Such vessels can rapidly change their position, retreat or advance, be replaced by fresh ones, or withdrawn altogether.

The means of transport of such vessels to the seat of war, although a secondary consideration, has been considered.

They might easily be placed in an outer shell of iron of a good form, which could be rigged complete, and so constructed as to give up its burden when arrived in the seas where it is to act—in fact, a ship of the class of small screw colliers, made to open at the bows and its contents floated out ready for action; but the gun-boat itself, when lightened of ammunition, and the gun lowered to the bottom as ballast, and fitted up with bulwarks, and a light movable iron chamber, forming a water-tight forecastle-deck reaching back, say 30 feet, and schooner-rigged, will, I undertake to say, make a very fair sea boat. Probably no compasses could be 'corrected' to be trusted to in such a mass of iron, but a compass fixed to the mizen mast, say at 30 feet from the deck, would be all that could be required.

Immediately abaft the hood or gun-chamber there would be a space under cover from shot where a companion and skylight could be fitted up when at sea, and through which light and air could be obtained at all times when fitted for service.

The funnel, if ever used when the vessel is not in action, would be removed for fighting, and the steam and smoke ejected through an oblique aperture right aft.

The only point to be determined by experiment is whether a moderate thickness of iron of the best quality will stand heavy shot at short range striking very obliquely, say at the worst at an angle of 30 degrees.

By the form of the proposed vessel, however, when placed in position at 250 yards of any of the Cronstadt forts, it could not be struck at 30 degrees, and probably 99 out of 100 shots that hit would graze at an angle of 10 to 20 degrees.

A small part only round the port, or what may be more correctly termed the muzzle-hole, could be struck with a direct blow.

There is every reason to believe that slabs of iron of good quality of 4 inches thick would stand against such grazing, provided they are put together without being weakened by holes and with some other precautions, and that sound forgings of 10 or 12 inches thick, if of sufficient weight in a single piece, would stand the direct blow. I do not believe that less than this would be safe against 68-lb., or, as we must expect to meet with, 120-lb. shot at short range, even when struck obliquely, and this thickness can be applied without requiring, with the gun ammunition, &c., more than 6 feet 9 inches, say 7 feet draught of water.

Another inch of thickness would require another foot of draught; but if it has been ascertained that the charts are correct, there would appear to be 10 to 15 feet of water close up to the principal detached forts, and it would be an immense advantage to take 9 feet draught of water, and have an unquestionably invulnerable skin.

If it were considered desirable to construct such a battery, it is now barely possible to do it in time for the coming season; but if possible, it could only be rendered so by ascertaining exactly the dimension and form of iron that each of the large makers could turn out with their present tools, and according to their present experience and habits, and to design the details to suit their existing means, sacrificing probably much that would render the result more perfect for the sake of rendering it possible to obtain anything in time. No doubt promises and even contracts could easily be obtained for making anything in any given time, and zealous and honest efforts afterwards made to effect what had been undertaken; but if the slightest attempt is made that involves new tools or new practices, promises and contracts will not effect impossibilities and the probability is that the short time still available will be lost.

While all the preparations shall be made on the assumption that the result is attainable and will be successful, trials must be made, without loss of time, on the several points to be determined—as to the resistance of the iron, &c. If they fail, the expense incurred up to that time in preparing for the whole work will not have been great. If they succeed, it is just possible that by great exertions but, above all, by judicious and methodical plans of proceeding, a complete battery might be launched ready for service in five months.

Lastly, I should observe that although the main feature of the plan is the resisting the effects of the enemy's shot by always exposing an oblique surface, yet the chances of fatal damage would be small if such vessels were to run the gauntlet, at night, through the deep channel, and get into the waters east of Cronstadt. Or if this is very desirable, as I should think it must be, nothing is easier than to lift the whole flotilla over the shoal water and launch them into the deep water beyond.

Mr. Brunel had matured these plans in September 1854, and they were then brought under the notice of the Admiralty; but no steps were taken to test the practicability of the scheme.

He was, however, induced to make a further representation to the Admiralty in the following July. He wrote,—

Having endeavoured ineffectually several times at the commencement of the war to impress upon members of the Government the great advantages that might be derived from the use of iron floating batteries or gun-boats, if properly constructed, I made another effort at the close of the last year's campaign, but early enough to have allowed of the construction of what I proposed before the opening of the Baltic in the present year and caused my plans to be submitted to the Admiralty through a friend. They were not approved, and I should judge from the answer I received that they were not understood, and I was never applied to for an explanation. I had no object in view but the public good, and I therefore kept the idea, such as it was, unpublished, believing the principle to be sound and good, and that the day would come when it might be usefully applied, and the more usefully to this country if not previously publicly discussed.

I have no other object to serve now; but after the clear proof that I was correct in the opinion I shared with so many other persons of the entire inability of any of the present floating ships, boats, or war engines to cope with any moderately constructed and well-armed land battery, I think it right once more, and this time more formally, to urge upon Government the consideration of the construction of armaments mechanically constructed and properly fitted for the special object. I beg to say that I do not mean a consideration in the ordinary mode, by the able and practical, but still executive officers of departments, whether engineers or ship-builders, because I believe that I must be myself, from my practical experience in this particular branch, at least as competent, if not more so, to judge in questions of mechanical construction, whether it be the forging or casting of a gun of large dimensions, or the construction of a vessel fit for navigation and capable of resisting shot, or the best mode of propelling such a vessel—on all which branches I have had much and tolerably successful practice; but I ask on public grounds for the deliberate consideration by men of judgment and experience in the military branch of the subject, such as the attack upon a fortified place, of which I cannot pretend to be a competent judge, and by men in a position to be able to express freely their independent opinions of the advantages that might be attained by the principles I propose, if capable of being successfully carried out.

Although the want of efficient gun-boats was then severely felt, this letter appears to have produced no effect upon the Board of Admiralty. But the friend who had originally brought the matter under the notice of the Board took the bold step of writing to Lord Palmerston, and acquainting him with what had passed.

Lord Palmerston at once saw the importance of investigating the subject, and sent for Mr. Brunel, who explained the plans to him. Lord Palmerston then asked him to see the officials at the Admiralty. Mr. Brunel did so; but great delay followed. It was, however, unimportant, as hostilities soon afterwards terminated, and there was no further need of gun-boats, good or bad.

This project did not exist only in the outline in which it is described in the memorandum given above. Mr. Brunel had worked out all the calculations of displacement, &c., and had made designs and models for the boat and its various appliances, and had been for some months in constant communication with Mr. W.G. Armstrong upon the form and construction of the gun.

This will be a fitting place to mention that in 1855 Captain Cowper Coles, C.B., showed Mr. Brunel the designs for his shot-proof raft, the principle of which was afterwards developed in the turret ship. Mr. Brunel gave Captain Coles the benefit of his advice on the various questions involved, and allowed him to use the services of his principal draughtsman, and to have the drawings got out in his office without expense.

Captain Coles, in a lecture which he delivered at the United Service Institution on June 29, 1860, warmly acknowledged the obligations he was under to Mr. Brunel for this act of kindness and generosity, and said that it had greatly encouraged him to persevere in bringing his plans into public notice.

### Renkioi Hospital

In February 1855, after the first winter of the allied armies in the Crimea, Mr. Brunel was asked by the War Department to undertake the design and construction of hospital buildings for the East.

He replied (on the same day that he received this application, February 16) that his 'time and best exertions would be, without any limitations, entirely at the service of Government.'

He was accordingly appointed as engineer, and proceeded to design and superintend the manufacture of the required buildings and all their internal arrangements.

They were sent out under his supervision, and erected at Renkioi, on the Dardanelles. All use for the buildings was ended with the conclusion of peace; but, for the seven months during which they were occupied, they added much to the comfort of more than thirteen hundred sick and wounded soldiers.

Many of the special arrangements adopted at Mr. Brunel's suggestion have been since brought into general use; and the success of these buildings was, to a considerable extent, influential in leading the Americans to construct similar hospitals during their civil war. These are now (1870) being copied in the German armies.

The history of the Renkioi hospital buildings is a striking instance of the zeal with which Mr. Brunel entered into any undertaking which had a claim upon his assistance, of the varied experience and fertility of invention which he could bring to bear upon any subject, however remote it might seem to be from his ordinary occupations, and of the minute personal attention he was accustomed to give to every detail, as the only certain means of ensuring success.

Mr. Brunel entered upon his duties on February 16, and reported to the War Office on March 5 that he had not lost any time nor spared any exertion or any means in his power to forward the important business he had undertaken. He stated that he availed himself freely of the advice and assistance of all persons to whom he could apply with any prospect of advantage; and he added, 'It is most gratifying to be able to state that from everybody I have received the most zealous and cordial assistance, and found it sufficient to mention the object of my enquiries to obtain immediately every assistance I could possibly require.'

An experimental ward was erected a few days afterwards on the premises of the Great Western Railway at Paddington, and was carefully criticised by competent persons; and, the plans having been approved of, specifications were made, with drawings of the various parts, and tenders were invited for the construction of the buildings.

The following paper gives a description of the buildings, and was written by Mr. Brunel in order to satisfy the curiosity of his friends[5]:—

March 1855.

The conditions that it was considered necessary to lay down in designing these buildings were,—

First. That they should be capable of adapting themselves to any plot of ground that might be selected, whatever its form, level, or inclination, within reasonable limits.

Secondly. That each set of buildings should be capable of being easily extended from one holding 500 patients to one for 1,000 or 1,500, or whatever might be the limit which sanitary or other conditions might prescribe.

Thirdly. That when erected they might be sure to contain every comfort which it would be possible under the circumstances to afford. And—

Fourthly. That they should be very portable, and of the cheapest construction.

The mode in which it has been sought to comply with these conditions is as follows:—

The whole hospital will consist of a number of separate buildings, each sufficiently large to admit of the most economical construction, but otherwise small and compact enough to be easily placed on ground with a considerable slope, without the necessity of placing the floor of any part below the level of the ground, or of having any considerable height of foundation to carry up under any other part.

These separate buildings have been made all of the same size and shape; so that, with an indefinite length of open corridor to connect the various parts, they may be arranged in any form, to suit the levels and shape of the ground.

Each building, except those designed for stores and general purposes, is made to contain in itself all that is absolutely essential for an independent hospital ward-room; so that, by the lengthening of the corridors, and the addition of any number of these buildings, the hospital may be extended to any degree.

To ensure the necessary comforts, and particularly to provide against the contingency of any cargo of materials not arriving on the spot in time, each building contains within itself two ward-rooms, one nurse's room, a small store-room, bath-room, and surgery, water-closets, lavatories, and ventilating apparatus.

The ward-room is made wide enough and high enough to ensure a good space of air to each bed, even if these should be unduly crowded. Each building contains two ward-rooms, intended for twenty-six beds each, which is found in practice to be a size of room admitting of proper control and supervision.

With respect to closets and lavatories, after examining and considering everything that has been done, both in hospitals of the best description and poor-houses of the cheapest construction, it was found that the requisite security for cleanliness and the greatest amount of economy of labour, and of consumption of water, could be obtained by a cheap description of water-closet designed for the purpose; and with the same object of diminishing the amount of labour and the waste of water, and securing cleanliness without depending upon the constant attention of assistants, fixed basins for lavatories and mechanical appliances for supplying and drawing off water were adopted.

As a protection against heat, experience in hot climates and experiments made expressly for the purpose satisfactorily proved that a covering of extremely thin and highly polished tin, which reflects all direct rays of heat, was the cheapest, lightest, and most effective protection, and every piece of woodwork not covered with tin is to be whitewashed externally. Internally the lime-wash has a slight tint of colour, to take off the glare.[6]

To secure ventilation in a hot climate with low buildings extending over a large area, and therefore incapable of being connected with any general system of ventilation, it was considered that forcing in fresh air by a small mechanical apparatus attached to each building would be the only effective means. Each ward-room is therefore furnished with a small fan, or rotatory air-pump, which, easily worked by one man, is found capable of supplying 1,000 to 1,500 cubic feet of air per minute, or 20 to 30 feet for each patient. This air is conveyed along the centre of the floors of each ward-room, and rising up under foot-boards placed under the tables, is found to flow over the floor to every part of the room.

Besides this mechanical supply of air, opening windows are provided along the whole length of the eaves, and spaces left immediately beneath the roof at the two gables, amply sufficient together to ventilate the rooms thoroughly if any breezes are stirring, without the help of the fan.

The light is admitted by a long range of narrow windows, immediately under the eaves, which protect them from the direct rays of the sun. These windows open, and are provided with shutters inside, which exclude the light, but admit the air.

By forcing the air into the room, instead of drawing it out, the entrance of bad air from the closets, drains, or any surrounding nuisances is prevented. The fan is placed at the opposite end to the closets and drains; and all the fans being in the open corridor, the workmen can be seen by a single sentry, and kept to their work.

The buildings, as now constructed, are adapted to protect the interior from external heat. Should winter come while they are still in use, the framework is adapted to receive an internal lining of boarding, and the interstices can be filled with a non-conductor.

Two buildings, of the same form and dimensions, are fitted up with every convenience as store-rooms and apothecaries' dispensaries.

An iron kitchen, slightly detached from the wooden buildings, fitted up with every contrivance capable of cooking for from 500 to 1,000 patients, is attached.

A similar building of iron is fitted up with all the machinery lately introduced in the baths and washhouses of London for washing and drying in the minimum space, and with the least amount of labour.

If an aggregate of buildings should be placed in one spot for more than 1,000 patients, a second kitchen would be added, but the single washhouse would be sufficient.

With each set of buildings is sent a pumping apparatus, a small general reservoir, and a sufficient length of main, with all its branches, to supply water to every detached building; and all the pipes and branches are of such construction as to admit of being put together without any soldering or cement. A system of drains is provided, formed of wooden trunks properly prepared, and of sufficient extent to form a complete and perfect system of drainage from every building to a safe distance from the general hospital.[7]

A number of small buildings, intended to be detached from the main body, are provided for residences for the officers and servants of the establishment, and for a small detachment of soldiers. A slaughter-house and store-yard and some other appurtenances are also provided, the extent of which depends on the circumstances of each case.

The construction of each building has been studied with very great care, so as to secure the minimum amount of material, the least possible amount of work in construction or erection, and the means of arranging all the parts in separate packages capable each of being carried by two men; and the result is that each building is the cheapest and lightest that has yet been constructed in proportion to the area covered.

For the transport of the materials to the spot selected, two sailing-vessels and three steamboats, capable of carrying one hospital for 1,000 men, which is the first about to be sent out, have been secured. In each vessel is sent a certain number of complete buildings, with every detail, including their proportion of water-pipes and drains, closets, lavatories, baths, &c., and a small amount of surplus material and tools; and in each of two separate vessels are sent a set of pumps and mains, and a kitchen and washhouse. So that by no accident, mistake, or confusion, short of the loss of several of the ships, can there fail to be a certain amount of hospital accommodation, provided with every comfort and essential.

The peculiar circumstances under which these establishments are likely to be placed have required not only peculiarities of construction, but these, in turn, have required numerous provisions and details specially designed for the case.

As all the buildings, except the kitchen and washhouse, are entirely constructed of wood, it is considered essential that no stove or fire-place of any description should be allowed in any part, except in the iron buildings; in these there is provision for an ample supply of hot water, but each ward-building is provided with a small boiler, heated by candles, which by experiment have been found amply sufficient for all that can be required. Candles are to be used exclusively for lighting, and lamps and lanterns have been constructed for the purpose.

A proper supply of fire-engines is provided, and other precautionary measures are adopted against fire.

The condition of portability requiring that the walls and roofs should be of the thinnest and slightest possible construction, protection against heat has been provided for in the manner before referred to, and good ventilation secured by mechanical means. But, in addition to this, there is a very simple provision made for passing the air over a considerable extent of water surface; which would not only cool it, but diminish the effect of excessive dryness, which is said to be occasionally in this climate more oppressive than even the temperature.

As the space in the wards is very liable to be encroached upon, and the beds crowded, portable baths have been designed, into which the more helpless patients can be lifted, and lowered, on a frame or sack, without requiring space for assistants to stand around, or with the bath placed only at the foot of the bed.

The kitchen and laundry have each required many special contrivances.

The instructions given to Mr. Brunton, the engineer, who has been sent out for the purpose of erecting these buildings, are, to commence by determining on his plan of arrangement to suit the peculiarities of the ground, and then to construct the complete system of drainage and to lay on the water supply before the buildings are rendered capable of receiving patients; and all the arrangements of the details are designed with the view of obtaining, as the first conditions, a perfect system of drainage, a good supply of water, free ventilation, and the most perfect cleanliness, quite independent of labour and of the continued attention of assistants; these conditions being assumed as essentials, preceding even the mere covering in of space and providing shelter for patients.

The cost of these buildings, delivered ready for shipment, will be from £18 to £22 per bed, allowing 1,000 cubic feet of space in each ward-room to each bed. If pressing emergency should lead to the beds being placed closer, and fifty per cent more patients should be introduced, it is believed that the perfect system of ventilation which is secured would render these hospitals very superior to any now in use for the army.

Of the cost above named, about £12 per bed is that due to the ward-rooms themselves, with all their conveniences attached, and the rest arises from the cost of the store-rooms, kitchen, machinery, residences, and appurtenances.

The cargo space required for their conveyance is about a ton and a half to a ton and three-quarters measurement per bed.

As the buildings were completed the work of transport was commenced; and twenty-three steamers and sailing-vessels were despatched, containing altogether about 11,500 tons measurement of materials and stores. The first vessel arrived out on May 7, 1855, and the last on December 5, in the same year.

Meanwhile the important question of the site for the hospital buildings was being determined by Dr. Parkes, the Medical Superintendent, with the assistance of Mr. Brunton, who was in constant communication with Mr. Brunel on the subject.

After visiting various places, Dr. Parkes finally selected a spot near the village of Renkioi, on the Dardanelles. In a report which he addressed to the Secretary of State for War upon the formation and general management of the hospital, he thus describes the nature of the site, and the means which were used in the formation of the hospital:—

The piece of land on which the hospital was placed was a shelving bank of a light, porous, sandy soil, resting on marl; it contained about 270 acres, stretched tongue-like into the waters of the Dardanelles, and was bounded inland by a low range of sandstone hills, which were themselves backed by rather lofty ranges of oolitic limestone, intersected by deep ravines. The tongue of land formed two bays, north and south, in both of which was good anchorage for ships, and as the wind blew almost always up or down the Dardanelles, i.e. from the north-east or south-west, one or other of these bays was comparatively calm in all winds except those which came infrequently from the west.

The position of the spot was on the Asiatic coast, nine miles from the mouth of the Dardanelles, in lat. 40° 2′, long. 26° 21′. It was the site of the port of an old Greek city, the ancient Ophrynium.

The extreme point of this tongue of land was about 10 feet above the sea, but from this point it rose regularly and gradually to about 100 feet above the sea. An admirable fall was thus given for drainage, and so gradual was the rise that the wooden houses were placed on the ground without terracing or excavation, whereby very great expense was saved. The extreme length from the point to a spot too steep for the erection of houses was about half a mile, and we were enabled thus to place down the centre of the tongue of land no less than thirty-four houses, capable of holding 1,500 sick, in one long line on either side of the central corridor, an arrangement which facilitated very greatly the laying of both water-pipes and drain-tubes. In fact, we were able to carry out the plan which Mr. Brunel had suggested as the best.

There was enough space on the tongue of land, on either side of this long central line, for two shorter parallel lines of seventeen houses each. These two lines were placed one to the north, and the other to the south of the large central hospital. Each was capable of containing 750 men, and one of them to the north was nearly completed when the declaration of peace put a stop to the works.

On the sides of the hills in rear were numerous small springs of excellent water, which were collected together and conveyed in earthenware pipes to a large reservoir, placed by Mr. Brunton 70 feet above the highest house, which was itself about 60 feet above the sea. From this reservoir the water was carried in iron pipes down the centre of the long corridor, and at every ward (which was placed at intervals at either side of the corridor) a leaden service pipe came off and led an abundant and never-ceasing supply into the ward cisterns, which supplied the baths, lavatories, and closets. By this arrangement all necessity for pumping water was avoided, and the sewers were able to be flushed very perfectly. The lavatories and closets were placed at the ends of the wards most remote from the corridor, and immediately outside them ran the two main sewers, which at their sea terminations were carried some distance into the Dardanelles.

The plan of the hospital may be at once understood by imagining a covered way, open at the sides, and 22 feet wide, running nearly east and west, and reaching for a length of more than a third of a mile, on either side of which stood, at intervals of 27 feet on the south side, and in most cases 94 feet on the north, the thirty-four houses, each of which, as already said, was 100 feet long, 40 feet wide, 12 feet high at the eaves, and 25 in the centre, and was capable of containing fifty patients, with an allowance of nearly 1,300 cubic feet of air for each man. Some portion of this space was occupied by the closets and some small rooms used as orderlies and bath-rooms. Thirty of these houses were used as wards; four were used as dispensaries and purveyor's stores. A drawing by Mr. Brunton, showing the arrangements of one of the wards, is attached.[8]

To the south of each division of ten houses was placed an iron kitchen, which afforded the necessary accommodation for preparing 500 diets.

At the inland extremity of the corridor were placed two iron laundries, the water from which (some 4,000 gallons daily) was passed into the sewers. Beyond the laundries were placed on either side the wooden houses of the medical and other officers, who were thus able to see down either side of this long line, and to preserve to a certain extent surveillance over the patients.

The two smaller hospitals were constructed on a similar plan, each range having, however, only one iron laundry inland, and one iron kitchen in the centre of the range.

About half a mile from the hospital, close to the sea in the south bay, three store-houses were erected, and a railway led from an adjacent jetty or pier by the side of these store-houses to the centre of the main hospital. Had the war continued, it would have been carried to the north pier and bay, and would also have had a branch running along the corridor of each hospital, so as to deposit the sick at the very doors of the wards into which they were to go.

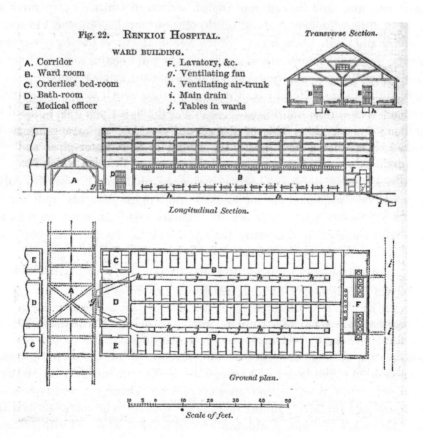

Fig. 22.   RENKIOI HOSPITAL.

WARD BUILDING.

*Transverse Section.*

A. Corridor
B. Ward room
C. Orderlies' bed-room
D. Bath-room
E. Medical officer

F. Lavatory, &c.
g. Ventilating fan
h. Ventilating air-trunk
i. Main drain
j. Tables in wards

*Longitudinal Section.*

*Ground plan.*

Scale of feet.

Nothing could exceed the simplicity of the whole arrangement; it was a repetition of similar parts throughout; and experience enables me to say, that nothing could be better adapted for a hospital than this system of isolated buildings, between every one of which was a large body of moving air, rendering ventilation easy, and communication of disease from ward to ward impossible.

The introduction of the covered way connecting the various houses was a happy idea. In the summer this corridor was left quite open at the sides, and formed a cool walk for the convalescents; while in winter we boarded up its north side, so that in the coldest blasts of the northern wind the men were protected, and were able to leave their wards and to take exercise. I need only further observe that, in order to secure perfect ventilation, not only were openings left under the eaves and in the gables of the buildings (which could be closed in cold weather), but air-shafts were placed under the floors through which 1,000 cubic feet of air per minute could be forced into the wards by fans placed in the corridor and worked by hand [fig. 22]. As the amount of wind at Renkioi was always considerable, we never had occasion to use these machines; but had the hospital been placed in a less airy situation, they would have been of the greatest use.

For the construction of this hospital every necessary part was sent out by Mr. Brunel. The houses were erected with great care by Mr. Brunton, assisted by Mr. Eassie, jun., and by eighteen English workmen (thirteen carpenters, one pipelayer, three plumbers, and one smith) sent out for this purpose. On account of the size and height of the houses (which were many times the size of the largest Crimean huts), the framework was obliged to be put together very carefully, and Mr. Brunton felt it necessary to employ none but the English workmen on this duty; consequently the erection of the houses took much longer time than we originally anticipated; but during the winter we had reason to be satisfied that Mr. Brunton had done wisely, for, in spite of the heavy winds we often had, no finished house was ever damaged, except in one or two instances to a very slight amount.

The erection of the houses was commenced on May 21, 1855. On July 12 I reported the hospital ready for 300 sick; on August 11 it was ready for 500, and on December 4 for 1,000 sick. By January 1856, viz. seven months after its commencement, it was ready for 1,500 sick; and when the works were discontinued, at the end of March 1856, we could, with a little pressure, have admitted 2,200 patients. In about three months more this immense establishment for 3,000 sick could have been finished and in full activity.

On the working of the system, Dr. Parkes says in his report:—

Although the hospital was ready for 300 patients on July 12, 1855, we were not called upon to receive sick till October 2. From that time till February 11, eleven ships arrived from Balaclava and Smyrna . . . After February 11, 1856, we received no more sick. The total number of military patients who were received from these ships was 1,244, and, in addition, 87 soldiers were admitted, either from the guard at Renkioi or Abydos, from transport ships which touched at Renkioi, or from the English soldiers attached to the Osmanli Horse stationed at the town of Dardanelles during the summer and autumn.

The total number of admissions was 1,331—

| Cured | 961 |
|---|---|
| Invalided | 320 |
| Deaths | 50 |

Besides the military patients, we admitted 77 civilians . . . The total number of patients actually treated was 1,408, the largest number at any one time 642 . . .

The anticipations we had formed of the health of the spot and of its adaptability for a hospital were quite confirmed by the experience of more than a year. The winter was mild, and the climate seemed especially adapted for pulmonary complaints, of which we had a large number. The changes of temperature, it is true, were very sudden and great; but as the men had warm wards, these changes were not felt, and there were few days in which the most delicate consumptive patient could not get out into the sheltered corridor for a short time during the day. The construction of the hospital was admirably adapted for men recovering from illnesses. As all the wards were on the ground, as soon as a man could crawl he could get into the air either in the cool and sheltered corridor or in the spaces round the hospital.

In April and May 1856 the greater number of the patients had been either discharged or invalided home, and . . . the medical and nursing staff was reduced more than one-half, and . . . in the middle of July the remaining staff was sent home.

All the stores which were likely to be used or to sell well in England were sent home, and everything else was sold on the ground. Major Chads, with twenty soldiers, and Mr. Brunton remained behind, to superintend the sale of the buildings, which took place on September 20.

1.  Reports on the Paris Universal Exhibition, 1867, vol. i. p. xxiv. 196.
2.  When, at the close of the Exhibition, Mr. Brunel was compelled, much against his will, to accept a pecuniary acknowledgment of his services, he spent the money in the erection of model cottages at Barton, a village near his property in Devonshire.
3.  See Report of the Select Committee of the House of Commons on Ordnance, 1863, Minutes of Evidence and Appendix (ordered to be printed July 23, 1863):—p.306, Report of Ordnance Select committee; p.44, Statement of Mr. Whitworth; p.56, Evidence of Mr. Whitworth (Q. 1329–1337); pp.58, 59, the same (Q.1385–1410); pp.402, 403, Papers delivered in by Mr. Isambard Brunel; p.110, Further Evidence of Mr. Whitworth (Q. 2545–2551); p. 112, the same (Q. 2602–2610); p. 548, Letter from Mr. Westley Richards to Mr. Isambard Brunel.

4. Captain Claxton, at Mr. Brunel's desire, went for a voyage in Messrs. Ruthven's vessel, the 'Enterprise,' in order to test her performances.

5. This appears to be the only instance in which Mr. Brunel printed an account of any of his works.

6. On this application of tin for the covering of the roof, Dr. Parkes observes (*Manual of Practical Hygiene,* London, 1869, p.317):—'In the Crimean War the roofs of Renkioi Hospital, on the Dardanelles, were covered with polished tin; it was found, however, somewhat difficult to place it so as to exclude rain, and the surface soon became tarnished. The thermometric experiments did not show a greater lessening of heat than 3° Fahr. below houses not tin-coated.'

7. At Renkioi, in Turkey, Mr. Brunel supplied square wooden sewers about 15 inches to the side; they were tarred inside, and acted most admirably without leakage for fifteen months, till the end of the war. The water-closets (Jennings's simple syphon), arranged with a small water-box below the cistern, to economise water, never got out of order, and in fact the drainage of the hospital was literally perfect. I have little doubt such well-tarred wooden sewers would last two or three years (Parkes's *Manual of Practical Hygiene,* p.635).

8. The woodcut, fig. 22, is a reduction of part of this drawing.

# VI

# MR. BRUNEL'S PROFESSIONAL OPINIONS AND PRACTICE

IT is proposed in the earlier part of this chapter to describe, principally by extracts from Mr. Brunel's correspondence, the position occupied by him in regard to the Companies which he served, and to the various classes of persons with whom he acted in the discharge of his duties as engineer to those Companies. These selections are followed by extracts relating to questions of general professional interest.

The first point to be considered is, his relations with the Companies which employed him as their engineer, and with the Directors who formed the governing body.

Mr. Brunel conceived that he was, by virtue of his appointment as engineer, the sole and confidential adviser of the Company in all matters relating to the construction and mechanical working of the undertaking. He did not permit any one to be associated with him in the supreme control over those matters which were in his department; and the moment he thought that confidence was no longer placed in him, he was prepared at every sacrifice to resign his office. But, as long as he was supported by the Directors, he thoroughly identified himself with their cause, and he never allowed considerations of health or convenience or pecuniary advantage to interfere with the performance of any service which he could render them. The fearless independence of his position, combined with his absolute devotion to the interests of his employers, was no doubt the secret of the immense influence he acquired, and of the affectionate esteem with which he was regarded by those whom he served.

*On the Direction of Railway Works*

March 4, 1845.

I have well considered the communication which you did me the honour of making on the part of the Government of His Majesty the King of Sardinia with reference to my undertaking the direction of the works of the proposed railway from Genoa to Alessandria, about to be executed by the Government itself . . .

In the first place I assume that if the direction of the works be confided to me as the engineer, the same degree of confidence will be placed in me, and the same authority will result from that confidence, as would be the case in England— that is to say, I should be the confidential adviser of the Government in all engineering questions connected with this railway, my communications would in all matters be made direct with the Government, and as long as I continued to be responsible for the direction of the works no other engineer would be consulted or allowed to interfere. Of course I claim no right to direct anything but that which has the sanction of the Government; but I should claim to be their sole adviser on all engineering points (connected with the construction of the railway), and to possess their entire confidence; and also that, if any portion of that full confidence were at any time withdrawn, the fact should be immediately communicated to me; when, after making every possible arrangement to prevent inconvenience to the Government, I might withdraw from the direction of the work. This is the position which an engineer of any standing occupies in this country, whether acting for the Government or for individuals; and I believe it to be as fully essential for the success of the proposed undertaking, and as necessary for the interests of His Majesty's Government as for my satisfaction, that I should be placed in a similar position.

The circumstance of my being a foreigner, of my being rarely present to meet objections, if any are raised, of the unavoidable frequency of real as well as apparent failures in works of such variety and so numerous as those which occur on this line, of the difficulties which always attend the introduction of novelties and everything connected with a railway, the rapid mode of its construction, the necessity which experience has proved of frequently adopting apparently hasty and hazardous methods to prevent the evil consequences of protracted delays,—all this will be novelty with you as it was years ago in England, all these circumstances combine to render it peculiarly essential to the satisfactory progress of the undertaking that it should be well known to all parties that full and entire confidence is placed in me by the Government.

It will of course also be necessary that all parties acting under me in the direction of the works should feel that their appointment or dismissal depends entirely upon me.

In return for the confidence thus placed in me and the authority given to me, I should of course know no interest but that of the Government. If the Government is willing to appoint me engineer according to this definition of my position, I shall feel pride in the appointment, and I shall devote my best energies to the accomplishment of one of the finest and most interesting works of the day . . .

*On the Position of Joint Engineer*

October 16, 1843.

The contents of your letter of yesterday take me quite by surprise; the expression you use of joint-engineership implies a view of our relative position diametrically opposed to the views which I have plainly and unequivocally expressed to you and to the Directors when such a thing as joint-engineership was proposed to and rejected by me . . .

You wind up your letter by saying 'we have accepted the duty of joint-engineers,' &c., and you add a postscript requesting me to lay your letter before the Directors: this I should have been obliged to do without any such request. I never accepted the duty of joint-engineer; I have always refused to do so. I thought I had made this very clear both to you and the Directors on several occasions; indeed I often feared that I expressed myself too strongly instead of leaving it capable of misapprehension.

### On the Position of Consulting Engineer
December 30, 1851.

I shall be happy to act in any capacity (subject to the exception I will further explain) which can be useful to your Company; but the exception I have to make is one which perhaps resolves itself merely into a question of name. The term 'Consulting Engineer' is a very vague one, and in practice has been too much used to mean a man who for a consideration sells his name, but nothing more. Now I never connect myself with an engineering work except as the Directing Engineer, who, under the Directors, has the sole responsibility and control of the engineering, and is therefore 'The Engineer;' and I have always objected to the term 'Consulting Engineer.'

In a railway the only works to be constructed are engineering works, and there can really be only one engineer; and in your case especially, where, as I apprehend, the contractor is part of the company, and has to be treated with consideration, and perhaps less vigorously, at all events differently from an ordinary contractor, considerable management and discretion will be required of your engineer, and a degree of responsibility which I would only undertake if sole engineer. Possibly this is what you meant, and that I alone see the distinction, but it is an important one with which you may not be so familiar as I am.

### On the Position of the Engineer in Relation to the Contractors
May 26, 1854.

I have in due course taken steps to prepare a report for the Directors on the state of the work, but you must not apply to the contractors for such reports. In the first place, it would lead to ridiculous contradictions, inasmuch as most likely my reports would differ materially from theirs; and also it would reverse the whole order of things. I must alone, so long at least as I am the professional adviser of the Company, be the medium of communication with contractors in all matters which the terms of the contract refer to me. I am very particular about the regularity of all these forms, because, although, while all goes smooth, they are of no consequence yet if, unfortunately, any little difficulties arise, then it is unpleasant and difficult to alter a previous course.

I would ask, then, that all official communications with the contractors should be made through me, and that, even as regards these, we should confer together before any formal resolution should at any time be passed, so as to be sure that it is properly worded, and that no awkward precedent is established.

You may have seen that a great appeal case before the Lords, affecting claims of some hundreds of thousands, has just been finally decided in favour of the Great Western Railway Company after fourteen years of litigation;[1] and this favourable decision was entirely obtained by carefully prepared specifications, and by my not having departed in any single case, in years of correspondence, from the letter and spirit of the contract, and particularly from the fact—strongly commented upon by Lords Cranworth and Brougham—that I had maintained my position of umpire between the Company and contractor. It is, then, as essential to the Company as to the contractor and to me that I should maintain that position.

### On the Relations between the Engineer and the Directors
### I

April 15, 1850.

You will remember that on the 2nd of November last I addressed you a letter on the subject of my acting as umpire in several cases of reference between the Company and contractors, who were raising heavy claims against the Company, in which I expressed my readiness to act in a very responsible, laborious, and thankless position . . .

I am also still acting, so far as my services can be useful, as your engineer; and in arranging with contractors, &c., there is even now much remaining to be done. As long as I enjoy your full confidence, it would be a great pleasure to me, although the profit might for the present be small, to continue your engineer, looking forward, as I do, at no distant period to the completion of the works; but the changes that have lately taken place in the Board, and the excitement which has been displayed by some of your proprietors in effecting those changes— such excitement as is too apt frequently to lead men to form most unjust and erroneous opinions of the trustworthiness of those who have been long engaged in their service—very naturally causes anxiety. I, and you who have been acting with me for so long, know that I have always advised you honestly, and I hope generally wisely, on such matters as I have been consulted upon, notwithstanding my known connection with other and sometimes conflicting interests. But new men may not have the same confidence in me, and I cannot afford to run the risk of any doubts being entertained upon the subject.

I must therefore beg of you to bring the matter before the Board formally, and if it is desired that I should continue to act as your engineer, that I may have it clearly understood that I do so with the full concurrence of the Board as now constituted, and particularly of the new Directors; and that to render the expression of their confidence clear, I wish them to understand that I am quite ready to tender them my resignation if they do not feel that confidence, and that such resignation cannot prejudice any of the cases under arbitration before me, as I undertook them upon the express understanding that it was as agreed umpire, and not as engineer to the Company.

I shall only add that by being engineer to a Company, I mean acting in perfect confidence with the Directors on matters generally, as I have been in the habit of doing.

## II

December 6, 1851.

I really cannot consent to forego in this case the rule I lay down for myself in my professional business, which is to yield as far as I comfortably can to the mere wishes of Directors in the mode in which I direct the works they may order, but not beyond this point—and if the Directors of any body claim the right to control the staff which I think necessary to carry on my work, I concede the right at once, and resign the direction of that staff. The staff for constructing engineering works is not a permanent establishment, which can be extended or restricted just as a Board of Directors may determine; it is much more of the character of a personal staff attached to the engineer-in-chief, and by means of which he is enabled to take great responsibility upon himself, both in respect of the works which he executes, and the settlement of large amounts of payments, in which settlement both the company and the contractors are very much at his mercy. It is the duty of an engineer of course to study economy as well as efficiency in his staff, and to pay attention to the wishes of the Directors so far as he can; but he must be the ultimate judge of the number and the persons of that staff. At least, that is my rule, and I cannot depart from it in this case. I should wish to state this as respectfully as possible to the Board, but at the same time to state it clearly.

## III

January 22, 1857.

. . . I have replied very fully to the observations of the Directors, but having done so I must again place my resignation in the hands of the Directors. A Board of Directors has a perfect right to dispense with the services of an engineer, or to lay down any rules of conduct they may think fit, and to engineer the works themselves, either as a body or by appointing one of their members, if they think fit; but if they desire to have the advice and responsibility of any respectable engineer—at all events, if they wish to have mine—they must place the usual amount of confidence in me; and as long as I am engineer they must leave me to conduct the engineering, and must act as if they assumed that I was more able to advise the Board upon all the usual practical questions of engineering than any one of the Directors. Admitting as I do the full right of a Board of Directors to determine whether they will have an engineer or not, if they have one they must trust him to do his own work. As a mere question of time it would be impossible that I could be constantly entering into long explanations of the most trifling character upon points on which I must be by far the most competent judge

—and in fact the competent and responsible judge, expressly employed, and paid professionally because I am so. If individual Directors are more competent than their engineer, let them assume the office and the responsibility; but do not have a divided responsibility. I am most happy at all times to receive any hints or suggestions from any body whether Directors or not, but I cannot undertake the labour of replying to them all, neither can I act as engineer to a Board of Directors who, either as a body or by one or two of their members, make a practice of taking upon themselves to judge of those details of which I must be assumed to be the competent judge, and thereby interfering with my proper duties.

If the engineer is not assumed to be worthy of trust both in respect of zeal in the interests of the Company, and very superior to any one of the Directors as regards experience, and ability to carry out what the Directors may determine upon, he is not fit to be the engineer; but if he is so, he must be left to carry out those details of his department of the business. These are the general principles upon which alone I can act.

It is, from the nature of the case, impossible by any extracts from Mr. Brunel's correspondence to give an adequate idea of the position which he occupied in relation to his assistants. There was so much personal intercourse between them that letter writing was but little resorted to.

His relations with them were of the most affectionate and intimate kind, and were maintained without any ostentation or outward show. He was in the habit of placing entire confidence in his subordinates as long as he considered that they deserved it; and, while he preserved his own proper position, he was always ready to shield them from the interference of others. The letters which are printed below are evidence of this. When, on the other hand, anything occurred to displease him in the conduct of his assistants, he was eager to give the offender a chance of retrieving his position; and he was always ready to help them in any difficulties.

### On Interference of Directors with the Assistant Engineers

#### I

January 19, 1842.

While I am upon the subject, and as I have referred to the impolicy of Directors taking notice of little things, and as I speak freely to you, I will mention that I have observed with pain on some occasions this tendency; and I will give one instance of what I must call most unwise interference. It was lately, and unfortunately at the same moment as this complaint, intimated that a pair of boxing gloves had been seen in one of the Company's offices, and that *the Directors had observed it.* Now I really do not know why a gentlemanly and industrious young man like —— should be subject to have his trifling actions remarked upon more than I myself, unless the observer gave him credit for a much more gentle temper than I possess; because I confess, if any man had taken upon himself to remark upon my having gone to the pantomime, which I always do at Christmas, no respect for Directors or any other officer would have restrained me. I will do my best to keep my team in order; but I cannot do it if my master sits by me, and amuses himself by touching them up with the whip.

## II

January 28, 1842.

I am much obliged to you for your letter. I am sorry to find, however, that the impression, a very erroneous one as I believe, remains upon your mind that the assistant engineers are predisposed to encourage, or at all events allow, improper conduct on the part either of contractors or the inferior agents of the Company . . .

From some experience in these matters it is that I have come to the conclusion that it is wise (however strange you may think the doctrine to be) to shut one's ears and eyes really and truly to everything which does not come forward in such a shape as to demand and admit of an enquiry; and it is for this reason also that I do entertain the opinion very strongly (in which you appear to differ from me), that it is not the interest, it is not wise, and therefore only it is not the duty of Directors to look after, or to see into, the smaller details of the conduct of an establishment which, being of a very temporary, changing, and uncertain character, cannot at the best be conducted with the discipline and regularity of a permanent establishment, in which the parties have their clearly defined and unchanging duties, and look forward to the permanent occupation of their places as their means of support.

At all events, when the Directors see anything they think desirable to correct or to modify, they can fully communicate it to me without the possibility of giving to me any soreness of feeling, which it is always desirable not to excite, even in the case of the lowest menial whose best services one wishes to have and use.

## III

December 12, 1851

With reference to your letter of December 11, stating that the Directors 'are satisfied that great irregularities have existed, and that they feel it to be their duty, and will not hesitate on any occasion, to represent to me any irregularity on the part of my staff that may come to their knowledge,' I am almost afraid, unless in a short note you may have failed to convey to me the meaning of the Directors, that they greatly misunderstand my feelings on the subject; my great desire, as great as, possibly greater than even that of the Directors, who cannot feel so personally responsible as I do for the efficiency of my staff—my great desire I say is to hear immediately from anybody, and particularly of course from a Director, of any supposed irregularity; and I should feel that I had ground of complaint even if any such report or any suspicion of any irregularity were not immediately communicated to me. The moment that the Directors could doubt my being as anxious as they can be to know and to remedy any irregularity, or that they should look upon me in such matters otherwise than as one of themselves, I should feel that I had lost their confidence, and could no longer carry on satisfactorily to myself my duties, and should therefore resign them. Such must, I beg, be our relative position as regards the future; and carrying out this principle as regards the past, I must beg of them to tell me explicitly what are the irregularities to which they refer as having been committed. I ought to be fully informed of such things—indeed, nothing ought to be suspected even without my knowing; for if *I* ought not to know, who ought?

As a matter of form, and to be strictly correct, I must guard myself against being supposed to mean that I could desire or approve of what the Directors I am sure would also disapprove of—namely, a system of fault-seeking—because in a very numerous staff or body of men, particularly where they have not the benefit of permanent situations, the perfection of regularity cannot be hoped for; what I principally seek and require of my assistants is an honest discharge of their duties, and any departure from this it is well known amongst them I never overlook. Have the goodness, therefore, to ascertain for me, and to let me know immediately, what these irregularities have been.

A few words may here be added on Mr. Brunel's practice in reference to taking pupils.

Although many of his assistants had been his pupils, he did not encourage young men to come to him with the object of learning their profession in his office. He never absolutely declined to take pupils; but he endeavoured, by fixing a high premium, to reduce the number of applicants.

He did not profess to do more for his pupils than to give them the opportunities of seeing work, afforded by his office, and the chance of being afterwards employed as his assistants. He attached much importance to private study of mathematics and other branches of science.

Passing on to the position assumed by Mr. Brunel in his relations to the profession at large;—it may be stated in a few words, that he was desirous on all occasions of promoting its welfare by encouraging friendly intercourse among its members, by healing strife, by suppressing as far as he could all cant or pretension, and by setting his face steadfastly against all attempts to fetter the freedom of invention or to lessen the independence of engineers by State patronage or control.

It may appear strange to affirm of one who was foremost in almost all the professional contests of his time, that he was zealous in healing strife; but it is nevertheless true that Mr. Brunel, while he was a bold and uncompromising advocate of his own schemes, was at the same time untiring in his exertions to limit the area of controversy, to confine it strictly within its proper bounds, and to divest it of all personality or of anything which could lead to unpleasant feeling or annoyance.

His endeavours to this end were greatly helped by the friendly relations which he maintained at all times with his professional brethren. He never allowed any divergence of opinion to interfere with private friendship; and, even in the height of controversy, he was glad to give, and ready to ask for, advice on matters connected with the scientific departments of civil engineering.

It is but seldom that extracts have been made from Mr. Brunel's private journals; but it may be permitted, in illustration of what has been said on this point, to give the following passage, written during the great contests of the year 1846.

May 5, 1846.

I am just returned from spending an evening with R. Stephenson. It is very delightful, in the midst of our incessant personal professional contests, carried to the extreme limit of fair opposition, to meet him on a perfectly friendly footing, and discuss engineering points . . . Again I cannot help recording the great pleasure I derive from these occasional though rare meetings.

Mr. Brunel's opinions on the working of the patent laws will be given below, as following more fitly after extracts from his correspondence relating to the position of his profession in regard to the Government; but, before entering upon that subject, a few words may be said in reference to a class of persons who formed a very large proportion of his correspondents—the class of 'Inventors.'

He used to receive numerous applications from persons who had invented, or who thought they had invented, some useful contrivance, from a locomotive which would save fifty per cent in fuel over those then in use, to a machine which, as Mr. Brunel assured its inventor, 'would not even have a tendency to move.' He was always ready to encourage inventions which seemed likely to produce good results, and to enquire into their merits, if they were patented; but not otherwise, lest it should be said that confidence had been placed in him.

The following is one of the many letters he had to write in answer to requests of this nature.

September 17, 1847

I could not have complied with your request of giving any opinion upon the merits of the invention. Simple as such an act may be, it too frequently involves one in controversy; and I never found, before I made the rule not to give opinions, that my advice was ever followed, if it was to discourage the inventor from further expense and trouble.

I should tell you in this case that the idea of dovetailing, which in your first letter I find was the principle of the invention, had long before been worked out in every shape and form that ingenuity or blundering could possibly give it.

Upon the important question how far, if at all, the practice of civil engineers should be subject to State control, Mr. Brunel held very decided views. He was strongly opposed both to any interference on the part of the State with the freedom of civil engineers in the conduct of their professional work, and to the recognition of merit by the bestowal of honours or rewards.

### On the Royal Co. on the Application of Iron to Railway Structures[2]

March 13, 1848.

I regret that the Commissioners should have done me the honour of requesting 'my opinion upon the enquiry referred to them;' because, as it is known to one or more of these Commissioners that I have expressed very strongly, both publicly and privately, my doubts of the advantage of such an enquiry, and my fears of its being, on the contrary, productive of much mischief, both to science and to the profession, it would expose me to a charge of weakness and of inconsistency if I were now to refrain from expressing those opinions, which otherwise I had no idea of intruding upon the Commissioners; and, indeed, I had hoped that, by making those opinions known to some of the members, I might have been passed over, and not invited to assist in the proceedings.

I shall be most happy to communicate, as I am at all times most anxious to do, any knowledge which I may obtain in the course of my practice, and such intercommunication of ideas and of experience amongst engineers I believe to be most useful; but the attempt to collect and re-issue as facts, with the stamp of authority, all that may be offered gratuitously to a Commission in the shape of evidence or opinions, to stamp with the same mark of value statements and facts, hasty opinions and well-considered and matured convictions, the good and the bad, the metal and the dross (and simple courtesy to the donors must prevent the Commissioners from attempting to draw distinctions which might appear invidious)—this, I believe, always has rendered, and always will render, such collections of miscalled evidence injurious instead of advantageous to science; and the facts or statements and opinions so collected will form generally, I believe, a lower average of information than that which is already in the possession, or at least within the reach, of those who have occasion to study the subject: for it is remarkable that in this particular enquiry the Commissioners can have no peculiar means of obtaining, and, as I believe, cannot hope to get better or more extended information than that possessed by any one of the principal engineers of the day; while they will be compelled to receive and to publish much that a prudent man, acting on his own responsibility, would either not have attended to, or would silently have rejected. This, however, is perhaps a negative evil, or, at most, one which cannot much affect the proceedings of the well informed in our profession; but the mischief which I anticipate is much more dangerous to the progress of science.

If the Commission is to enquire into the conditions 'to be observed,' it is to be presumed that they will give the result of their enquiries; or, in other words, that they will lay down, or at least suggest, 'rules' and 'conditions to be (hereafter) observed' in the construction of bridges, or, in other words, embarrass and shackle the progress of improvement to-morrow by recording and registering as law the prejudices or errors of to-day.

Nothing, I believe, has tended more to distinguish advantageously the profession of engineering in England and in America, nothing has conduced more to the great advance made in our profession and to our pre-eminence in the real practical application of the science, than the absence of all *règles de l'art*—a term which I fear is now going to be translated into English by the words 'conditions to be observed.' No man, however bold or however high he may stand in his profession, can resist the benumbing effect of rules laid down by authority. Occupied as leading men are, they could not afford the time, or trouble, or responsibility of constantly fighting against them—they would be compelled to abandon all idea of improving upon them; while incompetent men might commit the grossest blunder provided they followed the rules. For, in the simplest branch of construction, rules may be followed literally without any security as to the result. There is hardly a branch of engineering that could have been selected which in its present state is less capable of being made the subject of fixed laws or instructions than the application of iron to railway structures, and certainly there is no branch in which there is more room for improvement, or which offers so many different channels or directions for that improvement.

In the quality of the material, the workmanship, or the mode of manufacture, and in the application of it, there is every imaginable variety, there is room for almost any imaginable degree or nature of improvement; and unless the Commissioners are endowed with prophetic powers, it is impossible that they can now foresee what many be the result of changes in any one of these conditions . . .[3]

What rules or 'conditions to be observed' could be drawn up now that would not become, not merely worthless, but totally erroneous and misleading, under such improved circumstances? But above all, I fear—nay, I feel convinced— that any attempt to establish any rules, any publication of opinions which may create or guide public prejudice, any suggestions coming from authority, must close the door to improvement in any direction but that pointed out by the Commissioners, and must tend to lead and direct, and therefore to control and to limit, the number of the roads now open for advance.

I believe that nothing could tend more to arrest improvement than such assistance, and that any attempt to fix now, or at any given period, the conditions to be thereafter observed in the mode of construction of any specific work of art, and thus to dictate for the present and for the future the theory which is to be adopted as the correct one in any branch of engineering, is contrary to all sound philosophy, and will be productive of great mischief, in tending to check and to control the extent and direction of all improvements, and preventing that rapid advance in the useful application of science to mechanics which has resulted from the free exercise of engineering skill in this country, subjected as it ever is, under the present system, to the severe and unerring control and test of competing skill and of public opinion. Devoted as I am to my profession, I see with fear and regret that this tendency to legislate and to rule, which is the fashion of the day, is flowing in our direction.

I must repeat my regret that circumstances should have forced me to intrude these my opinions upon the Commissioners; but, for the reasons I have before given, the application to me, after the part I have taken, left me no alternative; but having expressed my opinions, and respectfully protested against the objects and proceedings of the Commissioners, I shall feel it my duty to attend to their summons, and afford any information in my power.

### On a Proposal to obtain the Recognition in England of Decorations Conferred at the Paris Exhibition of 1855

February 9, 1856.

I regretted to be under the necessity of declining to sign the memorial that was brought to me by a gentleman introducing himself with your card, without an opportunity of explaining to you my reasons; and it would be difficult to do so satisfactorily without an opportunity of personal explanation. In a few words, however, I will state that I disapprove strongly, and after full consideration, of any introduction into England of the system of distinctions conferred by Government upon individuals, whether engaged in professions, arts, or manufactures, whose merits can be so much better and more surely marked by public opinion. In countries where public opinion is not so searching and so powerful as in England, the evils of favouritism may be out-balanced by the advantages of some means of distinguishing men. I admit the possibility, though I doubt the fact; but I feel sure that the evils would be far greater than the advantages in England. The few cases of knighthood conferred in England generally follow public opinion, though I should not wish to see this system carried further. Such being my opinion, I could not consistently ask for my own letter of Chevalier de la Légion d'Honneur being recognised here.

On the question of the patent laws, Mr. Brunel held the opinion that the system of protecting inventions by means of letters patent was productive of immense evil. The prominent part which he took in all discussions upon this subject exposed him to much adverse criticism, which was perhaps the more freely bestowed, because it was felt that he was a very formidable opponent, not only from the force of his arguments, but also from the authority with which he spoke.

He was, from the necessity of his position as a civil engineer, himself an inventor; he had in his staff and among those with whom he acted many inventors; he did not, therefore, underrate the benefits conferred on science by those who, by inventing, add to its resources. He was continually being trammelled and thwarted in his various undertakings by patents, and he therefore could judge of their evil effects upon the progress of practical engineering; and, lastly, he had the best possible means of judging of their effects upon the inventors themselves, both from his opportunities of becoming acquainted with the fate of others, and from his own experience. His father, Sir Isambard Brunel, had taken out patents for most of his inventions, and, as Mr. Brunel stated before the House of Lords Committee of 1851, with very unfortunate results, especially in the case of the carbonic-gas engine (see Note B to Chapter I); where, if they had not been obliged to work secretly, in order to conceal the process before the patent was granted, they could have obtained valuable advice, which might either have led to an earlier abandonment of the project, or to its improvement in those points in which it failed.

Mr. Brunel drew up the following statement when asked to give evidence before the Select Committee of the House of Lords in 1851. His evidence will be found at p.246 of the Minutes of Evidence (ordered to be printed July 1, 1851).

*Memorandum for Evidence before the Select Committee of the House of Lords on the Patent Laws, 1851*

I have for many years had considerable experience of the operation of patents.

I have been engaged under my father in the working out of numerous inventions of his, and the taking out of patents on his account, also in advising others professionally with him, and by myself; and have been engaged in numerous questions of disputes resulting from patents; and I have had frequent occasion to use the patents and inventions of others. I have also had to introduce improvements of my own without patents, and to defend my use of them against patents.

I have thus for the last twenty-eight years been in the midst of everything connected with inventions, and in constant contact with the operation of the patent laws.

I have been behind the scenes the whole time.

The result has been that I have never taken out a patent myself; or ever thought of doing so; and I have gradually become convinced that the whole system of patents is, in the present advanced state of arts and science and manufactures, productive of immense evil.

I think that it does nothing of what it professes to do, and which I believe to be impracticable in the present state of things, but that, on the contrary, it impedes everything it means to encourage, and ruins the class it professes to protect, and that it is productive of immense mischief to the public.

I should wish to observe that my opinions are not formed from any theory, or from any consideration of what are or ought to be the laws of patents, or whether the details of such laws are capable of improvement or otherwise; but they are simply the result of a very long and tolerably intimate knowledge of the operation of the hope of protection held out, and the operation of that protection such as it can be when obtained; and these results do not, in my opinion, depend at all upon any question of whether patents are cheap or dear, whether they are granted sparingly or profusely, by a simple or by complicated machinery; it is the ruinous effects upon the class of inventors, of the false dreams and hopes excited by the system, and the injurious effect upon improvements of the greater or less degree of exclusive privilege which is attained, which I have had constantly before my eyes for so many years, and which must be increased by any real improvement of the patent laws.

I should, therefore, be an advocate for very cheap patents granted with great facility, to the poor illiterate workman, as well as to the rich manufacturer with his counsel and agents, and as well protected as legal ingenuity can devise.

If the system is good in principle it must bear extension; but I believe it could not stand a twelvemonth under such a test—every evil now inherent in the system would be greatly increased in quantity, and the absurdities which are now ascribed to errors of detail would all become so evident that the system would be abandoned by universal consent.

I believe, paradoxical as it may seem, that the privileges thus promised and granted to inventors are most injurious to them. To understand this, it must be known and borne in mind that useful inventions or improvements in the present day, certainly in nine hundred and ninety-nine cases out of a thousand, are not new discoveries, but generally slight modifications of what is already in use; judicious applications of known principles and of well-known and common parts of machinery, or of common substances, very often mere revivals or re-inventions of something which had many times previously been thought of, and perhaps tried, and failed from the want only of some substance or of some tool which has since been introduced.

I believe that the most useful and novel inventions and improvements of the present day are mere progressive steps in a highly wrought and highly advanced system, suggested by, and dependent on, other previous steps, their whole value and the means of their application probably dependent on the success of some or many other inventions, some old, some new. I think also that really good improvements are not the result of inspiration; they are not, strictly speaking, inventions, but more or less the results of an observing mind, brought to bear upon circumstances as they arise, with an intimate knowledge of what has already been done, or what might now be done, by means of the present improved state of things, and that in most cases they result from a demand which circumstances happen to create. The consequence is that most good things are being thought of by many persons at the same time; and if there were publicity and freedom of communication, instead of concealment and mystery, ten times or a hundred times the number of useful ideas would be generated by each man, and with less mental effort and far less expenditure of time and money.

In the present state of things, if a man thinks he has invented something, he immediately dreams of a patent, and of a fortune to be made by it. If he is a rich man he loses his money, and no great harm is done; but if he is a workman, and a poor man, his thoughts are divided between scheming at his machine in secret, and scheming at the mode of raising money to carry it out. He does not consult his fellow-workmen, or men engaged in the same pursuits, as to whether the same thing had ever been tried, why it had failed, what are the difficulties, or (what is most probable) whether something better is not already known, and waiting only the demand. In nine cases out of ten he devotes his time and money to the idea, instead of pursuing his legitimate and natural pursuits. In elaborating this idea his whole thoughts are turned to the means of making it different from what he may happen to know of similar ideas, so that he may secure a patent, rather than to an honest endeavour to obtain the most useful result. He does not make use of other good ideas which may be already patented or in use, even if he knows of them, because his sole object thenceforth is not improvement, but 'exclusive right.' After much time and money spent in experiments, he takes out a patent. I will assume that the mere patent costs him nothing, but the waste of time and money in elaborating his idea is generally considerable, and far more serious in its effects upon the man than the payment of the fees now demanded for a patent. When his patent is complete, and his invention published, the chances are, and ever will be, one hundred or one thousand to one that it is not worth a sixpence as an exclusive right which others will buy of him: every chance is against him.

*In the first place* it must be a good thing, it must be an improvement upon the very 'best thing' of the same sort: the chances are of course great against this.

*Secondly.*—It must be new, nothing of the sort must have been in use before— how very few things can be devised which will bear this test. I do not know of half-a-dozen clear cases of distinctly new inventions since I have been acquainted with machinery and science; and, judging from analogy, I cannot bring my mind to believe that these would bear the test of a strict and searching enquiry by interested parties. The chances are then immensely against his invention, if good enough to be disputed, proving to be new enough to stand as giving a claim to exclusiveness.

*Thirdly.*—It must not depend for its success upon the use of some other exclusive and privileged invention, or else of course it is of little saleable value, even if not an infringement upon the previous patent.

*Fourthly.*—There must be a demand existing or creatable for the article produced; or, like many other good things, it will be out of time, and drop accordingly.

*Fifthly.*—He must find some parties whose interest it is to encourage the introduction of the change, and who have the means of combating those interests which are embarked in previous monopolies.

Since all these conditions are necessary for success, it is not surprising that the result should be, as I am positive that it is in practice, that the aggregate of individual benefit derived by the exclusive privileges granted, is greatly below the aggregate expenditure of time and money involved in the production of the whole, taking the good and bad; but the proportion of the aggregate benefits as compared with the cost to the real inventor is still less.

It is known to all persons acquainted with the subject that, in nine cases out of ten of successful inventions, the patents are not beneficially enjoyed by the original inventor. And it always must be so. The mere original invention forms generally but a small part of the whole business or merit of bringing into useful operation any new thing. Judgment, a knowledge of the world, and of business and other qualities not particularly belonging to inventors, are just as requisite as mere ingenuity, although they are not the subject of protection. Capital and connection are also generally required. All these command, as they ought to do, a large share in the ultimate profits, but it is rare that an inventor at once finds such a partner. The invention generally changes hands once or twice, or oftener, till some chance brings it into operation; and ultimately it is (as must be admitted by all who know anything about these matters) very rarely, even in the case of good things, that the party who originated the subject of the patent has ultimately any large beneficial interest in it; and certainly, from these and all the other causes mentioned, it is an undoubted fact that inventors, on the whole, make a heavy annual loss, quite irrespective of mere patent fees, by inventions and patents, and that patents are not therefore a benefit to the present class of inventors, and particularly not to the poorer members of the class.

Without the hopes of any exclusive privileges, I believe that a clever man would produce many more good ideas, and derive much more easily some benefit from them. It is true that he will aim only at earning a few pounds instead of dreaming of thousands; but he will earn these few pounds frequently, and without interfering with his daily pursuits; on the contrary, he will make himself more useful.

An observing man sees what he thinks to be a mode of increasing the production of a certain machine or manufacturing operation in which he is engaged, or a better mode than that which he is acquainted with of producing some article. In all probability the same circumstances which led him to make the observation have attracted the attention of others before him; perhaps, at the same time, a little free communication with his fellow-workmen or with other manufacturers or men of science would show him that there were insuperable difficulties, and he would turn his attention at once to other things, or that there were better ways of doing the same thing, or, by pointing out difficulties, enable him to avoid useless investigations, or to make a change that would vastly improve his scheme; or he would communicate his ideas, instead of wasting his time in elaborating them, which very possibly others more acquainted with the particular branch would do much better than he; if he is a workman, his master would give him something for the idea, or if not, his value as a workman would soon become known. It is a great error to suppose that stupid men can live upon the clever man's brains if they are all left free scope in the use of their intellect; but if by artificial means an exclusive right or property in an idea can be secured, then of course the thief may steal the idea, and having registered his property in it, his inferiority of intellect is more than counterbalanced. Intelligent men who would always be suggesting improvements in a manufactory would soon become necessary, and would be valued accordingly. A manufacturer who was not surrounded by such assistants would stand no chance in the general competition, and what is necessary and valuable will in England fetch its price; and thus clever workmen will get well paid, and earn much more, and that more healthily, than the whole body of schemers now do.

The impediments thrown in the way of improvements by the existence of patents will hardly be credited by those who are not familiar with the operation of them. In the present state of things they create such barriers that it is almost wonderful that any improvements can be effected.

It will not be difficult to understand that, from the infinite number of patents that are now taken out, it is hardly possible to devise a mechanism or a chemical combination that does not, in some shape or other, form part of some previous invention or process. This would be the case even if patents were only taken out to secure real, or what are believed to be real, inventions. But the shoals of patents have brought into existence animals to feed upon them. There is a trade which nothing can destroy as long as patents last, and which must increase with the increase of patent whatever may be the mode of granting them. Patents are taken out even in very general terms, so as to embrace everything that can resemble some probable or imaginary improvement, and then, like a spider in his web, the patentee watches for his victims. Besides this, the honest but trading patentee, the more completely to secure a monopoly, often takes out several separate patents for nearly the same thing in different forms, some avowedly worthless. In doing this, without even intending it, he includes combinations, any beneficial application of which perhaps never crossed his mind, and which, in the shape in which he suggested them were good for nothing, but which nevertheless more or less prevent anybody else from touching them, even to make a good thing.

Again, the most respectable houses take out patents merely to secure a monopoly of some one form of article without much regard to the superiority of it. The result of all this is that it is almost impossible now to introduce the slightest real improvement in anything without infringing upon some patent, and exposing oneself to be proceeded against by some patentee.

The extent to which improvements are impeded by this state of things is hardly conceivable, except to those who, like myself, are daily suffering under it.

There is another very serious evil produced by the system. In taking out a patent, the necessity for avoiding all claim to anything that can be shown to have been patented or in common use before compels you in most cases to seek rather what part of it can best be patented than what it is that is good in the invention. Comparatively trivial points are frequently patented in order to secure the monopoly of that which, although it constitutes the merit of the invention, may not happen to be so new in every respect as to admit of a patent. When the patent is taken, other modes of carrying out the real invention are discovered; these modes are patented and the original patentee is obliged to lose the benefit of his invention or to buy up these new patents. There are instances where enormous sums have been spent in this manner to protect an original invention.

And, lastly, there is an evil which acts like a numbing disease on all improvements; a patentee frequently dares not himself introduce improvements in any apparatus or process which he has patented, for fear of thus injuring his patent; and I have seen numerous cases of very important inventions where all improvement has been thus checked and resisted by the very parties most capable of effecting them, and who would have brought great talent and zeal to the work if they had been free.

*Extract from Observations on the Patent Laws made by Mr. Brunel at a Meeting of the Society of Arts*

March 28, 1856.

He did not agree at all as to the advantages of patents. He quite agreed as to the desirability of protecting, as far as possible, a man's property, whether it was in the power of invention, or any other good thing that was within him, and still more would he protect in every possible way the property in inventions of those who possessed but little other property—the powers of the inventor and the ingenuity of the workman; but, having had some considerable experience with patentees, manufacturers, and workmen, he was of opinion that any practical benefits derived from the patent laws did not compensate for the injury inflicted. He believed, on the contrary, that both the inventors and the public greatly suffered from the attempt to protect inventions. He had had great experience on this subject, being compelled daily to examine inventions of various kinds, and having himself constantly to invent in the occupations in which he was engaged. Having, then, all his life, been connected with inventors and workmen, he had witnessed the injury, the waste of mind, the waste of time, the excitement of false hopes, the vast waste of money, caused by the patent laws, in fact, all the evils which generally resulted from the attempt to protect that which did not naturally admit of protection. He agreed as to the abstract desirability of protecting inventors in some way, provided it did not foster unhealthy invention, as he thought it desirable to protect every species of property that existed. He was disposed to encourage every step towards facilitating the obtaining patents; he hoped they would be made dirt cheap, as he thought that would be the most effectual way of destroying them altogether. Therefore, whenever he had been consulted on the subject of the patent laws, he had always advocated the rendering of patents as open and free and cheap as possible; in the first place, because he saw no reason for attaching a price to them, and next, because they would sooner arrive where the principle would be fully tested. We were already nearly arrived at that state of things when engineers were almost brought to a dead stand in their attempt to introduce improvements, from the excess of protection. He found that he could hardly introduce the slightest improvement in his own machinery without being stopped by a patent. Be could mention a striking instance, in which, a few months ago, wishing to introduce an improvement that he thought would have been valuable to the public in a large work on which he was engaged, he had no sooner entered upon it, with a willingness to incur considerable expense in the preliminary requirements and in the trial of it, than he was stopped by a patentee; but he was fortunate enough to find that another patent existed of the same thing, and a week after a third appeared. There was thus, fortunately, a probability that, by the destruction of all value in any of the patents, he might be able to continue the improvements he was desirous of introducing.

1.   The case of Ranger *v.* the Great Western Railway Company.
2.   See above, p.145.
3.   The remainder of this letter is printed above, p.145.

# XVII

# PRIVATE LIFE

UNDER any circumstances, and by whomsoever made, the attempt to describe Mr. Brunel's home life must fail to satisfy those who knew him, and who remember him in the midst of his family or among his friends.

But those who did not know him, except as a professional man, or who are only acquainted with his works, will expect to find in these pages some account of his private life, and of the manner in which he spent those brief intervals of relaxation which he permitted himself to enjoy.

Although Mr. Brunel was never an idle man, he was able, until he obtained business on his own account, to enjoy many amusements from which in after life he was completely debarred.

This arose partly from his work under his father being near his own home and his friends, and partly from the power he possessed, and which never deserted him, of being able to throw aside cares and anxieties and to join with the utmost zest in passing amusements.

The following letter, relating to this time, is written by one who was Mr. Brunel's constant companion during the period to which it refers:—

June 28, 1870.

Dear Isambard Brunel,—I will endeavour to supply you with some reminiscences of your father, before he became a public man, and was engrossed by the very severe labour of his profession.

The most striking feature in his character as a young man, and one which afterwards produced such great results, was an entire abnegation of self in his intercourse with his friends and associates.

His influence among them was unbounded, but never sought by him; it was the result of his love of fair play, of his uniform kindness and willingness to assist them, of the confidence he inspired in his judgment, and of the simplicity and high-mindedness of his character.

From 1824 to 1832 he joined his friends in every manly sport; and when, after his accident at the Tunnel, he was obliged to withdraw from more violent exercise, he was still ready to co-operate in the arrangements required to give effect to whatever was in hand.

Whether in boating, in picnic parties, or in private theatricals, he was always the life and soul of the party; for his skilful arrangements, as well as his never-failing invention and power of adaptation of whatever came to hand, made him the invariable leader in every amusement or sport in which he took part.

To ensure the success of his friends in a rowing match against time, from London to Oxford and back, in 1828, he designed and superintended the building of a four-oared boat, which, in length and in the proportion of its length to its breadth, far exceeded any boat of the kind which had then been seen on the Thames.

During that portion of the period to which these notes refer, when your father was engaged at the Tunnel works, the freshness and energy with which he joined in the amusements of his friends after many consecutive days and nights spent in the Tunnel—frequently he did not go to bed, I might almost say, for weeks together—surprised them all.

His power of doing without sleep for long intervals was most remarkable. He also possessed the power, which I have never seen equalled in any other man, of maintaining a calm and even temper, never showing irritation even when he was bearing an amount of mental and bodily fatigue which few could have sustained. His presence of mind and courage never failed him, and it was especially exhibited after the first irruption of water into the Tunnel; when he descended in the diving-bell to examine the extent of the disturbance of the bed of the river, and the injury, if any, which had been done to the brickwork.

The bell could not be lowered deep enough, and he dropped himself out of the bell, holding on by a rope, and ascertained by careful examination that the brick-work was uninjured.

He was several minutes in the water; and upon this fact being stated, many persons, and I think the officers of the Royal Humane Society, denied the possibility of his retaining his consciousness so long in the water, forgetting, which he did not, that his lungs were filled with air at two and a half atmospheres' pressure.

In 1830, he joined the Surrey Yeomanry and attended drill, and was out with the troop to which he belonged on several occasions.

In this capacity he was as popular as in every other; but his remarkable talent in obtaining personal influence, even among those with whom he was comparatively a stranger, was about this time most usefully exhibited during the election of his brother-in-law as member for Lambeth.

He made friends and conciliated opponents among all classes of electors—especially among working men, large bodies of whom he met on several occasions—and among all shades of politicians; and to his energy, good judgment and skilful arrangement of electioneering details, which were not then so well understood as they now are, very much of the success achieved was due.

No one, I believe, ever saw him out of temper or heard him utter an ill-natured word. He often said that spite and ill-nature were the most expensive luxuries in life; and his advice, then often sought, was given with that clearness and decision, and that absence of all prejudice, which characterised his opinions in after-life.

All his friends of his own age were attached to him in no ordinary degree, and they watched every step in his future career with pride and interest.

In fact, he was a joyous, open-hearted, considerate friend, willing to contribute to the pleasure and enjoyment of those about him; well knowing his own power, but never intruding it to the annoyance of others, unless he was thwarted or opposed by pretentious ignorance; and then, though at times decided and severe in his remarks, he generally preferred leaving such individuals to themselves, rather than, by noticing them, to give prominence to their deficiencies.

His appreciation of character was so exact, and his dislike to anything approaching to vulgarity in thought or action or to undue assumption was so decided, that to be his friend soon became a distinction; and the extent to which his society was sought, not only in private life, but in the scientific world, at this early period, marked strongly the distinguishing features of his mind and character.

In 1825 and 1826 he attended the morning lectures at the Royal Institution, and the eagerness and rapidity with which he followed the chemical discoveries which were then being made by Mr. Faraday, showed the facility with which he gained and retained scientific knowledge.

To write more would lead me to the events of a later period of his life, in the history of which you require no aid from me; nevertheless, I cannot refrain from adding a few words upon your father's personal and professional character, which was not, in my opinion, adequately appreciated by the public.

His professional friends before his death, and his private friends at all times, well knew the genius, the intense energy, and indefatigable industry with which every principle and detail of his profession was mastered; and both knew and valued the high moral tone which pervaded every act of his life.

The public, however, did not see him under the same circumstances.

Their imperfect acquaintance with his character arose in a great degree from his disregard of popular approbation, for he was never so satisfied with his own work as to feel himself entitled to receive praise in the adulatory style of modern writing, and he preferred to work quietly in his own sphere, and to rely on the intrinsic merits of his undertakings bringing their reward, rather than to court temporary popularity.

The rapidity with which he gained a high position as a civil engineer is the best evidence of his talents. He passed almost direct from boyhood to an equality with any one then in the profession—a position attained by the rapidity and accuracy with which he could apply theory to practice, and support his conclusions by mathematical demonstrations.

This knowledge, always used without ostentation, soon placed him above most of his contemporaries; and his intimate acquaintance with the strength and peculiarities of the various materials he had to employ, and of the best and most economical mode of applying them, impressed both directors and contractors with a degree of confidence in his estimates and opinions which no one had before possessed.

His power of observation was singularly accurate; he was not satisfied with a hasty or superficial examination, nor with the mere assertion of a fact; his mind required evidence of its correctness before he could receive and adopt it. I may illustrate this by a reference to the experiments he made with French mesmerists, and the pains he took to expose the farce of table-turning and its accompanying follies.

My object, however, by this addition to my note, is to dwell upon the fact that he left a mark upon his profession which cannot be obliterated. He set up a high standard of professional excellence, and endeavoured to impress on all who were associated with him, or under him professionally, that to attain the highest honours required the strictest integrity, sound mathematical knowledge, originality and accuracy of thought and expression, both in *viva voce* descriptions and in designs and working drawings, and a practical acquaintance with the durability and strength of materials, so as to know the best conditions under which each might be applied.

It was his excellence in these respects, when still young, which soon earned for him a great reputation as a witness before the Committees of the Houses of Parliament.

His calmness and unobtrusive manner, when under severe examination, or while attending public meetings, led many to think him cold, and regardless of the feelings or interests of those with whom he was associated; but nothing was further from his character, as every one knew who was engaged in the consultations upon the result of which future proceedings depended.

He was a prudent and cautious, but bold adviser, and a warm-hearted and generous friend.

Yours faithfully,<br>W. HAWES

Isambard Brunel, Esq.

The events of the year 1835 brought with them, not unnaturally, other changes. At the beginning of 1836, he removed to 18 Duke Street, Westminster, a large house looking on St. James's Park, and now (1870) the last in the street, next to the new India Office.

In July of the same year he married the eldest daughter of the late William Horsley, and granddaughter of Doctor Callcott. Of this marriage there was issue two sons and a daughter, all of whom survive him.

Although, as will be presently mentioned, he afterwards bought some property in Devonshire, the Duke Street house was always his home. He spent his life there, having his offices on the lower floors.

He had no wish to enter Parliament, although it had been more than once suggested to him to do so, and his work prevented his taking an active share, as an inhabitant of Westminster, in the concerns of his neighbourhood.

The only occasion on which he took a prominent part in local affairs was as a special constable in April 1848, when he acted as one of the two 'leaders' of the special constables in the district between Great George Street and Downing Street.

He was not without experience of the duties of a special constable, as he had been sworn in during the Bristol riots of 1830, and on that occasion saw active service. Happily, matters were better managed in London, and no actual collision took place between the constables, or the military, and the mob.

The extent to which Mr. Brunel kept his works in his own hands, and under his own superintendence, made it necessary for him to have a large amount of office accommodation; and the inconvenience of having branch offices in the streets near his house led him, in 1848, to enlarge his offices: with this object he added the adjoining house, 17 Duke Street, which he rebuilt. A large room on the ground floor, looking on the Park, was thenceforward his own office, and the room above was made the dining-room. It was decorated in the Elizabethan style, and was to have contained a collection of pictures illustrative of scenes in 'Shakespeare,' painted for him by the principal artists of the day. This project was never completely carried out, but several pictures (about ten in all) were painted and hung up, among them the 'Titania' of Sir Edwin Landseer. These subjects are again referred to in the following letter:—

February, 1870.

My dear Isambard,—You ask me to jot down for you any reminiscences I have of your father's love and feeling for art.

I remember with singular distinctness the first time I ever saw him, when I was a lad of fourteen, and had just obtained my studentship at the Royal Academy. He criticised with great keenness and judgment a drawing which I had with me, and at the same time gave me a lesson on paper straining. From that time till his death he was my most intimate friend. Being naturally imbued with artistic taste and perception of a very high order, his critical remarks were always of great value, and were made with an amount of good humour which softened their occasionally somewhat trying pungency. He had a remarkably accurate eye for proportion, as well as taste for form. This is evinced in every line to be found in his sketch books, and in all the architectural features of his various works.

So small an incident as the choice of colour in the original carriages of the Great Western Railway, and any decorative work called for on the line, gave public evidence of his taste in colour; but those who remember the gradual arrangement and fitting up of his house in Duke Street will want no assurance from me of your father's rare artistic feeling. He passed, I believe, the pleasantest of his leisure moments in decorating that house, and well do I remember our visits in search of rare furniture, china, bronzes, &c., with which he filled it, till it became one of the most remarkable and attractive houses in London. Its interest was greatly increased when he formed that magnificent dining-room, now, with the house of which it was a part, pulled down. This room, hung with pictures, with its richly carved fireplace, doorways, and ceiling, its silken hangings and Venetian mirrors, lighted up on one of the many festive gatherings frequent in that hospitable house, formed a scene which none will forget who had the privilege of taking part in it. When from time to time he went abroad, and especially in his visit to Venice in 1852, he added to his collection by purchases made with great judgment and skill. In buying pictures, your father evinced a taste often found in men of refined mind and feeling— viz. a repugnance to works, however excellent in themselves, where violent action was represented. He preferred pictures where the subject partook more of the suggestive than the positive, and where a considerable scope was left in which the imagination of the spectator might disport itself. This feeling was displayed in a great love of landscape art, and in the keenest appreciation of the beauties of nature. It is an interesting fact to record, and one which I often heard him mention, when his friends were admiring his beautiful grounds at Watcombe, that in the old posting days, when travelling on the cliff road between Teignmouth and Torquay, he constantly stopped the carriage to get out and admire the view which he had discovered from a field at Watcombe, little thinking then that it would ultimately be the site of his intended country home.

When your father and I went to Italy together in 1842, posting from Westminster to Rome and back again, I had ample opportunities of observing his love and enthusiasm for nature and art.

Overwhelmed as he was with work in England at the time, it was no easy matter for him to leave the country for a couple of months; and I remember that our starting at all was uncertain up to the last moment; and that, an hour before quitting London, it was only by a *coup de théâtre,* which he most adroitly performed, that he escaped the serving of a subpœna, the bearer of which had actually penetrated to the dining-room door in Duke Street.

We left London one evening in April 1842. During our journey we constantly passed several consecutive days and nights in the carriage; and I am sure there was not one of our waking hours in which some incident of interest did not occur.

I remember your father agreeing with me, that our experiences merely of post-boys and their various characteristics would be worthy of recording in detail— from Newman's two smart lads, who took us the first stage out of London, on to the genuine "postillon" (boots and all) we found at Calais; then to the wild young brigands (in appearance) who, inspired by the prospect of extra "buon mano," whirled us along the road from Civita Vecchia towards Rome, and winding up with the stolid German who rose slowly in his stirrups, and distracted us by a melancholy performance on the horn slung round him, and which no entreaty would induce him to give up.

We posted from Calais, *via* Paris, to Châlons-sur-Saône, marvelling the whole way whereabouts "La Belle France" was to be found; for a drearier and more utterly monotonous ride of something like 800 miles it is impossible to conceive. From Châlons we went down the river to Lyons, then onwards, visiting Nismes, and through Arles to Toulon.

From Toulon we went through Cannes and Nice and along the lovely Cornice road to Genoa. Your father was intensely delighted with this portion of the journey. Those wonderfully picturesque towns, with their roccoco churches looking like toys, and painted all over upon the principle of colour generally developed in that species of art, especially interested him. The streets were so narrow that it was sometimes doubtful whether the carriage could be squeezed through, and more than once it grazed the houses on either side as it passed on.

The work for which your father had come to Italy commenced at Genoa, and he was met there by a staff appointed by the Government to accompany him during his stay.

While at Genoa he came to me one morning and said, that, in consequence of some delay, he had a week in which to make complete holiday, and gave me the choice of Florence or Rome. I need scarcely say that I chose Rome, and for three days we were in the Eternal City, seeing more in that time than those to whom we related our proceedings could believe.

How well do I remember our entering Rome by the gate on the Civita Vecchia road, and standing up in the carriage to get our first view of St. Peter's, and, having seen it, the blank look of disappointment we turned on each other at the sight! But the interior of the great church as far exceeded our expectations as the exterior had fallen short of them.

We were back at Genoa to the minute your father had appointed; and the work being completed there, we went on to Turin. Here we were in time to be present at the Court balls and ceremonies consequent upon the marriage of the present King of Italy.

From Turin we proceeded to Milan.

At Milan your father parted from his staff, and completed the work he had undertaken as far as it was necessary to do so in Italy. From Milan, therefore, our journey home was one of uninterrupted enjoyment through those glorious Lombard towns to Venice, which happily we reached in a gondola from Mestre, and not by a railway viaduct; then through the Tyrol to Munich, and so down the Rhine to Belgium, reaching home from Antwerp.

Thus was completed an expedition in which there was neither hitch nor disagreeable adventure of any kind, and upon which I look back with unmixed pleasure.

The next and last time that your father and I journeyed on the Continent together was in April 1848, when he wished to see Paris in Republican garb, and asked me to accompany him.

We were there for some days, and, armed with cards of admission, on which our names were inscribed with the prefix of "Citoyen," heard and saw the various celebrities of the hour.

Affectionately yours,

J.C. HORSLEY

Isambard Brunel, Esq.

Within less than a year of Mr. Brunel's return from his visit to Italy, a strange accident happened to him, which placed his life in great jeopardy.

On April 3, 1843, he was amusing some children at his house by the exhibition of conjuring tricks, when, in pretending to pass a half-sovereign from his ear to his mouth, the coin he had placed in his mouth slipped down his throat. After few days he began to suffer from a troublesome cough, and on April 18 Sir Benjamin Brodie was consulted.

The nature of the accident and the course of treatment adopted are described in the following letter from Mr. Brunel's brother-in-law, the late Dr. Seth Thompson, which was published in the 'Times' newspaper of May 16, 1843:—

I shall be much obliged by your giving insertion to the following statement of the treatment pursued by Sir Benjamin Brodie in the case of Mr. Brunel, it being the wish of Mr. Brunel and his friends that the true facts should be known, as a just tribute to the skill of this eminent surgeon and as a guide to future practice. The accident happened on April 3; Sir B. Brodie was consulted on the 18th, and his opinion was that the half-sovereign had passed into the windpipe. The following day Mr. Brunel strengthened this opinion by a simple experiment. He bent his head and shoulders over a chair, and distinctly felt the coin drop towards the glottis; whilst raising himself a violent fit of coughing came on, which ceased after a few minutes. He repeated this a second time, with the same results. A consultation was held on the 22nd, at which it was decided that conclusive evidence existed of the half-sovereign having passed into the windpipe, that it was probably lodged at the bottom of the right bronchus, and that it was movable. It was determined that every effort should be made for its removal, and that for this purpose an apparatus should be constructed for inverting the body of the patient, in order that the weight of the coin might assist the natural effort to expel it by coughing. The first experiment was made on the 25th. The body of the patient being inverted, and the back gently struck with the hand between the shoulders, violent cough came on, but of so convulsive and alarming a nature that danger was apprehended, and the experiment was discontinued. On this occasion the coin was again moved from its situation, and slipped towards the glottis. On the 27th tracheotomy was performed by Sir B. Brodie, assisted by Mr. Aston Key, with the intention of extracting the coin by the forceps, if possible, or, in the event of this failing, with the expectation that the opening in the windpipe would facilitate a repetition of the experiment of the 22nd. On this occasion, and subsequently on May 2, the introduction of the forceps was attended with so much irritation, that it could not be persevered in without danger to life. On the 3rd another consultation was held, when Mr. Lawrence and Mr. Stanley entirely confirmed the views of Sir B. Brodie and Mr. Key, and it was agreed that the experiment of inversion should be repeated as soon as Mr. Brunel had recovered sufficient strength, the incision in the windpipe being kept open. On Saturday, the 13th, Mr. Brunel was again placed on the apparatus, the body inverted, and the back gently struck. After two or three coughs, he felt the coin quit its place on the right side of the chest, and in a few seconds it dropped from his mouth without exciting in its passage through the glottis any distress or inconvenience, the opening in the windpipe preventing any spasmodic action of the glottis.

In this remarkable case the following circumstances appear to be worthy of note—that a piece of gold remained in the air-tube for six weeks, quite movable and without exciting any inflammatory action, the breathing entirely undisturbed, and the only symptoms of its presence occasional uneasiness on the right side of the chest and frequent fits of coughing; that an accurate diagnosis was formed without being able to obtain any assistance from the stethoscope, although the chest was repeatedly and carefully examined; and also that, a fair trial having been given to the forceps, the application of this instrument to the removal of a body of this peculiar form from the bottom of the bronchus was proved to be attended with great risk to life; while the cautious and well-considered plan of treatment above detailed was attended with complete success, and without risk.

During the time that Mr. Brunel was in danger the public excitement was intense. His high professional position, the extraordinary nature of the accident, and the greatness of the loss, were the result to prove fatal, made his condition and the chances of his recovery an engrossing topic of conversation; and, when the news was spread that 'it is out,' the message needed no explanation.

That the result was successful was due, not only to the skill of the surgeons engaged, and to the anxious care with which those who nursed him left nothing undone to ensure his safety, but also to the remarkable coolness which Mr. Brunel himself displayed throughout. From the first he took part in the consultations which were held on his case, and assisted materially in determining the course of treatment which should be pursued.

The ten years which followed were the most prosperous in Mr. Brunel's life; he had attained to great eminence in his profession, and was still in the enjoyment of robust health. But the results of the gauge controversy and the fierce contests which followed it, and, above all, the failure of the Atmospheric System on the South Devon Railway, caused him grave anxiety and sorrow. Critics have erred greatly in representing him as a man who, in order to accomplish some vast design, thought but little of the distress which follows want of success in commercial enterprises. So far from its being true that Mr. Brunel was indifferent to the interests of his employers, his private journals show that throughout (to use his own words) 'the incessant warfare in which he was engaged' he was earnestly desiring peace and endeavouring to secure it, and that in times of difficulty, such as the trial of the Atmospheric System and the launch of the 'Great Eastern,' his chief thoughts were for those who would suffer through the failure of his plans.

In the midst of his professional occupations he was able occasionally, though rarely, to enjoy the society of his friends. After the session was over, in 1844 and 1845, he went to Italy on business, and in 1846 to Switzerland for a short holiday. In 1847 the South Devon Railway was occupying his attention, and he determined to take a house at Torquay. While there, the important character of his railway works in Devonshire and Cornwall led him to think of making a more permanent settlement in that part of the country.

After a good deal of hesitation between various places, he fixed upon a spot at Watcombe, about three miles from Torquay, on the Teignmouth turnpike road. He made his first purchase of land in the autumn of 1847; and from that time to within a year of his death the improvement of this property was his chief delight.

He had always a great love and appreciation of beautiful scenery, and in his choice of a place in which to plant and build he provided amply for his complete gratification.

The principal view, which, if the house had been built, would have been the view from the terrace, is one of the loveliest in that part of Devonshire. On one side is the sea, and on the other the range of Dartmoor, while in front is spread undulating country, bounded by the hills on the further side of Torbay, the bay itself looking like a lake, being shut in by the hills above Torquay.

When Mr. Brunel bought this property it consisted of fields divided by hedgerows; but, assisted by Mr. William Nesfield, he laid it out in plantations of choice trees. The occupation of arranging them gave him unfailing pleasure; and, although he could seldom spare more than a few days' holiday at a time, there can be little doubt that the happiest hours of his life were spent in walking about in the gardens with his wife and children, and discussing the condition and prospects of his favourite trees.[1]

He could not, of course, take a prominent part in the affairs of the parish, but he was always ready to assist in any work that had been taken in hand. He will be long remembered there by his friends in every rank of life.

In purchasing this property in Devonshire, Mr. Brunel had looked forward to retiring gradually from active professional life, 'to draw in and make room for others,' and to spend a greater portion of his time in the country.

It may well be questioned whether he would have been happy in giving up work while yet in middle life; but the wisdom of his resolve was not to be put to the test.

From the beginning of the year 1852 the 'Great Eastern' steam-ship began to occupy his time and thoughts. As the works progressed, he was more and more tied to London; and the large pecuniary investment he had made in the shares of the company caused him to hesitate before proceeding with the building of his house.

Thus the hopes he had formed for making his home in Devonshire faded gradually away, and were at length extinguished by the failure of his health.

Many things had happened in the earlier part of 1857 which gave him pleasure. In June he received, in company with Mr. Robert Stephenson, the honorary degree of Doctor in Civil Law from the University of Oxford.[2] In the summer he paid several visits to Devonshire, and at the beginning of September the floating of the first truss of the Saltash bridge was successfully accomplished.

The history of the launch of the 'Great Eastern,' which was commenced in November, has been already told. Throughout all the disappointments he then endured Mr. Brunel took comfort from the sympathy of valued friends, and from those higher sources of consolation on which it was his habit to rely. He paid for his exertions a heavy price, for they left him broken in health and already suffering from the disease of which in a little more than eighteen months afterwards he died.[3]

In May 1858 Mr. Brunel went to Vichy, and thence to Switzerland, returning home in the autumn by way of Holland. When at Lucerne he went up the Righi, and was so charmed with it that, instead of spending only a night there, he remained a week, working at the designs for the Eastern Bengal Railway.

It was on his return to England in September that the alarming nature of his illness was ascertained. After anxious consultation with Sir Benjamin Brodie and Dr. Bright, he was ordered to spend the winter in Egypt, in the hope that he might return in the March or April following in restored health.

He was very unwilling to be so long absent from England, especially as a new company had just been formed to finish the 'Great Eastern,' and the contracts for her completion were about to be let.

However, it was thought that very serious consequences might follow if he were at home; and in the beginning of December he left for Alexandria, with his wife and younger son.

Having stayed there a day or two, they went on to Cairo, where they found Mr. Robert Stephenson. He and Mr. Brunel dined together on Christmas Day.

On December 30 the journey up the Nile commenced. On January 21 they arrived at Thebes, and spent some days there. Mr. Brunel was able to ride about on a donkey, and made some sketches of the celebrated ruins in the neighbourhood.[4]

They reached Assouan on February 2, and made preparations for ascending the cataracts. They went as far as Dakkeh, and got back to Assouan on February 19.

The following letter from Mr. Brunel to his sister, Lady Hawes, describes some of the scenes through which he passed:—

Philæ, February 12, 1859.
I now write to you from a charming place; but Assouan, which I left to come here, is also beautiful, and I will speak of that first. It is strange that so little is said in the guide books of the picturesque beauty of these places. Approaching Assouan, you glide through a reef of rocks, large boulders of granite polished

by the action of the water charged with sand. You arrive at a charming bay or lake of perfectly still water and studded with these singular jet-black or red rock islands. In the distance you see a continuation of the river, with distant islands shut in by mountains, of beautiful colours, some a lilac sandstone, some the bright red yellow of the sands of the desert. Above the promontories the water excursions are delicious. You enter at once among the islands of the Cataracts, fantastic forms of granite heaps of boulders split and worn into singular shapes.

After spending a week at Assouan, with a trip by land to Philæ, I was so charmed with the appearance of the Cataracts as seen from the shore, and with the deliciously quiet repose of Philæ, that I determined to get a boat, and sleep a few nights there. We succeeded in hiring a country boat laden with dates, and emptied her, and fitted up her three cabins.[5] We put our cook and dragoman and provisions, &c., on board, and some men, and went up the Cataract. It was a most amusing affair, and most beautiful and curious scenery all the way. It is a long rapid of three miles, and perhaps one mile wide, full of rocky islands and isolated rocks. A bird's-eye view hardly shows a free passage, and some of the more rapid falls are between rocks not forty feet wide—in appearance not twenty. Although they do not drag the boats up perpendicular falls of three or four feet, as the travellers' books tell you, they really do drag the boats up rushes of water which, until I had seen it, and had then calculated the power required, I should imprudently have said could not be effected. We were dragged up at one place a gush of water, what might fairly be called a fall of about three feet, the water rushing past very formidably, and between rocks seemingly not more than wide enough to let our boat pass, and this only by some thirty-five men at three or four ropes, the men standing in the water and on the rocks in all directions, shouting, plunging into the water, swimming across the top or bottom of the fall, just as they wanted, then getting under the boat to push it off rocks, all with an immense expenditure of noise and apparent confusion and want of plan, yet on the whole properly and successfully. We were probably twenty or thirty minutes getting up this one, sometimes bumping hard on one rock, sometimes on another, and jammed hard first on one side and then on the other, the boat all the time on the fall with ropes all strained, sometimes going up a foot or two, sometimes losing it, till at last we crept to the top, and sailed quietly on in a perfectly smooth lake. These efforts up the different falls had been going on for nearly eight hours, and the relief from noise was delicious. We selected a quiet spot under the temples of Philæ . . . Our poultry-yard is on the sandbank, where fowls, pigeons, and turkeys are walking about loose, and, like all animals in this country, perfectly tame. Yes, they walk up and catch a pigeon to be killed when you like. In the midst of these and of the small birds which always walk and fly about us, have been walking for hours this morning three or four large eagles, who, with the politeness peculiar to animals here, pay no attention to our fowls, nor do they to the eagles. But here I am entering on the anomalies and contradictions of Egypt, which would fill volumes.

After leaving Egypt, Mr. Brunel went to Naples and Rome, where he spent Easter, and he returned to England in the middle of May.

When abroad, Mr. Brunel made sight-seeing a pleasure rather than a business; thus in Egypt he preferred to visit frequently the same places, and rather to enjoy that which he knew gave him pleasure, than to hurry about with the object of seeing all that was to be seen. At Philæ he stopped more than a week, and at Thebes he spent more time in a small outlying temple near Karnac than in the great ruin itself. So also at Rome he went frequently to the Colosseum, and he spent many hours in the interior of St. Peter's.

Shortly after his return to England he went to Plymouth, and over the Saltash bridge and other parts of the Cornwall Railway, which had been opened during his absence abroad.

Although it had by this time become certain that the disease under which he laboured had assumed a fatal character, he continued to give unremitting attention to his various professional duties; and in order to be nearer the 'Great Eastern,' he took a house at Sydenham, and removed there with his family in the beginning of August.

Almost every day he went to the great ship and superintended the preparations for getting her to sea. She was advertised to sail on September 6, and Mr. Brunel had intended going round in her to Weymouth.

He was on board early on the morning of the 5th, and his memorandum book has, under that date, an entry of some unfinished work which had to be looked after. Towards midday he felt symptoms of failing power, and went home to his house in Duke Street, when it became evident that he had been attacked with paralysis.

At one time it seemed possible that he might recover; but on the tenth day after his seizure, Thursday, September 15, all hope was taken away. In the afternoon he spoke to those who watched around him, calling them to him by their names; as evening closed in he gradually sank, and died at half-past ten, quietly and without pain.

The funeral was on September 20, at the Kensal Green Cemetery.

Along the road leading to the chapel many hundreds of his private and professional friends, his neighbours among the tradespeople of Westminster, the Council of the Institution of Civil Engineers, and the servants of the Great Western Railway Company, had assembled, and, with his family, followed his body to its place of burial, in the grave of his father and mother.[6]

It would be improper here to attempt to enter into a general criticism of Mr. Brunel's works, or to determine the position which he is entitled to occupy among civil engineers. That task has yet to be accomplished, and must be undertaken by those who can claim to be impartial judges. It has been the object of this book to provide, as far as possible, the materials on which a just judgment of his career can be based.

But it may be permitted, in conclusion, to place on record the following testimony to the high position held by Mr. Brunel in the esteem of his contemporaries.

On November 8, 1859, at the first meeting of the Institution of Civil Engineers after the death of Mr. Brunel and of Mr. Robert Stephenson, Mr. Joseph Locke, M.P., the President, rose and said—

I cannot permit the occasion of opening a new session to pass without alluding to the irreparable loss which the Institution has sustained by the death, during the recess, of its two most honoured and distinguished members.

In the midst of difficulties of no ordinary kind, with an ardour rarely equalled, and an application both of body and mind almost beyond the limit of physical endurance, in the full pursuit of a great and cherished idea, Brunel was suddenly struck down, before he had accomplished the task which his daring genius had set before him.

Following in the footsteps of his distinguished parent, Sir Isambard Brunel, his early career, even from its commencement, was remarkable for originality in the conception of the works confided to him. As his experience increased, his confidence in his own powers augmented; and the Great Western Railway, with its broad-gauge line, colossal engines, large carriages, and bold designs of every description, was carried onward, and ultimately embraced a wide district of the country.

The same feeling induced, in steam navigation, the construction of the "Great Western" steamer, the largest vessel of the time, until superseded by the "Great Britain," which was in its turn eclipsed by the "Great Eastern," the most gigantic experiment of the age.

The Great Ship was Brunel's peculiar child; he applied himself to it in a manner which could not fail to command respect; and, if he did not live to see its final and successful completion, he saw enough, in his later hours, to sustain him in the belief that his idea would ultimately become a triumphant reality.

The shock which the loss of Brunel created was yet felt, when we were startled by an announcement that another of our esteemed members had been summoned from us.[7]

. . . . . .

It is not my intention at this time to give even an outline of the works achieved by our two departed friends. Their lives and labours, however, are before us; and it will be our own fault if we fail to draw from them useful lessons for our own guidance. Man is not perfect, and it is not to be expected that he should be always successful; and, as in the midst of success we sometimes learn great truths before unknown to us, so also we often discover in failure the causes which frustrate our best directed efforts. Our two friends may probably form no exception to the general rule; but, judging by the position they had each secured, and by the universal respect and sympathy which the public has manifested for their loss, and remembering the brilliant ingenuity of argument, as well as the more homely appeals to their own long experience, often heard in this hall, we are well assured that they have not laboured in vain.

We, at least, who are benefited by their successes, who feel that our Institution has reason to be proud of its association with such names as Brunel and Stephenson, have a duty to perform; and that duty is, to honour their memory and emulate their example.

1. Although Mr. Brunel endeavoured, as far as possible, to lay aside work during his visit to Watcombe, he found occasions for the use of his mechanical knowledge. He prepared tools for transplanting young trees; he did not, however, succeed in making them flourish as well as before they were transplanted, though he attended to the work himself, and took care that a large amount of earth was moved with the tree. The bridge of rough poles across the turn pike road between Teignmouth and Torquay is as good of its kind as his larger works.

2. On June 10, 1830, when he was twenty-four years old, Mr. Brunel was elected a Fellow of the Royal Society. He subsequently became a member of most of the other scientific societies but he rarely attended any meetings, except those of the Institution of Civil Engineers.

3. At the beginning of 1858, on the expiration of Mr. Robert Stephenson's term of office as President of the Institution of Civil Engineers, Mr. Brunel came next in rotation for election; but his failing health, and the pressure of his professional duties, led him to request that he might not then be put in nomination.—See Inaugural Address of John Fowler, Esq., January 9, 1866.

4. A drawing of their Nile boat, the 'Florence,' which he made for his daughter, exhibits the same beautiful minuteness which appears in all his early sketches for the Clifton bridge and Great Western Railway.

5. Mr. Brunel's Nile boat, being of iron, could not safely go up the Cataracts.

6. A few weeks after Mr. Brunel's death, a meeting of his friends was held when it was determined to raise some memorial to him. A statue was made by the late Baron Marochetti, and a site for it promised by the First Commissioner of Works; but it has not yet been erected.

Mr. Brunel's family, by the permission of the Dean of Westminster, have placed a memorial window in the north aisle of the nave of Westminster Abbey.

Along the bottom of the window (which consists of two lights, each 23 feet 6 inches high and 4 feet wide, surmounted by a quatrefoil opening, 6 feet 6 inches across) is the Inscription, 'IN MEMORY OF ISAMBARD KINGDOM BRUNEL, CIVIL ENGINEER. BORN APRIL 9, 1806. DEPARTED THIS LIFE, SEPTEMBER 15, 1859.' Over this are four allegorical figures (two in each light): Fortitude, Justice, Faith, and Charity. The upper part of the window consists of six panels, divided by a pattern work of lilies and pomegranates. The panels contain subjects from the history of the Temple. The three subjects in the western light represent scenes from the Old Testament—viz. the Dedication of the Temple by Solomon, the Finding of the Book of the Law by Hilkiah, and the Laying the Foundations of the Second Temple. The subjects in the eastern light are from the New Testament—viz. Simeon Blessing the Infant Saviour,

Christ Disputing with the Doctors, and the Disciples pointing out to Christ the Buildings of the Temple. In the heads of each light are angels kneeling, and in the quatrefoil is a representation of Our Lord in Glory, surrounded by angels.

The work was placed in the hands of Mr. R. Norman Shaw, architect, who prepared the general design, arranged the scale of the various figures, and designed the ornamental pattern work. The figure subjects were drawn by Mr. Henry Holiday, and the whole design was executed in glass by Messrs. Heaton, Butler, & Bayne.

7.     Mr. Locke here spoke in feeling language of Mr. Robert Stephenson.

# Appendix I

## (SEE CHAPTER V ON THE BROAD GAUGE, P.83)

*Report to the Board of Directors of the Great Western Railway Company*
August 1838.

Gentlemen,—As the endeavour to obtain the opinions and reports of Mr. Walker, Mr. Stephenson, and Mr. Wood, prior to the next half-yearly meeting, has not been successful, I am anxious to record more fully than I have previously done, and to combine them into one report, my own views and opinions upon the success of the several plans which have been adopted at my recommendation in the formation and in the working of our line; and in justice to myself and to these plans, and indeed to enable others to arrive at any just conclusion as to the result which has been attained, or as to the probable ultimate success or advantages of the system, it is necessary that I should enter very fully, I fear even tediously, into a recapitulation of the circumstances, peculiar to this railway, which led to the consideration and the adoption of these plans, which some call innovations and wide deviations from the results of past experience, but the majority of which I will undertake to show are merely adaptations of those plans to our particular circumstances.

It will be necessary also that I should refer to all the numerous difficulties which we have had to encounter, which have necessarily prevented the perfect working of these plans in the first instance, but which have been overcome, or which are gradually and successively diminishing; and, finally, I am prepared to show that, notwithstanding the novelty of the circumstances, and the difficulties and delays which at the outset invariably attend any alteration, however necessary, or however desirable, from the accustomed mode of proceeding, and notwithstanding the violent prejudices excited against us, and the increased difficulties caused by these prejudices, the result is still such as to justify the attempt which has been made, and to show that in the main features, if not in all the details, the system hitherto followed is good, and ought to be pursued.

The peculiarity of the circumstances of this railway, to which I would more particularly refer, and which have frequently been mentioned, consists in the unusually favourable gradients and curves which we have been able to obtain.

With the capability of carrying the line upwards of fifty miles out of London on almost a dead level, and without any objectionable curve, and having beyond this, and for the whole distance to Bristol, excellent gradients, it was thought that unusually high speed might easily be attained, and that the very large extent of passenger traffic which such a line would certainly command would ensure a return for any advantages which could be offered to the public, either in increased speed or increased accommodations. With this view every possible attention was paid to the improvement of the line as originally laid down in the parliamentary plans. We ultimately succeeded in determining a maximum gradient of 4 feet per mile, which could be maintained for the unusual distance, before mentioned, of upwards of fifty miles from London, and also between Bristol and Bath, comprehending those parts of the line on which the principal portion of the passenger traffic will be carried. The attainment of high speed appeared to involve the question of the width of gauge, and on this point accordingly I expressed my opinion at a very early period.

It has been asserted that 4 feet 8 inches, the width adopted on the Liverpool and Manchester Railway, is exactly the proper width for all railways, and that to adopt any other dimension is to deviate from a positive rule which experience has proved correct; but such an assertion can be maintained by no reasoning. Admitting, for the sake of argument, that, under the particular circumstances in which it has been tried, 4 feet 8 inches has been proved the best possible dimension, the question would still remain—What are the best dimensions under the circumstances?

Although a breadth of 4 feet 8 inches has been found to create a certain resistance on curves of a certain radius, a greater breadth would produce only the same resistance on curves of greater radius. If carriages and engines, and more particularly if wheels and axles of a certain weight, have not been found inconvenient upon one railway, greater weights may be employed and the same results obtained on a railway with better gradients. To adopt a gauge of the same number of inches on the Great Western Railway as on the Grand Junction Railway, would in fact amount practically to the use of a different gauge in similar railways. The gauge which is well adapted to the one is not well adapted to the other, unless, indeed, some mysterious cause exists which has never yet been explained for the empirical law which would fix the gauge under all circumstances.

Fortunately this no longer requires to be argued, as too many authorities may now be quoted in support of a very considerable deviation from this prescribed width, and in every case this change has been an increase. I take it for granted that, in determining the dimensions in each case, due regard has been had to the curves and gradients of the line, which ought to form a most essential, if not the principal, condition.

In the Report of the Commissioners upon Irish Railways, the arguments are identically the same with those which I used when first addressing you on the subject in my Report of October 1835. The mechanical advantage to be gained by increasing the diameter of the carriage-wheels is pointed out, the necessity, to attain this, of increasing the width of way, the dimensions of the bridges, tunnels, and other principal works, not being materially affected by this; but, on the other hand, the circumstance which limits this increase being the curves on the line, and the increased proportional resistance on inclinations (and on this account it is stated to be almost solely applicable to very level lines); and, lastly, the increased expense, which could be justified only by a great traffic.

The whole is clearly argued in a general point of view, and then applied to the particular case, and the result of this application is the recommendation of the adoption of 6 feet 2 inches on the Irish railways. Thus, an increase in the breadth of way to attain one particular object—viz. the capability of increasing the diameter of the carriage-wheels without raising the bodies of the carriages— is admitted to be most desirable, but is limited by certain circumstances, namely, the gradients and curves of the line, and the extent of traffic.

Every argument here adduced, and every calculation made, would tend to the adoption of about 7 feet on the Great Western Railway.

The gradients of the lines laid down by the Irish Commission are considerably steeper than those of the London and Birmingham Railway, and four and five times the inclination of those on the Great Western Railway; the curves are by no means of very large radius, and indeed the Commissioners, after fixing the gauge of 6 feet 2 inches, express their opinion, that upon examination into the question of curves, with a view to economy, they do not find that the effect is so injurious as might have been anticipated, and imply therefore that curves, generally considered of small radius on our English lines, are not incompatible with the 6 feet 2 inch gauge; and, lastly, the traffic, instead of being unusually large, so as to justify any expense beyond that absolutely required, is such as to render assistance from Government necessary to ensure a return for the capital embarked. As compared with this, what are the circumstances in our case?

The object to be attained is the placing an ordinary coach body, which is upwards of 6 feet 6 inches in width, between the wheels. This necessarily involves a gauge of rail of about 6 feet 10½ inches to 6 feet 11 inches, but 7 feet allows of its being done easily; it allows, moreover, of a different arrangement of the body: it admits all sorts of carriages, stage-coaches, and carts to be carried between the wheels. And what are the limits in the case of the Great Western Railway, as compared to those on Irish railways? Gradients of one-fifth the inclination, very favourable curves, and probably the largest traffic in England.

I think it unnecessary to say another word to show that the Irish Commissioners would have arrived at 7 feet on the Great Western Railway by exactly the same train of argument that led them to adopt 6 feet 2 inches in the case then before them.

All these arguments were advanced by me in my first Report to you, and the subject was well considered. The circumstance of the Great Western Railway, and other principal railways likely to extend beyond it, having no connection with other lines then made, leaving us free from any prescribed dimension, the

7-feet gauge was ultimately determined upon. Many objections were certainly urged against it: the deviation from the established 4 feet 8 inches was then considered as the abandonment of the principle: this, however, was a mere assertion, unsupported even by plausible argument, and was gradually disused; but objections were still urged, that the original cost of construction of all the works connected with the formation of the line must be greatly increased; that the carriages must be so much stronger; that they would be proportionally heavier; that they would not run round the curves, and would be more liable to run off the rails; and particularly, that the increased length of the axles would render them liable to be broken: and these objections were not advanced as difficulties which, as existing in all railways, might be somewhat increased by the increase of gauge, but as peculiar to this, and fatal to the system.

With regard to the first objection, namely, the increased cost in the original construction of the line, if there be any, it is a question of calculation which is easily estimated, and was so estimated before the increased gauge was determined upon. Here, however, preconceived opinions have been allowed weight in lieu of arguments and calculations; cause and effect are mixed up, and without much consideration it was assumed at once that an increased gauge necessarily involved increased width of way, and dimensions of bridges, tunnels, &c.

Yet such is not the case within the limit we are now treating of: a 7-feet rail requires no wider bridge or tunnel than a 5-feet; the breadth is governed by a maximum width allowed for a loaded waggon, or the largest load to be carried on the railway, and the clear space to be allowed on either side beyond this.

On the Manchester and Liverpool Railway this total breadth is only 9 feet 10 inches, and the bridge and viaducts need only have been twice this, or 19 feet 8 inches; 9 feet 10 inches was found, however, rather too small, and in the London and Birmingham, with the same width of way, this was increased to 11 feet by widening the interval between the two rails.

In the space of 11 feet, allowed for each rail, a 7-feet gauge might be placed just as well as a 5-feet, leaving the bridges, tunnels, and viaducts exactly the same; but 11 feet was thought by some still too narrow: and when it is remembered that this barely allows a width of 10 feet for loads, whether of cotton, wool, agricultural produce, or other light goods, and which are liable also to be displaced in travelling, 13 feet (which has been fixed upon in the Great Western Railway, and which limits the maximum breadth, under any circumstances, to about 12 feet) will not be found excessive.

It is this which makes the minimum width, actually required under bridges and tunnels, 26 feet instead of 22 feet, and not the increased gauge.

The earthwork is slightly affected by the gauge, but only to the extent of 2 feet on the embankment, and not quite so much in the cuttings; but what, in practice, has been the result? The bridges over the railway on the London and Birmingham are 30 feet, and the width of viaducts 28 feet; on the Great Western Railway they are both 30 feet; no great expense is therefore incurred on these items, and certainly a very small one compared to the increased space gained, which, as I have stated, is from 10 to 12 feet. In the tunnels exists the greatest difference; on the London and Birmingham Railway, which I refer to as being the best and most analogous case to that of the Great Western Railway, the tunnels are 24 feet wide. On the Great Western Railway the constant width of 30 feet is maintained, more with a view of diminishing the objections to tunnels, and maintaining the same minimum space which hereafter may form a limit to the size and form of everything carried on the railway, than from such a width being absolutely necessary.

Without pretending to find fault with the dimensions fixed, and which have, no doubt, been well considered, upon the works on other lines, I may state that the principle which has governed has been to fix the minimum width, and to make all the works the same, considering it unnecessary to have a greater width between the parapet walls of a viaduct, which admits of being altered, than between the sides of a tunnel which cannot be altered.

The embankments on the London and Birmingham Railway are 26 feet, on the Great Western 30 feet, making an excess of about six and a half per cent on the actual quantity of earthwork.

The difference in the quantity of land required is under half an acre to a mile. On the whole, the increased dimensions from 10 to 12 feet will not cause any average increased expense in the construction of the works, and purchase of land, of above seven per cent—eight per cent having originally been assumed in my Report in 1835 as the excess to be provided for.

With respect to the weight of the carriages, although we have wheels of 4 feet diameter, instead of 3 feet, which, of course, involves an increased weight quite independent of the increase of width, and although the space allowed for each passenger is a trifle more, and the height of the body greater, yet the gross weight per passenger is somewhat less.

|                                             | Tons | cwt. | qrs. | lbs. |
|---------------------------------------------|------|------|------|------|
| A Birmingham first-class coach weighs       | 3    | 17   | 2    | 0    |
| Which with 18 passengers at 15 to the ton   | 1    | 4    | 0    | 0    |
|                                             | 5    | 1    | 2    | 0    |
| Or 631 lbs. per passenger                   |      |      |      |      |
| A Great Western first-class weighs          | 4    | 14   | 0    | 0    |
| And with 24 passengers                      | 1    | 12   | 0    | 0    |
|                                             | 6    | 6    | 0    | 0    |
| Or 588 lbs. per passenger                   |      |      |      |      |
| And our 6-wheeled first-class               | 6    | 11   | 0    | 0    |
| With 32 passengers                          | 2    | 2    | 2    | 0    |
|                                             | 8    | 13   | 2    | 0    |
| Or 600 lbs. per passenger                   |      |      |      |      |

Being an average of 594 lbs. on the two carriages.

This saving of weight does not arise from the increased width, and is notwithstanding the increased strength of the framing and the increased diameter and weight of the wheels; I have not weighed our second-class open carriages, but I should think the same proportion would exist.

As to the breaking of axles or running off the line, the practical result has been that, from some cause or other, we have been almost perfectly free from those very objections which have been felt so seriously on some other lines. Far from breaking any engine axles, not even a single cranked axle has been strained, although the engines have been subjected to rather severe trials. One of our largest having, a short time back, been sent along the line at night, when it was not expected, came in collision with some ballast waggons, and was thrown off the line nearly 6 feet; none of the axles were bent, or even strained in the least, although the front of the carriage, a piece of oak of very large scantling, was shattered. After ten weeks' running, one solitary instance has occurred of a carriage in a train getting off the line and dragging another with it, and which was not discovered till after running a mile and a half. As the carriage was in the middle of the train, and one end of the axle was thrown completely out of the axle guard, there must evidently have been some extraordinary cause—possibly a plank thrown across the railway by a blow from the carriage which preceded, and which might have produced the same effect on any railway; and at any rate it was a strong trial to the axle, which was not broken, but merely restored to its place, and the carriage sent on to London. The same mode of reasoning which has by some been used in favour of the 4 feet 8 inches gauge, if applied here, would prove that long axles are stronger than short, and wide rails best adapted for curves. All that I think proved, however, is this—that the increased tendency of the axles to break, or of the wheels to run off the rails, is so slight that it is more than counterbalanced by the increased steadiness from the width of the base, and the absence of those violent strains which arise from irregularity on the gauge and the harshness of the ordinary construction of rails. In fact, not one of the objections originally urged against the practical working of the wide gauge has been found to exist, while the object sought for is obtained, namely, the capability of increasing at any future period the diameter of the wheels, which cannot be done, however desirable it may hereafter be found, with the old width of rail. This may be said to be only prospective; but, in the meantime, contingent advantages are sensibly felt in the increased lateral steadiness of the carriages and engines, and the greater space which is afforded for the works of the locomotives. And here I wish particularly to call your attention to the fact that this prospective advantage—this absence of a most inconvenient limit to the reduction of the friction, which, with our gradients, forms four-fifths or eighty per cent of the total resistance—was the object sought for, and that, at the time of recommending it, I expressly stated as follows:—'I am not by any means prepared at present to recommend any particular size of wheel, or even any great increase of the present dimensions. I believe they *will* be materially increased; but my great object would be in every possible way to render each part *capable* of improvement, and to remove what appears an obstacle to any great progress in such a very important point as the diameter of the wheels, upon which the resistance, which governs the cost of transport and the speed that may be obtained, so materially depends.'

These advantages were considered important by you, they are now considered so by many others; and certainly everything which has occurred in the practical working of the line confirms me in my conviction that we have secured a most valuable power to the Great Western Railway, and that it would be folly to abandon it.

The next point I shall consider is the construction of the engines, the modifications in which, necessary to adapt them to higher speeds than usual, have, like the increased width of gauge, been condemned as innovations.

I shall not attempt to argue with those who consider any increase of speed unnecessary. The public will always prefer that conveyance which is most perfect, and speed within reasonable limits is a material ingredient in perfection in travelling.

A rate of thirty-five to forty miles an hour is not unfrequently attained at present on other railways in descending planes, or with light loads on a level, and is found practically to be attended with no inconvenience. To maintain such a speed with regularity on a level line, with moderate loads, is therefore quite practicable, and unquestionably desirable. With this view the engines were constructed, but nothing new was required or recommended by me.

A certain velocity of the piston is considered the most advantageous.

The engines intended for slow speeds have always had the driving wheels small in proportion to the length of stroke of the piston. The faster engines have had a different proportion; the wheels have been larger, or the strokes of the piston shorter. From the somewhat clamorous objections raised against the large wheels, and the construction of the Great Western Railway engines, and the opinions rather freely expressed of my judgment in directing this construction, it would naturally be supposed that some established principle had been departed from, and that I had recommended this departure.

The facts are, that a certain velocity of piston being found most advantageous, I fixed this velocity, so that the engines should be adapted to run thirty-five miles an hour, and capable of running forty—as the Manchester and Liverpool Railway engines are best calculated for twenty to twenty-five, but capable of running easily up to thirty and thirty-five miles per hour; and fixing also the load which the engine was to be capable of drawing, I left the form of construction and the proportions entirely to the manufacturers, stipulating merely that they should submit detail drawings to me for my approval. This was the substance of the circular, which, with your sanction, was sent to several of the most experienced manufacturers. Most of these manufacturers, of their own accord, and without previous communication with me, adopted the large wheels, as a necessary consequence of the speed required. The recommendation coming from such quarters, there can be no necessity for defending my opinion in its favour; neither have I now the slightest doubt of its correctness. As it has been supposed that the manufacturers may have been compelled or induced by me to adopt certain modes of construction, or certain dimensions, in other parts by a specification—a practice which has been adopted on some lines—and that these restrictions may have embarrassed them, I should wish to take this opportunity to state distinctly that such is not the case. I have indeed strongly recommended to their consideration the advantages of having very large and well-formed steam passages, which generally they have adopted, and with good results; and with this single exception, if it can be considered one, they have been left unfettered by me (perhaps too much so) and uninfluenced, except indeed by the prejudices and fears of those by whom they have been surrounded, which have by no means diminished the difficulties I have had to contend with.

The principal proportions of these engines being those which have been recommended by the most able experimentalists and writers and these having been adopted by the most experienced makers, it is difficult to understand who can constitute themselves objector; or what can be their objections.

Even if these engines had not been found effective, at least it must be admitted that the best and most liberal means had been adopted to procure them; but I am far from asking such an admission. The engines, I think, have proved to be well adapted to the particular task for which they were calculated—namely, high speeds—but circumstances prevent their being beneficially applied to this purpose at present, and they are, therefore, working under great disadvantages. An engine constructed expressly for a high velocity cannot, of course, be well adapted to exert great power at a low speed; neither can it be well adapted for stopping frequently and regaining its speed. But such was not the intention when these engines were made, neither will it be the case when the arrangements on the line are complete; in the meantime, our average rate of travelling is much greater than it was either on the Grand Junction or the Birmingham Railway within the same period of the opening. I have but one serious objection to make to our present engines, and for this, strange as it may seem, I feel that we are mainly indebted to those who have been most loud in their complaints—I refer to the unnecessary weight of the engines. There is nothing in the wide gauge which involves any considerable increased weight in the engine. An engine of the same power and capacity for speed, whether for a 4-feet 8-inch rail, or for a 7-feet rail, will have identically the same boiler, the same fire-box, the same cylinder and piston, and other working gear, the same side frames, and the same wheels; the axles and the cross-framing will alone differ, and upon these alone need there be any increase; but, if these were doubled in weight, the difference upon the whole engine would be immaterial. But the repeated assertion, frequently professing to come from experienced authorities, and repeated until it was supposed to be proved, that the increased gauge must require increased strength and great power, was not without its indirect effect upon the manufacturers. Unnecessary dimensions have been given to many parts, and the weight thereby increased— rather tending, as I believe, to diminish than to add to the strength of the whole.

I thought then, and I believe now, that it would have been unwise in this case to have resisted the general opinion, and taken upon myself the responsibility which belonged to the manufacturers; but I need not now hesitate to say that a very considerable reduction may be effected, and that no such unusual precautions are necessary to meet these anticipated strains and resistances—such being, in fact, imaginary. It cannot surprise anybody that, under such circumstances, attention was more occupied in endeavouring to meet these imaginary prejudiced objections, than in boldly taking advantage of the new circumstances, and that a piece of machinery constructed under such disadvantages was not likely to be a fair sample of what might be done. I am happy to say, however, that the result of the trials that have been made has entirely destroyed all credit in these alarmists with the manufacturer, and that we may hope in future to have the benefits of the free exercise of the intelligence and practical knowledge of engine-manufacturers.

The mode of laying the rails is the next point which I shall consider. It may appear strange that I should again in this case disclaim having attempted anything perfectly new; yet regard to truth compels me to do so. I have recommended, in the case of the Great Western, the principle of a continuous bearing of timber under the rail, instead of isolated supports—an old system recently revived, and as such I described it in my Report of January 1836; the result of many hundred miles laid in this manner in America, and of some detached portions of railways in England, was quite sufficient to prove that the system was attended with many advantages; but since we first adopted it these proofs have been multiplied—there need now be no apprehension. There are railways in full work, upon which the experiment has been tried sufficiently to prove beyond doubt, to those willing to be convinced, that a permanent way in continuous bearings of wood may be constructed, in which the motion will be much smoother, the noise less, and consequently—for they are effects produced by the same cause—the wear and tear of the machinery much less. Such a plan is certainly best adapted for high speeds, and this is the system recommended by me and adopted on our road. There are, no doubt, different modes of construction, and that which I have adopted as an improvement upon others may, on the contrary, be attended with disadvantages. For the system I will strenuously contend.

But I should be sorry to enter with any such determined feeling into a discussion of the merits of the particular mode of construction. I would refer to my last Report for the reasons which influenced me, and the objects I had in view in introducing the piling; that part which had been made under my own eye answered fully all my expectations. Here the piles did answer their purpose, and no inconvenience resulted from their use. The difficulties which we have since encountered, the bad state in which the line was for a considerable time, and which is only recently improved, have undoubtedly been aggravated, if not caused, by these piles; but not, as I believe, from a defect in the principle as applied in our case, where the line is mostly in cutting, or on the surface, but from defective execution; for, notwithstanding the determination to allow sufficient time for this most important operation, yet, to make up for previous delays and loss of time, it became necessary at last to force forward the work more rapidly than was at all consistent with due care in the execution; and during the whole of this period I was most unfortunately prevented by a serious accident from even seeing the work almost until the day of opening, when I ought to have personally superintended the whole. I do not mean that the work was neglected by those whose duty it was to supply my place—far from it; but in such a case, a new work cannot be properly directed except under the eye of the master. Following exactly the plan which had succeeded on the first piece completed, several serious faults were committed. A much greater density and firmness of packing is required than was previously supposed; the mode of packing adopted, and the material selected, in the first instance, have proved defective elsewhere; and over a great extent in the line, particularly in the clay cuttings, and where the work was at last most hurried, it has been badly executed. But many parts have stood well from the commencement; others are fast improving; and I have the satisfaction, although a very painful one, of seeing that if, in the first instance, a foundation of coarse gravel had been everywhere well rammed in before the timbers had been laid, and the packing formed upon this, we should, from the outset, have obtained as solid a road as we have now over a great part of the line. What we have been able to effect since the opening of the line has necessarily been a slow, expensive, and laborious operation. We have been compelled to open the ground, and excavate it to a depth of 18 inches under the longitudinal timbers, and this without interrupting the traffic: to remove the whole of the material thus obtained from off the line, and to replace it by coarse ballast; and not having the means of sufficiently consolidating this ballast by ramming while the timber is in its place, the packing has to be repeated once or twice after it has been compressed by the passing of the trains. This new packing, however, does stand, and in a few weeks I expect the line will be in a very different state from that in which it has been, or indeed now is. From what I have described as the result which can now be, and might have been, obtained from the commencement, it will be inferred that I am disposed still to defend the system of piling. I certainly could not abandon it from conviction of its inefficiency, for I see proofs of the contrary; and I feel that under similar circumstances I could now prevent the mischief which has occurred. Upon that portion of the line where the permanent way must next be formed, piling could not be resorted to, the ground being a solid hard chalk for many miles. I had intended, however, recommending the same principle, but in a different form, holding down the longitudinals by small iron rods driven into the chalk; but the same objection could not exist, because the chalk cannot yield under the timbers like clay, or even gravel. But I should wish most anxiously to avoid anything like an obstinate adherence to a plan, if the object which I believe essential can be obtained by other means, particularly when, that plan being my own, I may be somewhat prejudiced in its favour. I find that the system of piling involves considerable expenses in the first construction, and requires perhaps too great a perfection in the whole work, and that if the whole or a part of this cost were expended in increased scantling of timber and weight of metal, that a very solid continuous rail would be formed.

For this as a principle, as for the width of gauge, I am prepared to contend, and to stand or fall by it, believing it to be a most essential improvement, where high speeds are to be obtained. I strongly urge upon you not to hesitate upon these two main points, which, combined with what may be termed the natural advantages of the line, will eventually secure to you a superiority which, under other circumstances, cannot be obtained.

As regards the expense of forming the permanent way on this principle, I am quite prepared to maintain what I have on a former occasion advanced: that even on the system which we have adopted between London and Maidenhead, the total cost does not materially exceed that of a well-constructed line with stone blocks. I did not make in the outset an exact estimate of the cost of either mode; I was unable to obtain the cost which has actually been incurred on other lines; but a comparative estimate was made, and the result of that comparison led me to state that the one might exceed the other by £500 a mile. The actual cost of our permanent way appears, by the detailed account which has been made out, to have been above £9,000, including expenses of under-draining and forming the surfaces which cannot be included in the cost given in other cases, because that drainage (although I believe generally forming part of the plan) is not yet constructed. This sum includes the sidings at the stations, switches, joints, and other contingencies, and also the expenses incurred during the first month of working the line, and which, as I have before stated, consisted in removing and replacing work which had been improperly executed. These items will make a considerable reduction; and besides these, larger reductions may be effected in parts of the work which were new, and, from the circumstances naturally attending a first attempt, were not so economically conducted as they might be, or indeed, as they were towards the close of the works, when the different parts were let by contract.

Taking the prices at which the work was latterly actually executed, £8,000 per mile would be a liberal allowance for our future proceeding, even adopting the same system; and with a modified system, such as that suggested of simple longitudinal bearers of large scantling, and a rail of fifty-four pounds per yard, at the present high price of iron, the cost, calculated upon our actual past expenditure, would not exceed £7,400 per mile. This, I am aware, is a larger sum than that which has usually been assumed as the cost of the permanent way. I cannot prove that others have cost more, or even so much as this, as I have nothing but the published accounts to refer to; but this I can state, and prove if necessary, that rails and blocks, such as are now being adopted on the Manchester and Liverpool Railway, would upon our line cost at least as much.

The prime cost of rails and chairs delivered on the line would alone amount to half the money; and nothing is, perhaps, more certain than that the experience of other lines within the last two or three years has proved that this part of the construction of a railway is unavoidably much more expensive than was ever calculated for at the time our estimates were made.

I am, gentlemen, your obedient servant,

(Signed)                    I. K. BRUNEL

# Appendix II

(SEE CHAPTER IX ON THE 'GREAT BRITAIN' STEAM-SHIP, P.185)

*Report to the Directors of the Great Western Steam-ship Company*

October 1840.

GENTLEMEN,—I have now the pleasure to lay before you the result of the different experiments which I have made, and of the best consideration I have been enabled to give to the subject of the screw propeller.

The observations which I have to make are naturally divided under two principal heads, namely: first, the simple question of the applicability and efficiency of the screw considered merely as a means of propelling a vessel, compared with the ordinary paddle-wheel; and, secondly, the general advantages or disadvantages attending its use.

The consideration of the comparative efficiency of the screw as a means of propelling, of course embraces the whole question, not merely of the effect produced, but also that of the proportionate power absorbed in producing that effect.

With respect to the mere effect of a screw, the performance of the 'Archimedes' has proved, in a satisfactory and undeniable manner, that a screw acting against the water with a surface even much smaller than that offered by the paddle-boards of a well-proportioned paddlewheel, will propel the ship at a very fair speed, but at what expense of power this effect has been produced is not so evident.

I shall first examine into the principal cause of what amounts practically to a loss of power, and which is common in a greater or less degree to all modes of propelling a vessel by exerting a pressure against the water as against a fixed point.

The resistance, whether to the surface of a screw, or of a paddle-board, or of the blade of an oar, or any other propelling body, offered by the fluid against which it acts, is of course not perfect, and there is a certain amount of yielding, commonly called the slip, of the paddlewheel; the amount thus slipped causes a considerable waste of power, inasmuch as the full power of the engine is expended through the entire space passed over by the paddles or other propelling surface, while the useful effect produced is only equal to the same power expended over the space through which the vessel passes: this loss frequently amounts to one-quarter, and even one-third, of the whole power employed. To investigate theoretically the amount of slip due to any given form and quantity of surface, involves much more complicated calculations than have generally been applied, and would indeed require data which we hardly possess; but fortunately we have had the means of making experiments, the results of which enable us to determine the comparative slip of the paddle and of the screw, with sufficient accuracy for all practical purposes.

The screw in use on board the 'Archimedes' is 5 feet 9 inches diameter, with a pitch of 8 feet—that is to say, in making one revolution the thread of the screw advances 8 feet; the area of the screw, considered as a disc of the same diameter, or the extent of the surface of water which is acted upon in the direction of the axis of the vessel, is therefore about 26 feet, without deducting the section of the shaft-bearing, &c. The midship section of the vessel when I experimented upon her was, according to Mr. Patterson's estimate, 122 feet; the ratio of the resisting surface to the midship section being therefore as 1 to 4.7, which is a small proportion; and the form of the vessel is by no means peculiarly good as a steamboat. This proportion of propelling surface to midship section is much smaller—that is, the area of the screw is much less in proportion to the size of the vessel than is the area of paddle-boards immersed in steamboats generally.

The average paddle-board immersed and really effective is rather difficult to estimate, as allowances must be made for the disturbance of the water, when the wheel is in motion; but this average in the 'Great Western' measured perpendicularly—that is, allowing for the obliquity of the paddle—cannot be less than 180 to 200 feet, say only 180, while the midship section averages about 462 feet; the surface of paddle is therefore about ½.56 of the midship section.

I will now give the comparative effects of these different propelling surfaces in these two cases.

I have made very accurate experiments upon the comparative rate of the 'Archimedes,' and of the space passed through by the screw, and was enabled to determine this ratio with great certainty.

The average of a number of trials gave the following results:

Rate of ship, 50,867 feet per hour, or about 8⅓ knots.

Space passed through by screw due to the number of revolutions, 65,685 feet.

The average rate of vessel being to that of screw therefore as 1 to 1.2913.

In the performances of the 'Great Western,' upon an average of 20 voyages the ratio has been as 1 to 1.2997; but, separating from these 20 such voyages as were unusually short or long, and taking only such as, occupying 14, 15, or 16 days, may be considered as giving a fair average of the speed of the ship when not adversely affected by the wind or heavy seas, the average of these 13 voyages give 1 to 1.283; and leaving out again those of 16 days, and taking only 8 voyages of 14 and 15 days, the average gives a ratio of 1 to 1.27187.

    Of these, 5 voyages of 15 days give 1 to 1.29077
    and        3    "    14   "       1 to 1.23901

The last three, however, were short passages and homeward, when the currents and winds have been in favour, and consequently we may safely say that the ratio must be above 1 to 1.239; and after making every allowance for the effect of swell and other impediments (the experiments upon the 'Archimedes' being made in smooth water), the average of the 8 (5 of which were homeward voyages with favourable current and wind and the vessel in good trim), giving a ratio of 1 to 1.27, may be taken as a fair average.

The comparison between the 'Archimedes' and the 'Great Western' will therefore stand thus—

|  | Area of Propelling surface, the Midship Section being 1.0 | Difference of speed of vessel and Propelling surface, or amount of Slip, the ratio of Vessel being 1.0 |
|---|---|---|
| 'Archimedes,' screw | 0.203 | 0.2913 |
| 'Great Western,' paddle | 0.391 | 0.2708 |

Showing an amount of slip in the 'Great Western' very nearly equal to that of the 'Archimedes,' while the ratio of the propelling surface to the midship section in the case of the screw is little more than half that of the paddle-boards in the 'Great Western.'

In taking the average of the eight voyages of the 'Great Western' with favourable winds as I have done, I believe I have made full allowance for the different circumstances of smooth water and sea; but there is ample room in the above comparison to make even greater allowance for these circumstances, and still to leave a result which would prove that with similar areas the screw would meet with at least equal, if not a greater resistance, and consequently will slip as little or less than the ordinary paddle-board.

I subjoin a table also, taken from a well-known work on the steam-engine (Tredgold's), of the slip of a number of vessels, of which in every case the surface of paddle immersed is far greater in proportion to the midship section than that of the screw in the 'Archimedes.'

### Rate of Paddle, that of Ship being 1

| | | | |
|---|---|---|---|
| Medea | 1.595 | Monarch | 1.323 |
| Flamer | 1.483 | Magnet | 1.310 |
| Firebrand | 1.501 | Meteor | 1.490 |
| Columbine | 1.529 | Carron | 1.287 |
| Salamander | 1.200[1] | Average | 1.381 |
| Dee | 1.366 | | |
| Firefly | 1.364 | Great Western | 1.27 |
| Firebrand, as altered | 1.295 | Archimedes | 1.29 |
| Phito | 1.215 | | |

This list shows that the result in the 'Great Western,' with which ship I have made the comparison, is in itself a favourable one, and that compared with many others the 'Archimedes' would stand much better.

This apparent superiority of the screw over the paddle as regards the resistance offered to it by the water may at first appear startling, but there is a great mistake committed in assuming that the action of the screw is a very oblique action, tending rather to drive the water laterally with a rotatory motion than to push it steadily backwards.

Having witnessed and carefully observed the degree and the nature of the disturbance in the water caused by the screw, and comparing this with the violent displacement of the water by the action of paddle-boards, even under the most favourable circumstances, I no longer feel surprised.

The mass of water pushed backwards by the action of the screw appears to be very large, spreading from the screw probably in the form of an inverted cone, but there is little or no appearance of any rotatory motion, and the surface of the water is not put into rapid motion as in the case of the paddlewheel, which may be observed to impart a considerable velocity to the water, probably for a small depth only, but over a very large space.

As regards the oblique action also, a great mistake appears to have been generally made, and very naturally made, by most persons when first considering the working of the screw. It is generally assumed that the inclined plane formed by the thread of the screw strikes the particles of water at that angle and with the velocity of the revolution of the screw, but it is forgotten that the screw is moving forward with the ship, and therefore that the angle at which the water is struck by the plane is diminished by all that much that the ship with the screw advances—indeed, it is evident that if the ship advanced the whole amount of the pitch of the screw, the screw, oblique as it appears, and rapidly as it revolves, would not strike the water at all, but simply glide through.

The angle at which any given part of the screw does in fact strike the water is only equal to the difference between the angle to which that part of the screw is formed and the angle or direction in which it moves by the compound motion of the revolution of the screw and of the forward motion of the ship and screw; and, contrary to one's hastily imbibed notions of the action of the screw, this angle at which the plane of the screw is driven against the particles of water, is in such a screw as that of the 'Archimedes' very nearly equal over the greater portion of the surface, diminishing to nothing at the centre; and the motion imparted to the water, although perpendicular to the plane of the screw in point of direction, is small in extent or velocity, being also nearly the same over the whole surface of the screw, except close to the centre, where it is infinitely small.

In the 'Archimedes' screw, which appears to the eye so oblique, and the centre part of which would appear to act flat against the water, only causing it to revolve, the outer circumference being 18 feet and the slip 1 foot 8 inches, the angle at which this outer edge acts upon the water is only one in 11½.

The total amount of motion imparted to the water at right angles to the plane of the screw by one entire revolution even at the outer edge is not quite equal to the slip, being only 1.67 foot. The rotatory motion is still less, the total distance to which any particle of water is displaced laterally, or at right angles to the axis of the ship, by one entire revolution of the screw being at the outer edge only 0.69 foot, and the maximum distance being in any screw only equal to half the slip, and occurring at that part of the screw where the circumference is equal to the advance of the ship due to one revolution. This maximum of lateral motion is 0.9 foot, and takes place at 0.99 foot, or about 1 foot from the centre. In this mode of considering the direction at which the particles of water are acted upon by the plate of the screw I have taken no notice of the effect of the friction upon the surface of the screw, which, causing to be carried with it a film of water, will modify more or less according to the degree of smoothness of the surface the effect of the screw upon the water; and towards the centre this friction, however smooth the surface may be made, will gradually become equal to, and at last greater than, the propelling effect of that part of the screw; but this defect applies only to a very small portion of the whole area of the screw, and the absence of any very violent impulse to the water in a direction approaching to a right angle with the axis of the vessel, and which has always been assumed as an unavoidable evil in the screw, will account for the absence I have observed upon of any apparent rotatory motion.

I would not pretend, however, to advance these circumstances which I have observed, or these reasonings, as arguments whereon to found an opinion of the action of the screw, the facts as proved by the experiments are what I rely upon; but it is satisfactory to be able to account for the results by circumstances actually observed, and the reasons which suggest themselves.

The effect of a propelling surface in the form of a screw, and moving at a certain velocity, as compared with an equal surface moving at the same velocity but applied in the shape of paddle-boards, having been ascertained, it remains to determine the comparative power required to give motion to that surface.

The difficulty of determining this with any degree of accuracy from any experiments which we could make on board the 'Archimedes' was very great, but considering such results as I could obtain in conjunction with experiments which I have since made in our own works, and with the results upon steamboats recorded by others, and of those of experiments made by Colonel Beaufoy on the resistance of bodies in water, I think we may arrive at approximate conclusions sufficiently accurate for our purpose, and which may safely be relied upon.

In the case of the 'Archimedes' the engines were certainly not effective well-working engines, the proportions of the gearing or wheel-work between the engine and the screw was bad—such that the engine could not attain its proper speed—the friction of the gearing (which, whether it be a source of resistance necessarily attending the use of the screw or not, I shall consider afterwards) was very great, and the surface of the screw itself, which I had an opportunity of examining out of water, was so rough as necessarily to create very much more friction than would be caused by a tolerably smooth metallic surface. With all these sources of resistance, and under these unfavourable circumstances, the power calculated for the effective pressure on the piston and without deduction for friction or other causes, which, for the sake of distinction hereafter, I shall call the gross power, was about 145 horses, the speed of the vessel being about 8⅓ knots per hour, as actually measured by the land, and full 9 knots as measured with great care by heaving the common log, the midship section being, as before stated, 122 feet, and the lines of the vessel not so good as those of fast boats; comparing this with the gross power of the 'Great Western' engines when propelling that vessel at the same velocity, with the advantage of better lines and the other advantages arising from greater dimensions, there does not appear any such discrepancy as to indicate any loss of power by the use of the screw in the 'Archimedes'; on the contrary, the power expended in the 'Great Western' is actually as great as that in the 'Archimedes,' as compared with their relative midship sections—and if any great allowance is to be made for the circumstances which I have referred to of larger dimensions and better lines, there would appear to be actually less power expended in proportion to the dimensions and form of the 'Archimedes' than in the 'Great Western.'

The results obtained with the 'Great Western,' which as regards speed are similar to those of the 'Archimedes,' are necessarily taken from experiments made when she was rather deep, and the speed thereby reduced to 7.9 knots; but I have compared these with results reduced by calculations from experiments at higher speeds, and I find them agree satisfactorily—indeed, at the draft and consequent immersion of paddles when in this state, I consider the 'Great Western' as very nearly at her best as regards economy of power and effect produced. I should observe that the particular experiments from which the following calculations are deduced were made with the 'Great Western' in smooth water in the Severn. I have added also some calculations deduced from data given by Tredgold as to the performance of the 'Ruby,' a good boat with immense surface of paddle-board.

The comparison stands thus:

|  | GREAT WESTERN | ARCHIMEDES | RUBY |
|---|---|---|---|
| ACTUAL DIMENSIONS: |  |  |  |
| Midship section | 520 | 122 | 63 |
| Area of board immersed | 230 | — | 64 |
| Area of a disc of diameter of screw | — | 26 | — |
| RELATIVE DIMENSIONS AND POWER: |  |  |  |
| Area of propelling surface, midship section being = 1 | 0.442 | 0.213 | 1.016 |
| Gross power expended for one square foot of midship section | 1.023 | 1.026 | 0.976 |

The speed being the same, viz. 7.9 knots, the power expended is as nearly as possible the same in the three, and equal to one horse-power gross to one foot of midship section; while the relative propelling surface in the 'Archimedes' is equal to only half that of the 'Great Western,' and one-fifth that of the 'Ruby.' This *gross* horse-power, it will be observed, is *about* equal to one-half a nominal horse-power.

I have made several comparisons with recorded observations made on board the 'Great Western' at different times, and with experiments made in other vessels, and I find the same result; in estimating the powers used more particularly in some comparisons with the 'Great Western,' I have taken the mean pressure as ascertained on both sides of the piston, while in the 'Archimedes' I only obtained that on the top of the piston, which appears generally to be the best, and consequently the estimate is made unfavourably to the 'Archimedes.'

Such general results are all that I could obtain from the experiments on board the 'Archimedes,' but since that time I have made some experiments upon the friction of a plate of metal in water, and have compared these results with the experiments of Colonel Beaufoy, and the conclusion I have come to is that the power absorbed by friction in a well-made screw, apart from all question of the means adopted for working it, would not be such as to interfere with its beneficial application.

The resistance created by the screw itself arises principally from two sources— the resistance to the cutting edge and the tail-edge, and the friction of the surface in contact with the water. The amount of the first may of course be reduced to an unlimited extent by having a fine edge, and practically such edge ought to be much finer than that of the screw of the 'Archimedes.'

The friction upon the surface will of course materially depend upon the smoothness of that surface, and in the 'Archimedes' it was very rough, the iron being corroded at many places, with exfoliations and small holes—the corrosion arising apparently from the galvanic effect produced by the iron and the ship's copper.

The great number of revolutions required in the screw as compared with those of the paddlewheel, leads a person to assume, without much consideration, that a very high velocity is given to the cutting edges and to the surface of the screw, and consequently that great friction must be produced—this velocity is not, however, nearly so great as it at first appears.

In the present screw of the 'Archimedes' the velocity of the extreme point, following its oblique or spiral course, is only about three times that of the vessel, while the average velocity of either of these knife edges or of the surface is not twice that of the vessel.

Now without determining what the actual amount of these resistances may be, we can at once satisfy ourselves that it cannot be very considerable, by comparing it (which we have the means of doing) with the resistance caused by the cutwater and any given portion of the ship's bottom. The resistance of a knife-edge will be about as the square of the velocity, and if we assume the surface friction to increase in the ratio determined by Colonel Beaufoy—namely, at the 1.75th power, or as the 4th root of the 7th power of the velocity—then the resistance of the knife-edge will be equal to the resistance of a similar edge of about five and a-half times the length of the diameter of the screw moving at the same rate as the vessel, and the surface friction will be equal to that of a piece of the ship's bottom about six and five-eighth times the area of the screw—or, in the case of the 'Archimedes,' the additional power absorbed by the friction of the screw would be about equal to that absorbed by the friction of little more than twice the space of the dead wood which had been cut out to receive the screw—while the knife-edges would be about equivalent to three knife-edges immersed in the water, of the same depth as the ship's stem.

The actual amount of power absorbed in driving the 'Archimedes' screw was probably about twenty horse-power gross, or from ten to twelve nominal horse-power; but I have no doubt that a screw of similar diameter and in good condition would not absorb half that power: and this amount may be still further, and very much reduced, by increasing the relative size of the screw to that of the ship, and thereby reducing the slip, and proportionately reducing the number of revolutions required.

The great extent to which this is capable of being carried will at once be seen when I state that if the ship's progress were made to be 7 feet instead of 6 feet to each revolution of the screw, which a very slight increase of diameter and pitch of screw would effect, the power absorbed in driving the screw would be diminished in the ratio of the $6^2 \sqrt[4]{6^7}$ to $7^2 \sqrt[4]{7^7}$—that is, as $6^{15/4}$ to $7^{15/4}$, or about as 3 to 2.

I must repeat here the observation I have previously made, and remind you that these calculations are not introduced as *proving*, but merely as *explaining*, that which appears to me proved by the general results of the experiments on the 'Archimedes'—namely, that the effect produced was, considering all the circumstances, fully proportionate to the power expended, while the experiments and calculations which I have since made also satisfy me that these results may be very much improved upon.

As regards the first of the two heads under which I stated that I proposed to consider the subject, namely, the mere efficiency of the screw as a propeller, I think but one conclusion can be drawn from the results of the experiments quoted, and that is, that as compared with the ordinary paddlewheel of sea-going steamers, the screw is, both as regards the effect produced, and the proportionate power required to obtain that effect, an efficient propeller.

I limit the comparison to the ordinary paddlewheels of sea-going steamers, first, because those are the circumstances which *we* have alone to consider; and, secondly, because it is *possible*, by increasing the diameter and breadth of the paddles, which, for the attainment of an adequate object is practicable to any extent in a mere river boat, to render the action of the common paddle all but perfect, and probably more effective than any other propeller.

In considering the advantages and disadvantages likely to attend the use of the screw propeller, I will, commencing with the latter, consider such objections as have been advanced by others, as well as those which may have occurred to myself.

The only objections, however, which I think worth consideration are:—

First. The necessity of a peculiar form of vessel.

Secondly. The situation of the screw under water, and consequently to a certain extent unseen and inaccessible, and the liability to injury from its position from grounding or in other ways.

Thirdly. The probability of its being lifted out of water when the ship pitched deep.

Fourthly. The difficulty of getting up the required number of revolutions, and the great defects of the mode employed in the 'Archimedes,' and the shaking caused by the machinery.

As regards the form of vessel, undoubtedly a shallow boat, intended for shallow waters, would be very unfit for the application of the screw, which would probably require a greater depth of water than the whole draft of the vessel; but I see no defect or difficulty of this description in the vessel now under consideration, nor can I anticipate any in any vessel this Company is likely to be interested in; a clean run is the most essential condition, and I should suppose no ship was ever built in which this principle of form was carried to a greater extent than in our new iron ship. Her present form I believe to be excellent for the screw, and with a very slight dropping of the keel towards the stern, which can easily be done now without any expense, assisted by the different trim, which, as I shall presently show, will be effected by the use of the screw, the required draft of water will be attained.

It may, perhaps, be as well to mention here, that the diameter of the screw, if in the same ratio to the midship section as in the 'Archimedes,' would be only 12 feet 3 inches, my friend Captain Claxton having made a mistake upon this point in calculating it at 16 feet, and that if increased only to 14 feet 4 inches, the diminution referred to in a former part of my Report of one-third in the power lost in working the screw would be effected; considering the speed we wish to attain, probably 15 feet 6 inches would be a good diameter. Upon the whole I think the vessel is as well fitted for a screw as she is for paddles, and much better adapted for either than the 'Archimedes;' but if originally intended for a screw, possibly some trifling modification in the form and construction, principally of the keel near the stern, might have been introduced which would have rendered the whole a more perfect job than she would now be if altered—but the absence of this would in no way lessen the efficiency of the screw, and I cannot think that any alteration we might now be obliged to make would exceed in cost the sum of £200.

Secondly, the inaccessibility of the screw and liability to damage; this appears to me the objection most plausible, but I cannot say that I attach much weight to it, particularly in the case of a vessel intended for long voyages and across the ocean. During the whole passage in deep water I consider the screw far less exposed to injury than a paddlewheel, and that the chances of injury are so remote that even if it were quite inaccessible it would still be altogether safer than paddles, which are so much exposed; but it is by no means inaccessible, the screw may be rendered stationary at any time or during any weather, when it would be barely safe to stop the engines with common paddles, and when it would be very difficult to do anything to the paddles even if the engines were stopped, while the whole of the screw, bearings, &c., may easily be examined and felt from above, and, if necessary, men sent down with common diving jackets and hoods to replace bearings, or attach tackle to move the screw, or clear away any obstacle entangled in it. When in port I still think the chances of injury very remote; an inspection of our model will satisfy you that from the form and size of her midship section the vessel cannot lay in any position in which the screw would touch the ground, while at that time the whole screw may be very easily examined and replaced without any necessity for going into dock.

Thirdly, the probability of its being lifted out of water when the ship pitches deep.

This appears at first to be a very natural and an unavoidable objection, but the result of observations proves that the motion of vessels, of steamers at least, is not such as to cause the apprehended difficulty. Among the observations made on board the 'Great Western' steam-ship by Mr. Berkeley Claxton, under my direction, were measurements of the angles of rolling and pitching, and from these it was evident that the vessel never pitches to so great an angle as that to which she rises; such a result might indeed have been anticipated by considering the form of the vessel forward and aft, and the circumstance that a steamer is almost invariably meeting or passing the seas, or, if overtaken by them, is still going at a good rate, which reduces the relative speed of the sea; consequently, although the vessel may be frequently thrown up very violently forward, yet the stern, which has no displacement under water, settles down quietly and heavily upon the surface; or, considering it in another way, the variation of displacement at the stern is very rapid, falling off almost to nothing at a few feet below the water-line, and spreading out to a great extent at a few feet above, whilst forward the difference of displacement is comparatively small, the centre of motion, therefore, is thrown very far aft, and while the bows, which are also opposed to the first shock, are thrown alternately high out of water or plunged deeply into it, the stern floats nearly steady, the vessel resting on its broad counter nearly as the centre of motion: whatever may be the explanation such is the operation, not only as measured by instruments, but more particularly as observed since, practically.

In the 'Great Western' the whole cutwater and, it is said, a considerable length of keel, is frequently seen out of water from the bow-sprit, while astern it is very doubtful whether more than half the stern part was ever seen; marks have been made by my direction on the rudder to observe this; as yet the 9-foot mark is the lowest seen, and this occurring rarely, and for very short intervals.

In the 'Archimedes,' during a voyage performed in her by Mr. Guppy from Bristol to Liverpool, and during which they were exposed on more than one occasion to violent pitching, the screw (which can be watched from the deck) never was uncovered; and Mr. Smith and others on board the 'Archimedes,' whose whole conduct was such as to inspire unusual confidence in all information obtained from them, assured me that such was always the case.

In the voyage to Oporto and back, in which I sent Mr. Berkeley Claxton, he made the same observation; these facts, in conjunction with previous and subsequent observations on board the 'Great Western,' convince me that nothing is to be feared on this head; but even if the screw were occasionally to be partly exposed, I know of no evil consequences likely to ensue, as I shall clearly point out when referring to the *advantages* of this propeller over the common paddle.

Fourthly, the difficulty of getting up the required number of revolutions, and the great defects of the mode employed in the 'Archimedes.'

Upon this point certainly the 'Archimedes' offers but a miserable example, and the result is almost enough to prejudice the mind of any person against the whole scheme; the proportions of the gearing, as I have before stated, are so bad the engines appear, even to the eye, to labour ineffectually to get up their speed. The required speed of the screw is not nearly attained, while the noise and tremor caused by the machinery is such as to render the vessel uninhabitable, and perfectly unfit for passengers, I should almost say for a crew. I never attached much importance to these circumstances, because I felt convinced that such a mere mechanical difficulty would by some means be overcome, if, as I confess I did not then at all anticipate, the screw itself should prove efficient.

The most simple and effectual means of overcoming all objections on these heads always appeared to me to be by the use of straps instead of gearing; and all my experience, and I have seen a great deal of the working of machinery by straps and ropes in the numerous works executed by my father, led me to the conclusion that there existed no difficulty whatever in sending the necessary power through a rope or hemp strap, but I was hardly prepared to find the result so entirely satisfactory as it has proved to be.

In an experiment made in your works at the yard, I have sent through two small whale lines, a power equal to about one-thirtieth of that which would be required in the strap if used in the new ship, and this without any slip or straining of the rope which would be injurious in practice, and without any peculiar means of ensuring adhesion to the drums; so that we have ascertained beyond doubt that sixty such whale lines upon a drum of only 4 feet 3 inches diameter is adequate to our wants, but if we suppose seventy lines of superior manufacture to that used in the experiments with a perfect mode of tightening and working upon a drum of 6 feet diameter, all of which can easily be had, it will ensure the perfect and easy working of a mode of obtaining the required number of revolutions of the screw without noise or tremor. The strap in question would be only about 3 feet or 3 feet 3 inches broad, easily replaced piecemeal, and even, if necessary, without stopping the engines.

All the difficulties enumerated under the fifth head may be considered as entirely overcome, or rather as ceasing to exist; and so far from the working of the screw involving difficulties and unavoidable friction, noise, or tremor, it may be worked with unquestionable and perfect facility, and as compared even with the best-made paddles in smooth water, the whole machine will be noiseless.

It is almost unnecessary that I should say that the screw, apart from the gearing in use on board the 'Archimedes,' *cannot* and *does not* produce the slightest tremor or noise—it was with some difficulty, and at least only by attentively listening, that the revolutions of the screw could be counted, even when disconnected and free from the noise of the engine or gearing and the vessel being towed, and then only from some defect in the bearings or the shaft of the screw causing a slight beat.

In thus answering the objections supposed to have been urged against the use of the screw, I may probably have appeared to see everything in a favourable light; unhesitatingly I admit that it is so, and that both formerly when I was completely sceptical as to the mere efficiency of the screw as a propeller, and since my doubts on that head have been removed, I always felt that upon all other points the screw possessed every superiority that could be desired over a common paddlewheel for a sea-going vessel.

I shall now proceed to point out the principal advantages peculiar to the use of the screw; they are—

First. A considerable saving of weight, and that principally top weight.

Secondly. The admitting of a better and simpler form of vessel, having greater stiffness with the same quantity of material, and offering less resistance to head wind and seas, and affording more available space within.

Thirdly. The operation of the screw being unaffected by the trim or the rolling of the vessel, and allowing of the free use of sails, with the capability of entirely disconnecting the screw or of varying the multiplying motion so as to adapt the power of the engine to the circumstance either of strong adverse winds or scudding.

Fourthly. Perfect regularity of motion and freedom from the possibility of violent shocks to the engines.

Fifthly. The singularly increased power of steering given to the vessel—and

Sixthly. The great reduction in the breadth of beam.

I have gone into some detail in calculating the weights of the parts which are not common to the two systems, and I find that the difference, or actual diminution in weight in favour of the screw as applied to our new ship, is upwards of ninety-five tons; but that a much greater weight even than this is transposed from the top of the ship to the bottom—no less a mass than one hundred and sixty tons is removed from the level of the paddle-shaft or from about 10 feet above the water-line, and replaced by sixty-five tons at about 7 feet below the water-line; not only is buoyancy, and consequently proportionate space for cargo, gained to the extent of the difference, but the relief to the labouring of the vessel in bad weather from the change of position must be immense. If the reverse were under consideration, if in a vessel fitted for sea, however stiff in trim or form, it were suggested to remove sixty-five tons of her ballast, and to place one hundred and sixty tons upon her deck, and thus navigate her across the Atlantic in all weathers, it would probably be considered, not merely as highly dangerous, but as actually impossible. Although such an opinion as that it would be impracticable we now know would be incorrect, yet the extent of the beneficial change is much more striking when considered in this way. As regards the trim of the ship, about one hundred and forty-five tons would be removed from nearly the centre of flotation, and the balance of fifty tons added and distributed over the after part, principally quite aft.

I have not calculated the exact effect of this upon her trim; it would only bring her down by the stern, and this is a defect which there seems, as we too well know, never any difficulty in remedying.

Secondly. The simplifying and improving the form of the ship—both as regards strength and mass exposed to the wind and sea.

The necessity of contracting the midships of a steamer, and making her completely wall-sided, and forming a sort of recess to receive the paddles, interferes considerably with the framing of the ship. In a wooden ship of the size of our new one the whole beam of the ship would have to be contracted in order to carry the planking through in a direct line and obtain the requisite fore and aft tie as has been done in the 'Great Western'; in the new ship the almost infinite resources afforded by the material used, enabled us to expand the sides and obtain breadth of beam for cabin room, both before and abaft the paddles, and contract the sides at the paddles as seen upon the plan; but in order to strengthen this part, so evidently weak by form, much contrivance and much material was required. By dispensing with paddles, the best form of ship is left free to be adopted; perfect lines may be preserved, more equal strength obtained with increased space, and the whole mass of paddle-boxes and their accompanying sponsons and deck-houses swept away, and the resistance of these huge wings to head winds or seas entirely avoided.

The space gained by avoiding the contraction is calculated by Mr. Patterson to amount to two hundred tons measurement; this would be entirely gained, and would not even involve increased dues or tolls, as it would be added to the engine-room; it would therefore perhaps counterbalance any loss of room caused by the shaft conveying the movement from the engine to the screw, but I believe this nominal increase at one part would not be so great, while in fact the ship would really be more compact, and, though to a very small extent, a smaller ship, as the sponsons would be removed.

The third point of advantage named is perhaps the most important. With paddles, the action is materially affected by the depth of immersion; when the vessel is deep, and consequently the paddles deep, their action is impeded, a greater part of the power of the engine is absorbed in driving the paddle, the speed of the engine is reduced and the effect diminished; when too light also the paddles do not take sufficient hold of the water, the amount of slip increases and power is wasted; in rolling the same effects are produced, and thus at those times when the greatest effect is required, namely, with deep immersion or in bad weather to overcome the increased resistance offered to the vessel, the propelling power is least effective, and Captain Hoskins actually estimates this loss as occasionally equal to two-thirds the whole power.

The bad effects of one paddle being immersed too deeply, and the other not sufficiently, also prevents the free use of the sails; and it must often occur that the impediment thus offered to the working of the paddles more than counterbalances the good effects of a tolerably fair wind. With the screw the effect is constant, at least unaffected by the position or motion of the ship, whether deep or light the screw acts nearly the same, and as to rolling or heeling over the screw would work equally well (as long as it be immersed) if the vessel were on her beam-ends or bottom upwards.

The screw therefore leaves the ship free to be used as a sailing-vessel to any extent that other circumstances will admit of, and as long as the sails draw there can generally be no doubt that the wind is assisting the ship. The screw may also be thrown in and out of gear at any time and during any weather, either in case of accident to the engines, or in the event of her scudding before a gale of wind, when the engine would be useless; this last, however, I do not consider a probable occurrence, particularly if another arrangement of which the screw is susceptible be taken advantage of.

If the motion be conveyed by a strap, as I have recommended, there is no difficulty in having two or even three drums on the screw-shaft of different diameters, and thus when the resistance to the ship is very much increased by strong head winds, deep draught, and other causes, to use the slow motion and obtain an increased propelling force, or when, on the contrary, the vessel is running before the wind to use the quick motion—by which, in both cases, a great increase of speed would be attained.

This is in fact obtaining at once, and by simple means all those advantages, and to a much fuller extent, which are aimed at in the reefing-paddles.

Fourthly, great regularity of motion is naturally consequent upon the screw being unaffected by the rolling of the ship, and upon its being immersed and not exposed therefore to blows from the sea, and except in the case of its being lifted out of water, the resistance is perfectly uniform and perfectly smooth.

An engine could not have a work less capable of causing any jar or shock as to the effect; even if lifted partially out of water the variation of resistance would be as easy or soft, to use a mechanical term, as possible, while the extent of the variation could never approach to that to which paddles continually expose an engine. A heavy sea or a deep plunge will occasionally bring the engines nearly to a stand; while at other moments, if the engineers are to be believed, the paddles are left free and the engines run away at a fearful speed. I am inclined to think this description of the effects somewhat exaggerated; but certainly the screw cannot by possibility be exposed to the same variations as the paddles—it cannot be stopped by the action of the sea, indeed, being wholly immersed, the resistance cannot be increased at all, while under no circumstances can it be relieved to the extent to which paddles are, which may both on some rare occasions be quite out of the water; and therefore whether the resistance of the screw is so constant as I believe it to be, or not, yet as compared with that offered by paddles, it is certainly all but perfectly constant.

Fifthly, the effect upon the steerage is singular, the mass of water put into motion by the thrust of the screw is thrown directly upon the rudder, and the consequence is not only that when the ship is going at any given rate, the rudder is passing through the water at a greater rate, and consequently is more sensible, and acts more powerfully upon the ship; but even when the ship has no way, but the screw is at work, the rudder is acted upon by water moving perhaps at two or three knots per hour, and the vessel is still under command—this must be a most important power to possess in a ship, and must materially diminish many of the greatest dangers arising from a strong head wind and sea, and at the same time and under the same circumstances must increase the speed by improving the steerage.

And lastly, her diminished breadth of beam. Important as this alteration would be to any vessel, it is peculiarly so as connected with Bristol; the total breadth, including paddle-boxes, would be at least 78 feet; with the screw, and taking all the increased beam that might be convenient, it would be under 50—very nearly 30 feet of difference. One of the principal objections to her coming up the river would be removed, and the dock gates might easily be made to receive her.[2]

There are many other points upon which comparisons may be drawn, but I am not aware that any very important differences exist.

As regards first cost I believe there would be little difference—if any, it would be in favour of the screw; as a reduction of ninety-five tons of iron can hardly fail to cause some saving, although some portion of the substituted machinery may be more costly per ton.

As regards wear and tear I can have no doubt that some considerable saving would be effected; the paddles are a constant source of trouble and expense, and seem never to be capable of being kept in good repair; indeed, they are huge and comparatively light frameworks subjected to extraordinary and constantly repeated shocks, each arm receiving direct about 260,000 very sharp blows per voyage, independently of the more violent shocks from heavy seas, while the screw can be subjected to no such constant source of mischief.

From all that I have said it must be evident to you, gentlemen, that my opinion is strong and decided in favour of the advantage of employing the screw in the new ship; it certainly is so. I am fully aware of the responsibility I take upon myself by giving this advice, I am also fully sensible of the large amount we have at stake, and I have not forgotten the nature and tone of the observations which have on more occasions than one been so freely made by individuals upon the course we have hitherto pursued; although, and I have pleasure in referring to the fact, this course has in every instance where results have been obtained proved successful; but my conviction of the wisdom, I may almost say the necessity, of our adopting the improvement I now recommend is too strong, and I feel it is too well founded, for me to hesitate or to shrink from the responsibility.

I think I have hardly advanced an opinion which I have not supported, and in most cases preceded, by a statement of facts, leaving no doubt as to the correctness and safety of relying on these opinions; still it would be too much to hope that my mode of laying before you these facts which I have collected and the opinions I have formed could produce as strong a conviction in your minds as the consideration of them has in my own; but if you bear in mind that the actual results of the fair and full trial of the 'Archimedes' for several months has completely established the fact of the efficiency of the screw as a propeller; that the experiments I have made, as well as the general and apparent results of her working, have equally satisfactorily explained the fact of the power required being no greater in proportion to the effect produced than in the 'Great Western' steam-ship, and many other good steamboats; and that these results are satisfactorily explained by theory, you cannot fail to draw the same conclusion that I have done as to the general question of at least the equal efficiency of the screw.

As to the comparisons I have drawn between the general and what I may call the indirect advantages of the one mode of propelling over the other, they seem to me so evident that I am disposed to apologise to you for having occupied your time in pointing them out, and we have the satisfaction of knowing that they are now very generally admitted, particularly by practical men.

In conclusion, I must observe that much more detailed information and recorded results than appear on the face of this Report have been required to enable me to form correct comparisons, and to reduce to calculation and to actual figures and amounts many results observed; and that it would have been impossible for me to have given you such clear and positive facts on many most important points without the very detailed observations made and recorded by Mr. Berkeley Claxton in the several voyages of the 'Great Western,' and also in one on board the 'Archimedes.'

The information obtained from these logs has been, and may still be, of the greatest importance to us in our future working, and I have much pleasure in adding that the manner in which my directions were carried out was highly creditable to Mr. Berkeley Claxton, who, I think, has conferred a great benefit on the Company by his labours. I have to express also my thanks to my friends Captain Claxton and Mr. Guppy for their assistance in the various experiments which have been made, and in working out the results.

I am, Gentlemen,

Yours very faithfully,

(Signed)                                                    I. K. BRUNEL.

1. There appears, from the table, to be an immense extent of paddle-board to this vessel, if the table be correct—greater than the midship section.

2. In a Report by Mr. Brunel on improvements in the port of Bristol, dated December 26, 1839, the following passage occurs: 'A great change is unquestionably about to take place in the carriage of merchandise by sea, a change similar, though possibly not so striking, as that which has so suddenly been effected by railways in land carriage. To ensure the speed which the passenger traffic demands, great size in the vessels is required; in the course of a very few years we shall find Atlantic steamers desirous of taking 500 or 600 tons of cargo to make up their draught.'

# INDEX